INTRODUCTORY

HORTICULTURE

INTRODUCTORY HORTICULTURE

DELMAR PUBLISHERS
COPYRIGHT ©1979
BY LITTON EDUCATIONAL PUBLISHING, INC.

LIBRARY OF CONGRESS CATALOG CARD NUMBER: 77-81006
ISBN: 0-8273-1644-5

Printed in the United States of America
Published Simultaneously in Canada by
Delmar Publishers, A Division of
Van Nostrand Reinhold, Ltd.

H. Edward Reiley
& Carroll L. Shry, Jr.

DELMAR PUBLISHERS • ALBANY, NEW YORK 12205
A DIVISION OF LITTON EDUCATIONAL PUBLISHING, INC.

Preface

Agriculture programs have been enjoying a surge of popularity in recent years. One of the areas of most rapid growth is the study of horticulture. This newfound interest is due in a large part to an increasing awareness on the part of the general public, as people look inwardly at a need to be closer to nature and outwardly at a world in which efficient methods of food production and conservation of land are increasingly important.

This awareness and the resulting growth of the field have generated a need for a broad-based first level text which explores the basic principles of horticulture and methods of practical application of these principles. INTRODUCTORY HORTICULTURE, the first text in Delmar's new Agriculture Series, was written to fill this need.

Foremost in the text is an examination of fundamental horticulture principles, from careers in the field to the cultivation of specific crops. Special care has been taken to treat highly technical subjects, such as plant propagation and taxonomy, in a comprehensive yet understandable manner. A section on pesticides includes up-to-date information on the subject, with special emphasis on personal safety and the protection of human beings and the environment. Certain topics, such as the creation of holiday arrangements, the construction of terrariums and bonsai plantings, and the cultivation of houseplants, have been added especially for student appeal. The units in the text are arranged for timeliness of topic presentation. For example, if a typical two-semester program is followed, the student will cover the section on holiday arrangements in November or December and the material on small fruit and vegetable gardens in the spring.

The format of the text complements the authors' direct approach to the subject matter. A general objective and list of competencies at the beginning of each unit tell what the student should accomplish before proceeding to the next unit. Review material and activities, also included in each unit, allow the students to put into practice skills they learn through study of the text material. Numerous line drawings, photographs, and charts clarify and summarize text content. A glossary at the end of the book lists terms with which the horticulture student should be familiar.

An extensive instructor's guide accompanies the text. A pretest and post-test are included so that individual comprehension of subject matter may be gauged before and after study of the text. Also contained in the guide are answers to review questions, additional student activities, suggestions for instructor demonstrations, supplemental reading lists and when appropriate, special activities for the advanced student and masters for overhead transparencies.

The authors of INTRODUCTORY HORTICULTURE, H. Edward Reiley and Carroll L. Shry, are both involved in agriculture education in Maryland. Mr. Reiley is presently Supervisor of Vocational Education for the Frederick County Board of Education. He received a Bachelor of Science degree in Horticulture and a Master of Science degree in Agriculture Economics from the University of Maryland. Before working in a supervisory capacity, Mr. Reiley taught agriculture in secondary school and continuing education classes. He is a member of various professional organizations, including the American Vocational Association, the American Horticultural Society, and the American Rhododendron Society. In his spare time, Mr. Reiley breeds new varieties of rhododendrons and azalea hybrids.

Carroll L. Shry is presently a horticulture instructor at Linganore High School in Frederick County, Maryland. He holds a Bachelor of Science degree in Agriculture Education from the University of Maryland and a Master of Science degree in Horticulture from West Virginia University. He also has taught agriculture-related subjects at the community college and continuing education levels. Mr. Shry had the distinction of being named Maryland's Conservation Teacher of the Year in 1975.

Contents

Section 1
The
Horticulture
Industry

unit 1
Exploring the Horticulture Field

OBJECTIVE

To obtain an overview of the horticulture industry and jobs available in the field so that a career choice can be made.

COMPETENCIES TO BE DEVELOPED

After studying this unit, the student will be able to

- discuss the scope, size, and economic importance of the horticulture industry in the United States.

- describe five careers in the horticulture industry, including working conditions and educational and personal requirements.

- use the Job Analysis Chart, Self-Analysis Table, and job descriptions to make a career choice in horticulture.

- survey the local job market and select a job that is compatible with student interests and preferences. List skills needed for the job. Describe how to prepare for the job.

The word *horticulture* is derived from a Latin word meaning "garden cultivation" (*hortus* [garden] + *culture* [cultivation]). Over the years, the horticulture industry has developed far beyond this simple definition, however. Today, horticulture includes the cultivation, processing, and sale of fruits, nuts, vegetables, ornamental plants, and flowers. As an industry, it contributes over 16 billion dollars to the United State's economy every year.

The horticulture industry consists of four major divisions:

Pomology. The science and practice of growing, harvesting, handling, storing, processing, and marketing tree fruits.

Job Title	Does it entail year-round work?	Are there regular hours?	Is most of the work outdoors or indoors?	Does it offer variety?	Is the work in one place?	Are there fringe benefits?	Does the job involve working with others?	What are the educational requirements?	Is there an opportunity for promotion?
greenhouse worker	yes	Generally, but some overtime is usually required.	indoors	yes	yes	yes	Yes, to some extent.	high school diploma with a course in agriculture or horticulture	yes
nursery worker	in many cases	Yes, but there are peak seasons.	mostly outdoors	yes	yes	some	yes	high school diploma with a course in agriculture or horticulture	yes
garden center employee	yes	Yes, but there are peak seasons.	both indoors and outdoors	yes	yes	some	yes	high school diploma with a course in agriculture or horticulture	yes
golf course employee	no	Yes, during the golfing season.	outdoors	yes	yes	not to a great extent	yes	high school diploma with a course in agriculture or horticulture	yes
assistant grounds keeper	yes	Yes, some overtime is required on occasion.	mostly outdoors	yes	yes	There may be some.	not necessarily	high school diploma with a course in agriculture or horticulture	yes
park employee	yes	yes	outdoors	yes	yes	not to a great extent	yes	high school diploma with a course in agriculture or horticulture	yes
vegetable grower	depends upon grower	no, seasonal	outdoors	yes	yes	some	yes	high school diploma with a course in agriculture or horticulture	yes
orchard employee	depends upon grower	no, seasonal	outdoors	yes	yes	some	yes	high school diploma with a course in agriculture or horticulture	yes
employee of small fruit grower	depends upon grower	no, seasonal	outdoors	yes	yes	some	yes	high school diploma with a course in agriculture or horticulture	yes
employee of floral design shop	yes	yes, with some overtime	indoors	no	yes	yes	yes	high school diploma with a course in agriculture or horticulture	yes

Fig. 1-1 Job analysis chart. (Adapted from the Job Profile Chart, Ohio State University.)

Olericulture. The science and practice of growing, harvesting, storing, processing, and marketing vegetables.

Floriculture. The science and practice of growing, harvesting, storing, designing, and marketing flowering plants.

Ornamental and Landscape Horticulture. The science and practice of propagating, growing, maintaining, and using grasses, annual plants, shrubs, and trees in the landscape.

Additional specialized areas include:

Seed Production. The science and practice of producing, processing, and selling high quality seed crops for use in growing a wide range of horticultural crops that grow true from seed.

Related Occupations. Those areas that provide products and services necessary in the production of horticultural crops. Such resources include goods such as fertilizers, pesticides, containers, machinery, and services such as education and research.

JOBS IN HORTICULTURE

There are various sources of employment for individuals trained in horticultural practices. These businesses include greenhouses, nurseries, garden centers, golf courses, parks, orchards, floral design shops, grounds maintenance operations, and vegetable and fruit growers. Figure 1-1 is a job analysis chart which explains positions in these areas of employment. The following job descriptions, adapted from the *Handbook of Agricultural Occupations* (published by the United States Department of Agriculture), provide a more detailed account of the most common positions in the industry.

Job Title: Greenhouse Employee

Job Description. A greenhouse employee grows plants in a heated glass or plastic greenhouse, figure 1-2. The plants that are grown may be vegetables or flowers. The work involved may include the propagation of trees or ornamental shrubs from seed or cuttings. Greenhouse activities include the production of out-of-season vegetables such as tomatoes, cucumbers, or leaf lettuce; the growing of cut flowers, potted plants, and bedding plants in preparation for sale; and the starting of plants for outdoor nursery beds. When a greenhouse is part of a nursery operation, a worker may have the

Fig. 1-2 A greenhouse worker placing plants in the greenhouse.
(R. Kreh, Photographer)

duties of a nursery employee as well as those of a greenhouse worker. Employees in greenhouses screen, mix, and sterilize soil and place it in containers for growing plants. They sow seed, start cuttings, and transplant seedlings and plants. They water, weed, thin, prune, fertilize, and spray growing plants. They are also responsible for maintaining the greenhouse structure and equipment.

Working Conditions. Most of the work of a greenhouse employee is done indoors. Outdoor jobs are usually done during favorable weather. This work requires a great deal of walking, stooping, and bending over plant or seedbeds. Hands and clothing may be stained from soil and plant sap or juices.

A job in a greenhouse involves a great deal of manual labor, but the labor is not usually difficult. Working hours are usually regular, with work generally steady throughout the year. Seasonal demands sometimes require working overtime.

Personal and Educational Qualifications. For individuals to really enjoy greenhouse work, they must be interested in watching plants grow and develop. Good health is important, but certain types of physical disabilities will not prevent an interested individual from entering the field. A high school education with a course in vocational agriculture or horticulture is preferred for one interested in this vocation.

Closely related to greenhouse work is flower arranging, figure 1-3.

Job Title: Nursery Employee

Job Description. The nursery employee grows seedlings and plants for landscaping, fruit production, and replanting in forests, figure 1-4. The employee may work in one of several different types of nurseries. Some nurseries specialize in producing fruit trees and

Fig. 1-3 Flower arranging is an important part of the florist trade. (Richard Kreh, Photographer)

small fruit transplants; some in ornamental trees and shrubs; and some in forest replanting materials. Some nursery employees operate greenhouses and produce their own seedlings and plants from cuttings. Some produce planting stock of two or more major lines of plants, trees, or shrubs.

Individuals employed in nurseries prepare seedbeds; plant seedlings; prepare cuttings for rooting; and weed, cultivate, water, and prune plants. They also perform other cultural practices such as spraying and grafting. They dig, grade, and pack plants for shipment. They may cut, lift, and lay sod. They also transplant shrubs and trees, and, in a tree nursery, gather and process forest tree seeds. Aiding in the maintenance and repair of buildings and equipment is also usually a part of the job.

Working Conditions. Most of the nursery worker's time is spent outdoors. If a greenhouse is a major part of the employer's business, the employee will naturally spend a considerable amount of time in the greenhouse. Planting and cultivating must be accomplished when weather conditions are good, but much outdoor work at a nursery can be done in rather bad weather.

Most of the work is considered manual labor; some of it involves heavy lifting. Working hours are regular. Employment may be for the entire year, depending upon the skills of the worker.

Personal and Educational Qualifications. The nursery employee should enjoy working with plants and being outdoors. A truck driver's license may be required. The ability to identify plants and the knowledge of how plants are used in the landscape are very desirable. A high school education with a course in agriculture or horticulture is preferred.

Job Title: Garden Center Employee

Job Description. The employee of a garden center has many jobs, including caring for ornamental plants; moving plants and supplies into selling areas; arranging plants and supplies for display purposes; and selling the various products handled by the center. A garden center may be a part of a large retail store, a part of a nursery or greenhouse operation, or a retail store operated independently of other business.

Fig. 1-4 Nursery worker checking seedlings. In just one of these beds, there are more than 30,000 seedlings. This is enough to plant 30 to 60 acres, depending on how far apart the trees are planted. (Courtesy U.S. Forest Service)

The work of garden center employees includes cleaning, stocking, and arranging garden supplies on shelves and counters and in windows. They water, spray, and trim ornamental plants and control environmental conditions. They unload and unpack supplies as they arrive from wholesalers, make deliveries, and load orders onto trucks and into customers' cars. They also give information and advice to customers concerning plants and lawns and their care.

Working Conditions. Garden centers are built and arranged to attract customers. As a result, a garden center employee usually works in a clean, pleasant, and comfortable environment. Parts of the work area are usually heated during cold weather. Other areas are unheated but protected from rain, snow, and wind. Some of the work may be outdoors during the season in which ornamental shrubs and trees are sold. The work is fairly regular, but has seasonal peaks. Some garden centers close completely or operate with only a small crew during the winter months.

Personal and Educational Qualifications. Employees of garden centers should enjoy working with people and caring for ornamental plants. They must be good salespeople and must therefore be able to talk easily and in a friendly way. A high school education with a course in vocational agriculture or horticulture is desirable.

Job Title: Grounds Maintenance Employee

Job Description. A grounds maintenance employee cares for the area surrounding an industry, business, church, school, airport, apartment building, private estate, cemetery, or shopping center. These employees plant and care for lawns and ornamental plants. The work entails mowing grass, reseeding, controlling weeds, and planting and spraying ornamental plants. They also rake grounds and dispose of leaves and other refuse. A year-round job is provided through maintenance and repair of walks, driveways, and equipment. The work may involve making minor repairs to buildings and providing for snow removal.

Working Conditions. The grounds maintenance employee works outdoors and deals mainly with ornamental plant materials. The work does provide variety, however. Most of the work is manual labor, but is not extremely strenuous. The environment in which the employee works is usually very pleasant, although certain jobs must be done under a variety of weather conditions. The work is steady throughout the year. Employees work regular hours, but there are some peaks in the work load.

Personal and Educational Qualifications. The grounds maintenance employee should not mind working alone and should enjoy working with plants, tools, and small garden equipment.

Job Title: Golf Course Employee

Job Description. Golf course employees are responsible for the overall maintenance of golf courses. They care for the turf on both the greens and fairways. They install and use irrigation and drainage equipment; clean and maintain sand traps; change the locations of cups; and aerate the soil. They may also prune shrubs and trees, replace soil as needed, and repair equipment and buildings.

Working Conditions. The work of the golf course employee is usually done outdoors. Most of it is done during fairly good weather. The workday usually consists of eight hours. In the southern part of the United States, employment is steady throughout the year. In sections of the country which have cold winters, employment is from March or April through October or November.

Personal And Educational Qualifications. Because a great deal of walking is done in golf course maintenance, the employee should have good health, but certain physical handicaps will not interfere with one's success. The employee should enjoy working outdoors and be able to get along well with others. A high school education with a course in agriculture or horticulture is desirable.

Job Title: City, State, or National Park Employee

Job Description. The park employee does that work which is necessary for the proper maintenance of parks. This includes maintaining trees, shrubs, flowers, and lawns that make up the planting area. The city park worker deals mainly with formal flower beds and lawn areas, while the state or national park worker is usually concerned with the care and maintenance of natural woodlands or forests.

The work of the park employee includes mowing grass, trimming the edges of walks and driveways; planting, pruning, and caring for trees, shrubs, hedges, lawns, and flower beds; controlling insects, diseases, and weeds; and caring for the soil. It also includes such jobs as the removal of trash and snow, maintenance of swimming pools, care of boating facilities, general maintenance of buildings and equipment, and repair of roadways and drives.

Working Conditions. The park employee is outdoors a great deal of the time. Most of the work is manual labor, and is accomplished in a healthy, pleasant environment. At times, park improvement jobs must be done in bad weather conditions. This occupation usually gives year-round employment, and working hours are regular. In certain types of state park work, there may be peak periods.

Personal and Educational Qualifications. A high school education with a course in vocational agriculture or horticulture is desirable.

SELECTING A JOB

To simplify the task of choosing a specific career in the horticulture industry, first study the questions posed in figure 1-5, Self-Analysis Table. When the questions have been answered, match the information with the job analysis chart found in figure 1-1, page 2. This procedure may be followed for jobs other than

In a career, do you desire a job that involves:

year-round work?
regular working hours?
a variety of tasks?
working in one place?
the security of fringe benefits?
working alone?
working with others?
specific educational requirements?
the guarantee of promotion?

Fig. 1-5 Self-analysis table.

those listed in the job analysis chart. Simply collect the information concerning the job and match it with your self-analysis.

Career opportunities may also be broken down according to the type of work they involve: production, management, and sales, figure 1-6, page 8. If outside work is preferred, an area in production might be considered. If inside work without the company of other people is appealing, a professional career, such as research, should be considered.

There are other factors to consider before selecting a career:

• Will the job be challenging?
• What types of duties does the work involve?
• What skills does the job require?
• Are jobs in the field available?

Using figures 1-1 and 1-5, check several jobs listed to determine which come closest to satisfying your personal job expectations. Research the job selected. If the job is available in your community or work area, arrange to discuss your interest with someone involved in that career, such as the manager or owner

Jobs that require work with the hands, tools, and equipment. Much of the work is done outside.

greenhouse worker   nursery grower
propagator   bedder
tree surgeon   pruner
turf worker   grounds keeper
greens keeper   landscape gardener
orchardist   small fruit grower
vegetable crop grower   farm chemicals sales worker

Jobs that may require advanced training, planning, and paperwork. Much of the work is done inside.

landscape architect   consultant
teacher   researcher
plant breeder   office supervisor
sales person   plant disease specialist
greenhouse manager   nursery manager
garden center manager   pest control specialist
pesticide specialist

Note: This represents only a partial list. It is suggested that a survey of jobs available in the local job market be made. These jobs may then be analyzed using the process mentioned earlier.

Fig. 1-6 Job list.

of a local business. A visit to the class by the owner or a visit to the business may be arranged.

THE FUTURE OF THE HORTICULTURE INDUSTRY

The horticulture industry is currently in a state of rapid growth. As people become more concerned about their surroundings, world resources, nature, conservation, and controlling pollution, horticulture becomes more important. More plants will be used to beautify surroundings, to conserve soil and control pollutants, and to protect and feed wildlife. As food costs increase, more families will plant gardens. The future appears bright for employment which involves working with plants or providing goods and services related to the horticulture business.

Production of Horticultural Crops

As the population increases and people have more leisure time, more gardening of all types will be done. The demand for fruits, vegetables, flowers, ornamental plants, and landscaping services will grow. Increase in the demand for vegetable plants by home gardeners means a substantial increase in the production business.

Service Industries

As more horticultural goods are produced, more and more seeds, fertilizers, insecticides, fungicides, growth-regulating chemicals, hand tools, and equipment are also needed. These service industries generally seek to employ persons with a background in horticulture.

Marketing

More people are engaged in marketing and distributing horticultural products than in their production. As the volume of products increases, so does the need for employees in marketing with a background in horticulture.

Inspection

Government agencies employ inspectors of both fresh and processed horticultural crops and products. Private processing firms interested in a quality product also employ field inspectors to spot disease and insect outbreaks as well as to determine when the crop is at peak quality for harvesting.

Teaching and Extension Work

More and more students need to be trained in horticulture at secondary and post-secondary school levels. This requires more teachers to fill the need. A quickly changing technology also requires a field force of specialists, such as extension agents and county agricultural agents, to work with producers. The home-owners and those who garden as a hobby also require the advice of professionals. Educators often use the news media as a mass communication method to reach the public more easily.

Research

As demands for greater production are placed on the horticulture industry, researchers play an important role, figure 1-7. Better varieties of plants must be developed; more effective insect and disease control established, both through resistant varieties and through biological and chemical control; and better ways of growing, harvesting, preserving, packaging, storing, and marketing must be developed.

The horticulture industry has met the challenge in the past through the development of new products and procedures such as:

- hybrid varieties
- better greenhouse climate control
- mist propagation
- automatic watering and fertilizing
- control of the blooming date of plants by regulation of the length of darkness or light
- plant growth regulators

Fig. 1-7 A researcher checks results of a chemical growth retardant. These plants are all the same age, but were subjected to applications of varying strengths. (USDA photo by Murray Lemmon)

- rooting hormones
- weed control
- resistant varieties of plants
- extended storage life of foods
- mechanical harvesting
- dwarfing rootstock for tree fruits
- better methods of packaging and processing products such as concentrated juices and frozen foods
- improved cultural practices

There is every reason to believe the challenge will be met in the future. You can be part of this exciting industry.

STUDENT ACTIVITIES

1. Using a telephone book, make a list of horticulture businesses in the local area.

2. Select one business in the community and interview the owner as well as several employees. Report to the class on jobs available and the type of duties involved in each job.

SELF-EVALUATION

Select the best answer from the choices offered to complete the statement or answer the question.

1. The science and practice of growing and harvesting tree fruits, small fruits, and tree nuts is called

   a. ornamental and landscape horticulture.   c. olericulture.
   b. floriculture.   d. pomology.

2. The science and practice of growing vegetable crops is called

   a. ornamental and landscape horticulture.   c. olericulture.
   b. floriculture.   d. pomology.

3. The science and practice of growing and harvesting flowering plants is called

   a. ornamental and landscape horticulture.   c. olericulture.
   b. floriculture.   d. pomology.

4. The science and practice of propagating, growing, maintaining, and using grasses, annual plants, shrubs, and trees in the landscape is called

   a. ornamental and landscape horticulture.   c. olericulture.
   b. floriculture.   d. pomology.

5. A greenhouse employee works

   a. mostly outdoors.   c. with nursery stock only.
   b. mostly indoors.   d. alone.

6. The greenhouse worker has a job which consists of

a. very irregular and seasonal work.
b. very strenuous manual labor.
c. regular hours which are steady throughout the year.
d. constant outside work.

7. Nursery workers

a. spend a great deal of time outdoors.
b. spend most working hours in the greenhouse.
c. can always depend on steady work.
d. do not need to be able to identify plants.

8. To find out what jobs are available locally, the student should

a. make a survey of local horticultural businesses.
b. check with the department of labor.
c. check with a local employment agency.
d. all of the above

9. A garden center employee should

a. enjoy meeting people.
b. enjoy a job that is mostly paperwork.
c. enjoy working alone.
d. not be concerned about finishing high school.

10. A grounds maintenance employee should

a. enjoy working outdoors and working alone.
b. enjoy mowing grass as part of the job.
c. enjoy repairing equipment and walks.
d. all of the above

11. As the population grows and more people build homes, the horticulture industry will

    a. remain at the present level of employment.
    b. grow also as the demand for plants increases.
    c. get smaller, since home owners will grow their own plants.
    d. none of the above

12. Which one of the positions listed below is never related to horticulture?

    a. researcher
    b. teacher
    c. marketing specialist
    d. stock broker

13. Most of the jobs in horticulture require

    a. a high school diploma.
    b. excellent physical condition.
    c. an ability to meet people.
    d. a person who enjoys being alone.

14. Three jobs in horticulture that require outdoor work and that are done with hand-operated tools and equipment are

    a. tree surgeon, researcher, landscape gardener.
    b. teacher, orchardist, turf worker.
    c. tree surgeon, turf worker, small fruit grower.
    d. nursery worker, pruner, plant breeder.

15. Horticultural salespeople should have a background in horticulture because

    a. it helps them to know the product they are selling.
    b. it helps them to talk to growers.
    c. it helps them to know which items to offer for sale.
    d. all of the above

16. The word *horticulture*, a Latin word, means

    a. "grower of crops."
    b. "plant cultivator."
    c. "lover of plants."
    d. "garden cultivation."

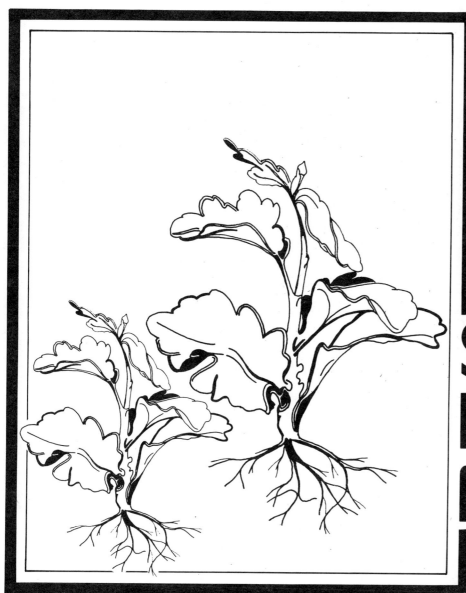

Section 2
How
Plants Grow

OBJECTIVE

To recognize the main parts of a plant and describe the function of each.

COMPETENCIES TO BE DEVELOPED

After studying this unit, the student will be able to

- note two contributions of plants to the life cycle on earth.
- list and describe the purpose of the four main parts of plants.
- explain the process of photosynthesis.
- explain the major structural difference between dicot and monocot stems, and how the stems grow.
- describe the process of pollination.

unit 2
Parts of the Plant and Their Functions

Materials List

a complete plant with roots, stem, leaves, and flower or seeds

a plant with a fibrous root system

a plant with a tap root system

a large flower for disection

transparencies of plant parts

overhead projector

knives for disecting flower parts

THE IMPORTANCE OF PLANTS

The importance of plant life on earth cannot be overemphasized. Without plants, life on earth could not exist. Directly or indirectly, plants are the primary source of food for humans and animals. Whether people eat plants or eat animals that feed on plants, plant life is vital as a food source.

Plants play another essential role by producing oxygen. Without oxygen, life on earth could not exist. Plants are the major producers of oxygen on this planet. All plant life, from the smallest plankton in the ocean to the giant redwood tree, works to produce oxygen.

In addition to supplying food and oxygen, plants help to keep us cool, renew the air, slow down the wind, hold soil in place, provide a home for wildlife, beautify our surroundings, perfume the air, and furnish building materials and fuel.

PARTS OF A PLANT

Most plants are made up of four basic parts: leaves, stems, roots, and flowers, which later become fruit or seeds, figure 2-1.

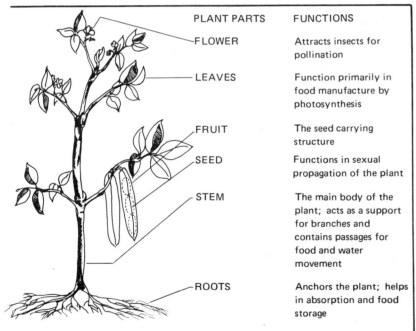

PLANT PARTS	FUNCTIONS
FLOWER	Attracts insects for pollination
LEAVES	Function primarily in food manufacture by photosynthesis
FRUIT	The seed carrying structure
SEED	Functions in sexual propagation of the plant
STEM	The main body of the plant; acts as a support for branches and contains passages for food and water movement
ROOTS	Anchors the plant; helps in absorption and food storage

Fig. 2-1 Parts of the plant and their functions

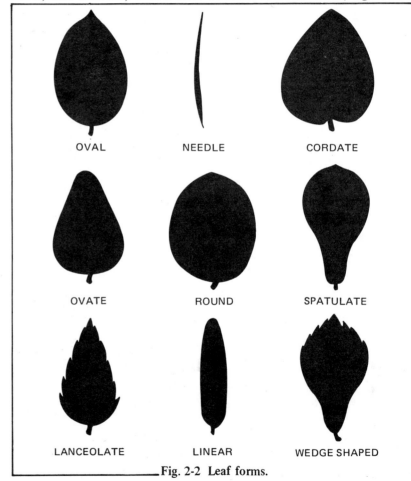

OVAL   NEEDLE   CORDATE

OVATE   ROUND   SPATULATE

LANCEOLATE   LINEAR   WEDGE SHAPED

Fig. 2-2 Leaf forms.

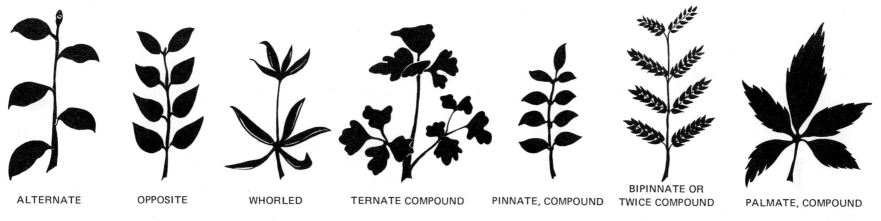

| ALTERNATE | OPPOSITE | WHORLED | TERNATE COMPOUND | PINNATE, COMPOUND | BIPINNATE OR TWICE COMPOUND | PALMATE, COMPOUND |

Fig. 2-3 Leaf arrangements.

Leaves

Leaves are the food factory of the plant, producing all food that is used by the plant and stored for later use by the plant or by animals.

Leaves vary a great deal in shape and size. Most leaves are flat. Some, such as the leaves of pine trees, are needlelike, while others, such as onion leaves, are cylindrical. The shape and size of leaves helps to identify plants, figure 2-2.

The arrangement of leaves on plants also differs. Some plants have leaves which alternate on the stem; some are positioned opposite one another. Others are *whorled* (arranged in a circle around the stem), figure 2-3.

External Leaf Structure. Leaves consist of the *petiole*, or leaf stalk, and the *blade,* a larger, usually flat part of the leaf, figure 2-4. Notice that the leaf blade has veins and a midrib. The

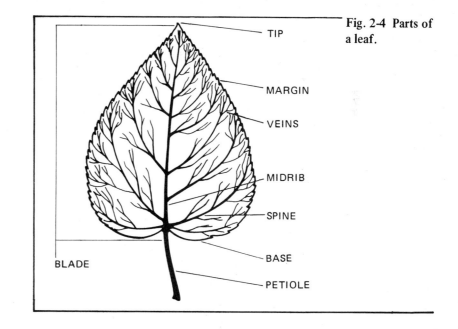

Fig. 2-4 Parts of a leaf.

TIP

MARGIN

VEINS

MIDRIB

SPINE

BASE

PETIOLE

BLADE

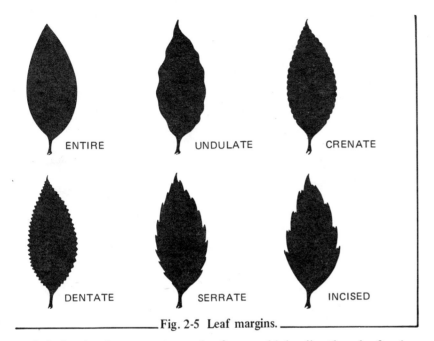

Fig. 2-5 Leaf margins.

ENTIRE UNDULATE CRENATE

DENTATE SERRATE INCISED

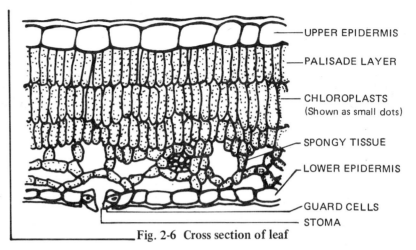

UPPER EPIDERMIS

PALISADE LAYER

CHLOROPLASTS
(Shown as small dots)

SPONGY TISSUE

LOWER EPIDERMIS

GUARD CELLS

STOMA

Fig. 2-6 Cross section of leaf

midrib is the large center vein from which all other leaf veins extend.

The *veins* of the leaf form its structural framework. Most leaves have one of the forms shown in figure 2-2. Different leaf *margins* (edges of plant leaves) are shown in figure 2-5. Awareness of the different leaf margins assists in plant identification.

Internal Leaf Structure. Internally, leaves have specialized cells which perform very important tasks, figure 2-6. The skin of the leaf, called the *epidermis*, is a single layer of cells. Its chief function is to protect the leaf from loss of too much moisture. There are special cells in the leaf skin known as *guard cells*. These cells open and close a small space or pore on the underside of the leaf called a *stoma* to allow the leaf to breathe and *transpire*, or give off moisture.

In the center of the leaf are food-making cells which contain *chloroplasts*. The green color of the chloroplasts, which gives green leaves their color, comes from the chlorophyll they contain. These cells, through a process called photosynthesis, manufacture food. *Photosynthesis* is the process by which carbon dioxide and water in the presence of light are converted to sugar and oxygen. This is the beginning of the food chain for all living things on earth. The chemical process is explained here in chemist's shorthand:

$$\text{LIGHT ENERGY}$$

$$6\,CO_2 \;+\; 6\,H_2O \;\;=\;\; C_6H_{12}O_6 \;+\; 6\,O_2$$

$$\underset{\text{dioxide}}{\text{carbon}} \;+\; \text{water} \;=\; \text{sugar} \;+\; \text{oxygen}$$

$$\text{CHLOROPHYLL}$$

The oxygen which results from the photosynthesis process is used directly by animals. It is the vital ingredient in all forms of *oxidation* (combining with oxygen), such as burning, rusting, and rotting.

Food manufactured in the leaves moves downward through the stem to the roots. It is then used by the plant or stored in the stem or root in the form of sugar, starch, or protein. The leaves themselves are also used as food for various animals, including human beings. They are often the most nutritious part of the plant.

Stems

Stems have two main functions: (1) the movement of materials, such as the movement of water and minerals from roots upward to the leaves, and the movement of manufactured food from the leaves down to the roots and (2) the support of the leaves and reproductive structures (flowers and fruit or seeds). Stems are also used for food storage, as in the Irish potato, and for reproduction methods which involve stem cuttings or grafting. Green stems manufacture food just as leaves do.

External Stem Structure. The outside of the stem consists of *lenticels*, or breathing pores, bud scale scars, and leaf scars. *Bud scale scars* indicate where a terminal bud has been located a previous year. The distance between two scars represents one year of growth. *Leaf scars* show where leaves were attached, figure 2-7.

Some plants, such as the Irish potato and the gladiolus, have stems which differ greatly from the stems of other plants. The stems in these cases are used for food storage and plant reproduction. A detailed discussion of this type of stem and how it differs from others is given in Section 5, Plant Propagation.

Internal Stem Structure. The stem of woody plants is composed of bark called *phloem* and wood called *xylem,* figure 2-8, page 20.

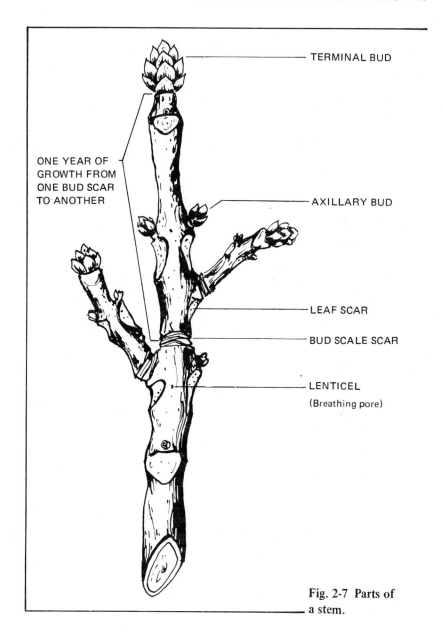

TERMINAL BUD

ONE YEAR OF GROWTH FROM ONE BUD SCAR TO ANOTHER

AXILLARY BUD

LEAF SCAR

BUD SCALE SCAR

LENTICEL
(Breathing pore)

Fig. 2-7 Parts of a stem.

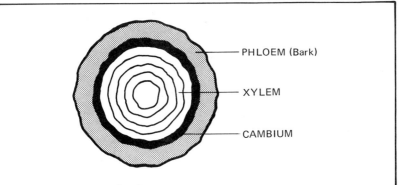

Fig. 2-8 Cross section of a dicot stem

PHLOEM (Bark)

XYLEM

CAMBIUM

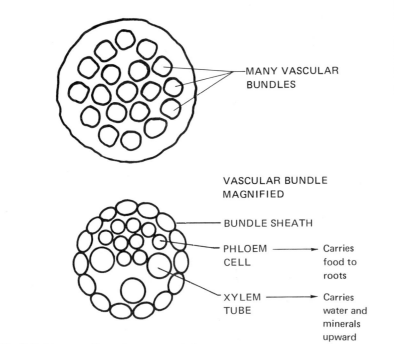

MANY VASCULAR BUNDLES

VASCULAR BUNDLE MAGNIFIED

BUNDLE SHEATH

PHLOEM CELL → Carries food to roots

XYLEM TUBE → Carries water and minerals upward

Fig. 2-9 Cross-section of a monocot stem

Water and minerals travel up the sapwood, or xylem, and manufactured food travels down the bark, or phloem. These two layers are separated by the *cambium* which produces all new cells. This stem structure is typical of *dicots*, plants which have two seed leaves. Dicot stems may continue to grow because the cambium builds new phloem cells on the outside and new xylem cells on the inside. Trees are good examples of dicots that have stems which continue to grow.

Some plants, such as grasses, have a different stem structure, figure 2-9. These plants are referred to as *monocots* because they have only one cotyledon (seed leaf). Corn is an example of a monocot. Monocots have vascular bundles which contain both phloem and xylem tissue in each small bundle. (See figure 2-9.) Notice that the monocot stem has no cambium. All cells are formed in the initial stage of stem growth and merely enlarge to create the size of the mature stem. There is no further enlargement of stem size by formation of new cells.

Economic Importance. The stems of some plants, such as the Irish potato and asparagus, are used as food. Others are used as building materials, such as the lumber from tree trunks.

Roots

Roots are usually underground and, therefore, are not easily visible. Roots function to

- anchor the plant and hold it upright.
- absorb water and minerals from the soil and conduct them to the stem.
- store large quantities of plant food.
- propagate or reproduce some plants.

The first three functions are essential to all plants.

Structure. The internal structure of a root is much like that of a stem. Older roots of shrubs and trees have phloem (corklike bark) on the outside, a cambium layer, and xylem (wood) inside, just as stems do. The **p**hloem carries manufactured food down to the root for food and storage and the xylem carries water and minerals up to the stem.

The external structure of the root is very different from that of the stem. Whereas the stem has a terminal bud which initiates new growth, roots have a root cap. Just behind the root cap are many root hairs. Side roots of increasing size form as the root grows older, figure 2-10. The *root cap* protects the root as it grows and pushes through the soil in search of moisture and nutrients. The *root hairs* absorb moisture and minerals which are conducted to the larger roots and to the stem of the plant.

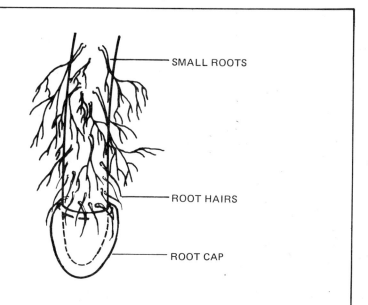

SMALL ROOTS

ROOT HAIRS

ROOT CAP

Fig. 2-10 External view of young root (greatly enlarged)

In addition to their function within the plant itself, many roots are important as cash crops for food. Vegetables such as carrots, beets, radishes, and sweet potatoes are all roots. Roots also serve in the process of propagation. Plants with tuberous roots, such as the dahlia, peony, and sweet potato, are propagated by separating the root clump or by rooting sprouts from the root, as is the case with the sweet potato.

Type of Root System. The ease with which nursery stock is transplanted or moved depends to a great extent on the type of root system that the plant possesses.

Fibrous Root System v. Tap Root System. Plants with fibrous root systems are much easier to transplant than plants which have tap root systems. As illustrated in figure 2-11, when a plant with a fibrous root system and a plant with a tap root system are dug from the same depth in the ground, a greater percentage of the fibrous root system is saved. The roots are shorter, smaller, and more compact.

The tap root system has longer and fewer roots. Because of this, much of the root system is cut off when a plant is dug. The ends of the roots which are lost in the cutting contain many

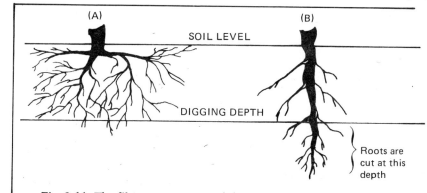

(A) (B)

SOIL LEVEL

DIGGING DEPTH

Roots are cut at this depth

Fig. 2-11 The fibrous root system (A) and the tap root system (B)

root hairs necessary in the absorption of water and minerals from the soil. The larger roots serve only to conduct and store water, nutrients, and food. If too many of the small roots are lost, the plant may not be able to replenish the moisture lost by the leaves, and the plant will dry out and die. Heavy pruning of the top of the plant may prevent this from happening.

FLOWERS, FRUITS, AND SEEDS

To most people, flowers are something of beauty meant to be seen and enjoyed. Those same people usually think of fruits and seeds as healthful foods. The parts that are admired and enjoyed by human beings, however, have an entirely different purpose for the plant.

The beauty of the flower, for example, is necessary to attract insects. In their visits for nectar or pollen, these insects fertilize the flower by means of a process called *pollination*. This is the beginning of fruit and seed formation. The fruits and seeds are made attractive to animals and birds so that they are collected, eaten, and spread. This, in turn, reproduces the plant. Some seeds have sticky coats and are carried on the animal's fur to be dropped in a new location. These devices and actions are not for the convenience of human beings or animals, but to ensure the continued existence of the plant itself.

Parts of the Flower

Flowers differ in such features as size, shape, and color, but generally have the same basic parts. These basic parts are necessary for the production of seed.

Seed is the most common way plants reproduce in nature. Seeds are produced by a sexual process with a male and a female parent involved. A *complete flower* has both male and female

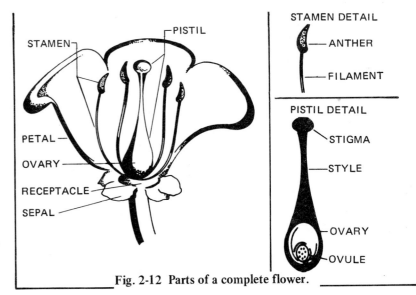

Fig. 2-12 **Parts of a complete flower.**

parts and only one parent is necessary if the plant is *self-fruitful*, or can pollinate itself. The following is a description of a *complete flower*, figure 2-12. The complete flower contains four main parts: sepals, petals, stamen, and pistil.

The sepal is the green leaflike part of the flower which covers and protects the flower bud before it opens.

The petals are actually leaves, but are generally known as the most striking part of the flower. The bright colors which are usually present on flowers act to attract insects for pollination.

The stamens make up the male reproductive part of the flower. The stamen consists of a short stalk called a *filament* and a sac-like structure on top of the filament called an *anther*. The anther contains pollen, which is the male sex cell.

The pistil, located in the center of the flower, is the female part of the flower. It produces the female sex cells, the eggs. These eggs, if fertilized, become seeds. The pistil has three main parts. These include a sticky *stigma* on top to catch pollen and a *style*, a tube which leads to the third part, the *ovary*. The egg cells develop in the ovary. After fertilization, the ovary grows to become a fruit or a seed coat.

The flower is constructed so that insects attracted to it for nectar must first climb over the anther and brush pollen on the hairy surface of their bodies. As they climb onto the center of the flower for nectar, part of the pollen is brushed onto the stigma of the pistil. This allows the fertilization process to begin. The pollen grain sprouts like a seed and sends a long stalk down the style to the ovary and the egg cells. The pollen sperm cell then fertilizes the egg cells or *ovules*, and seeds begin to develop. The ovary enlarges into a seed coat or fruit.

An *incomplete flower* has only the male parts or the female parts of the flower, but not both. Thus, a male flower has sepals, petals, and stamen, but no pistil. A female flower has sepals, petals, and pistil, but no stamen.

Flowers play a very important role in the florist and nursery businesses. Many plants are grown solely for the beautiful flowers they produce. Plants may grow flowers only to attract insects for pollination, but people grow them for their beauty and economic value. The floriculture industry (that part of the horticulture industry that deals with flowers) is a multimillion dollar business.

STUDENT ACTIVITIES

1. Dissect a flower. Sketch and label all parts.

2. In a notebook, draw and label a cross section of the four main plant parts.

3. Cut a cross section of a dicot stem and monocot stem and draw a sketch, noting the differences.

> **Caution: Exercise care when handling knives or saws.**

4. Obtain two plants common to the area and label a sketch of a leaf from each plant. Note the shape or form, type of margin, and arrangement of leaves. See text for information.

5. Start a monocot seed (corn) and a dicot seed (bean). As they develop, note that the monocot has one seed leaf and the dicot has two seed leaves.

SELF-EVALUATION

A.

1. Write the formula for photosynthesis in chemist's shorthand and in words.

2. Label the lettered parts of the plant shown below.

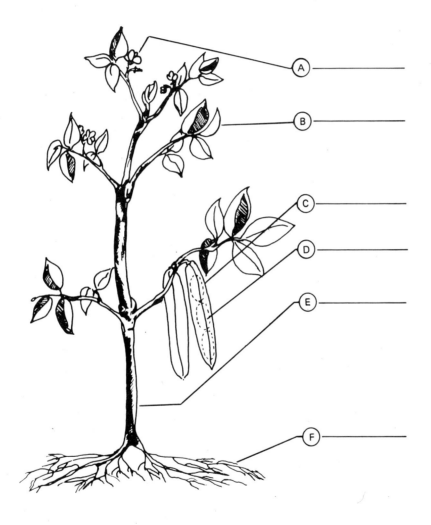

B. The column on the left contains a statement associated with plants and their functions. Select the correct term from the right-hand list and match it with the proper statement on the left.

1. A specialized stem that stores food and is used in plant propagation.
2. Water and minerals travel up this part of the plant stem (the wood of the stem).
3. Manufactured food travels down this part of the plant stem (the bark of the stem).
4. The part of the stem found in dicots but not monocots, which allows dicot plants to continue to produce new cells after initial cell division.
5. Plants with these root systems are easier to transplant than plants with tap root systems.
6. The flower is important to the plant only if it is fertilized and produces this plant part.

a. seed
b. phloem
c. Irish potato
d. food
e. xylem
f. fibrous
g. cambium
h. anchor
i. fertilize

C. Select the best answer from the choices offered to complete the statement or answer the question.

1. The two most important things that human beings receive from plants are

a. shade and food.
b. oxygen and food.
c. beauty and food.
d. food and clothing.

2. Plants also accomplish other purposes. They

a. keep us cool and clean the air.
b. slow the wind, prevent erosion, and provide beauty.
c. perfume the air and furnish fuel and building materials.
d. all of the above

3. The four basic parts of a plant are the

   a. leaves, stems, roots, and flowers which produce seeds.
   b. flowers, leaves, pollen, and fruit.
   c. sepal, pistil, stamen, and ovary.
   d. none of the above

4. Two main functions of stems are

   a. to store food and move water up to the leaves.
   b. movement of materials and support of plant parts.
   c. to manufacture food and store it for future use.
   d. to furnish food for human beings and other animals.

5. The stems of dicots, which have a cambium layer of cells,

   a. produce all new cells quickly, then cease growing.
   b. continue to produce new cells and grow larger.
   c. are always green.
   d. are cold resistant.

6. Monocot stems have the xylem and phloem

   a. on each side of a cambium layer.
   b. in small bundles scattered through the stem.
   c. side by side.
   d. near the outside of the stem.

7. The xylem, or wood, of a stem

   a. conducts manufactured food down to the roots.
   b. is very hard.
   c. is green in color.
   d. conducts water and minerals up to the leaves.

8. Which of the following is not a function of roots?

   a. storage of food   c. anchoring of plants
   b. absorption of water   d. manufacture of food

9. Which three of the following vegetables are roots?

   a. carrots, beets, sweet potatoes   c. Irish potatoes, carrots, beans
   b. celery, Irish potatoes, radishes   d. kale, Irish potatoes, beets

10. The green color of leaves is caused by tiny particles in the food-producing cells called

   a. guard cells.   c. chloroplasts.
   b. epidermis.   d. starch.

11. Green plants are able to manufacture food only in the presence of

   a. light.   c. water.
   b. carbon dioxide.   d. all of these.

12. The four main parts of a flower are the

   a. pollen, ovary, pistil, and stamen.   c. sepal, pistil, ovary, and stigma.
   b. sepal, petal, stamen, and pistil.   d. none of these

13. Which of the following is provided for a plant by its leaves?

   a. shade   c. an air-conditioning unit
   b. a food factory   d. a breathing aparatus

14. Plants are easier to transplant if they have a

   a. tap root system.   c. fibrous root system.
   b. large root system.   d. small root system.

15. An example of a plant with a specialized root that stores food and is used to propagate the plant is the

   a. Irish potato.   c. radish.
   b. carrot.   d. sweet potato.

16. The major function of root hairs on roots is to

    a. grow into larger roots.
    b. absorb water and minerals from the soil.
    c. protect the root as it pushes through the soil.
    d. keep the root warm.

17. Pollination is a sexual process in which pollen is deposited on the stigma of the plant. It starts the process of fertilization and

    a. growth of the pollen tube.
    b. seed formation.
    c. production of a fruit or seed coat.
    d. all of the above

18. The stamen is

    a. the male part of the flower.
    b. the part of the flower that produces pollen.
    c. the part of the flower that holds the anther.
    d. all of the above

19. The pistil is

    a. the female part of the flower.
    b. the male part of the flower.
    c. the showy part of the flower.
    d. the pollen-producing part of the flower.

20. When fertilized, the eggs in the ovary grow into

    a. fruit.
    b. seed cases.
    c. seeds.
    d. flowers.

unit 3
Environmental Requirements for Good Plant Growth

OBJECTIVE

To explain the basic needs of plants and the various factors that make up their environment.

COMPETENCIES TO BE DEVELOPED

After studying this unit, the student will be able to

- list four factors that affect the roots of plants.
- describe the differences between clayey, sandy, and loamy soils and identify a sample of each.
- explain three ways to improve soil drainage and two ways to increase moisture retention of soil.
- explain what is meant by the pH value of soil.
- compose a balanced fertilizer program for one plant that is grown commercially in the area.
- list four aboveground requirements for good plant growth.
- list the three major plant food elements and two functions of each.

Materials List

fertilizer bag of garden fertilizer and lawn fertilizer

fertilizer bag or container of water soluble fertilizer with minor elements

samples of clayey, sandy, and loamy soils

complete plant

three half-gallon jars

four potted plants for light and temperature experiments

To properly grow into profit-making crops, plants require a certain environment. This environment can be divided into two parts: the underground environment in which the roots live and grow, and the aboveground environment in which the visible part of the plant exists. Figure 3-1 illustrates the two environmental systems of trees and how they relate to tree growth.

CROWN

Trees increase each year in height and spread of branches by adding a new growth of twigs. This new growth comes from young cells in the buds at the ends of the twigs.

TRUNK

The tree trunk supports the crown and produces the bulk of the useful wood.

ROOTS

Roots anchor the tree; absorb water, dissolved minerals and nitrogen necessary for the living cells which make the food; and help hold the soil against erosion. A layer of growth cells at the root tips makes new root tissue throughout the growing season.

FIRE RUINS TIMBER

Disease and insects enter through fire scars

Diseased or decayed wood

Insect damage

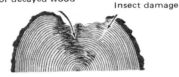

THINNING INCREASES GROWTH

35 years growth before thinning   16 years growth after thinning

INNER BARK (Phloem) carries food made in the leaves down to the branches, trunk, and roots.

OUTER BARK protects tree from injuries.

CAMBIUM (a layer of cells between bark and wood) is where growth in diameter occurs. It forms annual rings of new wood inside and new bark outside.

SAPWOOD (Xylem) carries sap from roots to leaves.

HEARTWOOD (was sapwood, now inactive) gives strength.

PHOTOSYNTHESIS

Leaves are the most important chemical factories in the world. Without their basic product, sugar, there would be no food for man or animal, no wood for shelter, no humus for the soil, no coal for fuel.

Inside each leaf, millions of green-colored, microscopic "synthetic chemists" (chloroplasts) manufacture sugar. They trap radiant energy from sunlight for power. Their raw materials are carbon dioxide from the air and water from the soil. Oxygen, a byproduct, is released. This fundamental energy-storing, sugar-making process is called photosynthesis.

What happens to this leaf-made sugar in a tree? With the aid of "chemical specialists" (enzymes),

every living cell — from root tips to crown top — goes to work on the sugar. New products result. Each enzyme does a certain job, working with split-second timing and in harmony with the others. In general, they break down sugar and recombine it with nitrogen and minerals to form other substances.

ENZYMES

— Change some sugar to other foods such as starches, fats, oils, and proteins, which help form fruits, nuts, and seeds.

— Convert some sugar to cell-wall substances such as cellulose, wood, and bark.

— Make some of the sugar into other substances which find special uses in industry. Some of these are rosin and turpentine from southern pines; syrup from maples; chewing gum from chicle trees and spruces; tannin from hemlocks, oaks, and chestnuts.

— Use some of the sugar directly for energy in the growing parts of the tree — its buds, cambium layer, and root tips.

TRANSPIRATION

Transpiration is the release of water-vapor from living plants. Most of it occurs through the pores (stomates) on the underside of the leaves. Air also passes in and out.

Fig. 3-1 The growth of a tree (Courtesy United States Department of Agriculture)

In the commercial production of plants, individuals must control the plant environment to obtain the *optimum* (best) return for the investment made, figure 3-2.

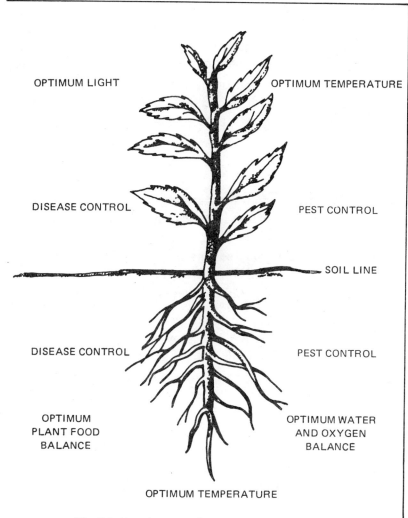

OPTIMUM LIGHT

OPTIMUM TEMPERATURE

DISEASE CONTROL

PEST CONTROL

SOIL LINE

DISEASE CONTROL

PEST CONTROL

OPTIMUM PLANT FOOD BALANCE

OPTIMUM WATER AND OXYGEN BALANCE

OPTIMUM TEMPERATURE

Fig. 3-2 Requirements for optimum plant growth

THE UNDERGROUND ENVIRONMENT

Although some plants require a more specialized underground environment, there are certain factors that affect the growth and development of all plants. The *medium* (soil or soil substitute) in which plants are grown is a very important factor. (*Note*: The plural of medium is *media* or *mediums*.) Through their roots which anchor them in the soil, plants take in air, moisture, and minerals — all vital to plant life. Many times, plant food is added to the soil to encourage better growth.

Soil

Soil is made up of sand, silt, clay, organic matter, living organisms, and pore spaces which hold water and air. Soils are classified according to the percentage of sand, silt, and clay they contain.

Soil particles vary greatly in size. A sand particle is much larger than a silt particle. Clay particles are by far the smallest. These clay particles hold moisture and plant food elements much more effectively than larger particles. A certain amount of clay in all soil is important for this reason.

Soils vary greatly in general composition, depending on their origin. Some soils were formed as a result of rock breaking down over thousands of years; others developed as certain materials were deposited by water. A normal soil profile consists of three layers: (1) topsoil, (2) subsoil, and (3) soil bedrock, or if rock is not present, lower subsoil, figure 3-3, page 32. Topsoil represents the depth normally plowed or tilled, and contains the most organic matter or decaying plant parts. Deep-rooting plants send roots down into the subsoil, which is a well-defined layer immediately below the topsoil. If the soil is well drained, roots penetrate deeper into the subsoil since oxygen is available at greater depths.

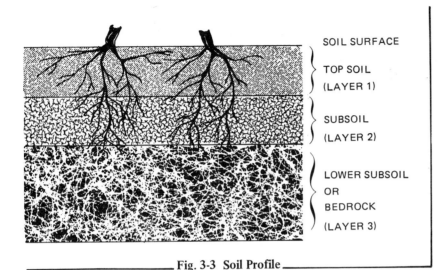

SOIL SURFACE

TOP SOIL
(LAYER 1)

SUBSOIL
(LAYER 2)

LOWER SUBSOIL
OR
BEDROCK
(LAYER 3)

Fig. 3-3 Soil Profile

Roots may penetrate until rock, hard clay, or water prevents further growth.

The natural structure of soils is more important to the gardener, fruit grower, and the nursery worker who plants outside than to the greenhouse operator or nursery worker who grows plants in containers. The worker who grows plants in containers can add ingredients to the soil to change its structure, moisture-holding ability, drainage ability, or fertileness. In fact, many plants are grown in completely artificial mixes which contain no soil.

Sandy Soil. *Sandy* (or *light*) soils include soils in which silt and clay make up less than 20 percent of the material by weight. These soils drain well, but have little capacity to hold moisture and plant food.

Clayey Soil. To be classified as a clayey soil, a soil must contain at least 30 percent clay. Such a soil is known as a heavy soil.

Heavy soils have relatively poor drainage and aeration capabilities. (To *aerate* is to supply with air.) Because of this, heavy soils tend to hold more moisture than is good for plants. However, this type of soil also holds fertilizer and plant food well, which can be beneficial to plant growth.

Loamy Soil. This is the most desirable soil for general use. *Loam* is a mixture of approximately equal parts of sand, silt, and clay. If it has more sand than silt or clay, it is known as a *sandy loam*; more clay, a *clayey loam*; more silt, a *silty loam*. The texture triangle shown in figure 3-4 is helpful in determining the names of soils.

Soil Improvement

Soils used for outdoor plant growth may be improved through increased drainage, irrigation methods, and the addition of organic matter and plant food in the form of fertilizers.

Since it would be very expensive to change the percentage of sand, silt, or clay in a soil to improve the soil's drainage, aeration, or moisture-holding capacity, other practices must be used.

Drainage. Drainage can be improved by changing the soil structure. One way in which this is done is by adding organic matter to encourage earthworms. Their tunnels and castings result in better soil structure through *aggregation*, the clinging together of soil particles in large crumblike shapes. Lime and gypsum (calcium sulfate) also aid in soil aggregation and improve structure in some cases. In effect, aggregation increases the size of soil particles much as if sand were added; that is, by the formation of larger spaces between particles. Another method is the use of tile drains to remove water from the soil. Raising planting beds and placing ditches between the beds are also methods which are used to control moisture in soil.

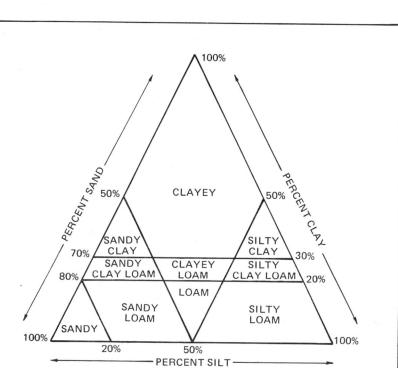

Fig. 3-4 Texture Triangle. This triangle may be used to determine the textural name of a soil by mechanical analysis (actually measuring the percent of sand, silt, and clay present in a soil). After the percentage of silt and clay are determined, these amounts are plotted on the soil triangle. This is done by projecting lines inward from the point on each side of the triangle which represents the percentage of that particular type of soil. The line drawn from the silt side of the triangle is placed parallel to the sand side of the triangle. The line projected from the clay line runs parallel to the silt line. The location of the point at which these two lines intersect indicates the name of the soil. The name of the section of the triangle in which the point is located is the name of the soil.

Moisture Retention. Often, sandy soils are not able to hold sufficient water for plants; the large particles and pore spaces encourage too much drainage. Adding organic matter improves the water-holding capacity of soils. Organic material holds many times its own weight in water. For example, peat moss holds fifteen times its weight in water. Organic matter also holds plant food effectively and allows the slow release of the food for plant use as the organic matter decomposes in the soil. Animal manure, green manure, peat moss, and sawdust are good sources of organic matter. (*Green manure* is a green herbaceous crop plowed under to improve soil.) *Mulches* such as compost or wood chips, are placed on the surface of the soil to help retain soil moisture by reducing runoff, thereby allowing more rainwater to be absorbed into the soil. Mulches also keep soil cool, thus reducing evaporation loss. Irrigation methods are used to add water when rains do not supply enough.

Fertilizers should be used when necessary and according to recommendations of a soil test. Obtain directions on how to take a soil sample from a local extension agent or horticulture instructor. Send samples to the local soil test laboratory, asking for recommendations for the crop being grown.

Beyond these practices, the grower using field or nursery areas to cultivate crops has little control over the soil structure or its moisture-holding capacity.

Diseases. Soils harbor certain diseases such as root rot and wilt. Nematodes, which are tiny animals, and insects may also damage roots. Resistant varieties of crops or natural controls such as crop rotation must be used to satisfactorily control these problems. Chemicals are sometimes used as a last resort to protect plants against diseases and insects.

Soil pasturization is necessary any time soil or sand is used in a planting media. The soil or sand should be heated thoroughly to 180°F for 30 minutes before being mixed with other materials.

Nutritional Deficiencies

Nutritional or plant food deficiencies often show up on the leaves of plants. Yellow or pale green leaves indicate a nitrogen deficiency. A phosphorus deficiency shows up as a purple color on the underside of the leaf. By the time these symptoms appear, damage has already been done to the plant. A soil test should have been used earlier to determine plant needs.

Planting Media Mixes

More and more growers are using planting mixes that contain little, if any, soil. In greenhouse operations and nurseries where plants are grown in flats, pots, or other containers, this is both convenient and economical.

Advantages

There are several advantages to using soilless media.

- The mix is uniform; that is, it does not vary in fertility, acidity (pH), or texture.
- The mixes are *sterile* (containing no disease organisms, insects, or weed seeds).
- Soilless mixes are lighter in weight, and thus easier to handle and ship.
- Good moisture retention and drainage are possible through the proper combination of ingredients.

Disadvantages

Artificial mixes have some disadvantages.

- Since they are very lightweight, containers kept outdoors may be blown over.
- Since the mineral content in most mixes is low, minor plant food elements necessary to plant life may be missing. (examples: iron, sulfur, manganese, zinc, and calcium)
- Plants transplanted from mixes to soils often hesitate to extend roots into such a different growing medium. A problem develops when the growing media remaining on the roots fails to blend with the soil into which the plant is transplanted. Moisture and other necessary nutrients are then unable to pass from the new soil to the roots. In addition, if the soil is heavy in clay, the plant's roots may never venture beyond the original root ball into the clayey soil. This situation may result in the death of the plant from lack of moisture. (The rhododendron and azalea, which are often grown in peat moss or sphagnum moss, are good examples of plants with this transplanting problem.) The problem may be solved by adding 50 percent soil to the growing mix, or by exposing the root ends to the soil when the plant is transplanted. Mixing sphagnum moss into the soil at the permanent planting site also makes the transition easier.

Content of Mixes. Soilless mixes contain various combinations of the following materials.

Perlite, a gray-white material of volcanic origin, is usually used to improve aeration of the media. Horticultural grade perlite should be used because it has larger particles and provides better drainage and aeration than the finer builder's perlite. Sponge "ROK," an aluminum silicate, is used in the same manner as perlite.

Sphagnum moss is the dehydrated remains of acid bog plants, used in shredded form for seeds. Most horticulturists use shredded sphagnum moss to cover seeds because it is relatively sterile and lightweight, controls disease well, and has an excellent moisture-holding capacity.

Peat moss is partially decomposed vegetation that has been preserved underwater. The peat is collected from marshes, bogs, or swamps. It has a very high moisture-holding capacity, contains about 1 percent nitrogen, and is low in phosphorus and potassium.

Vermiculite is a very light, expanded mineral with a neutral pH (neither acidic nor alkaline). (When vermiculite is heated, the moisture in the mineral becomes steam and causes it to expand.) It has a very high moisture-holding capacity.

Limestone refers to ground natural limestone, also known as calcium carbonate ($CaCO_3$).

Tree bark is usually the bark of pine or oak trees broken into small pieces and used in planting mixes.

Slow-release fertilizers are fertilizers containing plant food which is gradually made available to plants over a period of time.

Mixes may contain two or more of the above ingredients. For plants requiring excellent drainage, coarse materials, such as bark or perlite should make up a high percentage of the mix. Most mixes contain an organic material such as sphagnum or peat moss for moisture retention and a material such as perlite or bark for drainage and aeration. One of the most popular commercial mixes is 50 percent shredded sphagnum moss and 50 percent vermiculite, with a slow-release plant food added. Mixes are sold bagged and ready for use.

Soilless mixes may also be used as soil conditioners by digging them into the soil in varying amounts as needed. The mixes add organic matter which increases moisture-holding capacity and improves soil drainage.

If soil is used as part of the planting mix, be sure it is sterilized or pasteurized. A mixture of one-third soil, one-third peat, shaving, or leaf mold, and one-third sand is suitable for most plants.

Plant Food and Fertilizers

Plants may be grown in soil, in soilless mixes, or in a combination of the two. In any case, plant food must be readily available to the plant. Fertilizer must be added as the plant requires food and the plant food elements must be in a water soluble form to be available to the plant.

Plant food may be divided into two groups: (1) *major elements,* nitrogen, phosphorus, and potassium (stated as *total nitrogen, available phosphoric acid,* and *soluble potash* when listed as fertilizer content), and (2) minor elements, calcium, magnesium, sulfur, iron, manganese, boron, copper, and zinc. The elements are listed as major and minor according to the relative amounts of each element necessary for good plant growth. Plants require relatively large amounts of the major elements and relatively smaller amounts of the minor elements. Other minor elements may also be needed in artificial mixes.

Commercial fertilizers and plant foods show the percentage or pounds per hundred weight of the three major elements in large numbers on the bag or container. If the container has the numbers 5-10-5 on the label, the mix is 5 percent elemental nitrogen, 10 percent available phosphoric acid, and 5 percent water soluble potash. The other 80 percent is filler material which makes it easier to spread the plant food evenly. The three main plant food elements, nitrogen, phosphorus, and potassium, are always listed in that order. It is easy to see how much actual

Fig. 3-5 The fertilizer bag shown on the left is sold as a garden fertilizer. The analysis, in the top left-hand corner, indicates that the mix is 10 percent nitrogen, 10 percent available phosphoric acid, and 10 percent soluble potash, with a total of 30 percent actual plant food. The other 70 percent is filler, which aids in spreading the fertilizer. The bag on the right, a lawn fertilizer, has a high percentage of nitrogen compared with the other elements. Notice that the fertilizer on the left contains the same percentage of all three elements. Grasses require more nitrogen than phosphorus or potash; this accounts for the high percentage of nitrogen in the mix on the right. (Richard Kreh, Photographer)

plant food is being purchased on a percentage basis by reading the label, figure 3-5.

Minor elements may also be listed on the container. In parts of the country that have acidic soils, the reaction of acid content to the soil may be expressed in a statement such as "Acid equivalent to _____ pounds of limestone."

> Read the label on a plant food container. Record the information on ingredients you find.

Just as human beings require a balanced diet, plants need a balance of food for best growth. Only a soil test can determine the amount of various plant food elements needed. If shortages exist, the plant will often show symptoms of deficiencies.

Nitrogen. Nitrogen is generally purchased in one of four forms:

- *nitrate of soda* ($NaNO_3$), which is highly soluble and quickly available. It also tends to lower soil acidity.

- *ammonium nitrate* (NH_4NO_3), which is not as soluble and available over a longer period of time.

- *ammonium sulfate* ($(NH_4)_2SO_4$) which becomes available more slowly and leaves the soil more acidic. It is good for plants that grow well in very acidic soil.

- *urea formaldehyde*, which is an organic form of nitrogen and is more slowly available than the inorganic forms.

Of the three major plant food elements, nitrogen has the most noticeable effect on plants, with the effects showing soonest. Nitrogen encourages aboveground vegetative growth and gives a dark green color to leaves. It tends to produce soft, tender growth, a good quality for crops such as lettuce to possess. The tender growth makes the plant better tasting. Nitrogen also seems to regulate the use of the other major elements.

Because the addition of nitrogen quickly produces a visible effect, there is often a tendency to overuse it. Too much importance is often placed on this element without regard for a balanced plant food program.

Too much nitrogen may: (1) lower the plant's resistance to disease; (2) weaken the stem because of long soft growth; (3) lower the quality of fruits, causing them to be too soft to ship; and (4) delay maturity or hardness of tissue and thus increase winter damage to plants.

Not enough nitrogen results in a plant being: (1) yellow or light green in color and (2) stunted in root and top growth.

Nitrogen is lost from soil very easily by *leaching* (washing out). It is very soluble in water and is not held by the soil particles because of the charges of the particles involved. Soil particles have a negative charge; nitrogen also has a negative charge. Since like charges repel, nitrogen particles are not held in the soil. However, organic matter does tend to hold insoluble nitrogen which is released slowly into the soil. Nitrogen should not be used in excess for two reasons: (1) it is quickly lost from the soil through leaching, especially in sandy soils which lose water faster, and (2) it can damage plants if applied in too great an amount.

Phosphorus. Phosphorus is present to some extent in all soils. Unlike nitrogen, it is held tightly by soil particles and therefore is not easily leached from soil. However, because it may not be in the water soluble form, it is usually not available to plants in the amount needed. This means that additional phosphate fertilizers should be applied. Whether or not additional phosphorus is necessary can be determined by the use of a soil test.

Phosphorus affects plants in several ways.

- It encourages plant cell division.

- Flowers and seeds do not form without it.

- It hastens maturity, thereby offsetting the quick growth caused by nitrogen.

- It encourages root growth and the development of strong root systems.

- It makes potash (potassium) more easily available.

- It increases the plant's resistance to disease.

- It improves the quality of grain, root, and fruit crops.

Since phosphorus is held very tightly by soil particles, it does not usually cause damage to the field grown plants if excessive amounts are applied. However, container grown plants can be damaged by excesses of any soluble fertilizer element since it increases the soluble salt content present in the media. Fertilizers which are high in soluble salts *dehydrate* (dry out) plant roots by pulling water from the roots.

Insufficient phosphorus results in (1) purple coloring on the undersurface of leaves; (2) reduced flower, fruit, and seed production; (3) susceptibility to cold injury; (4) susceptibility to plant diseases; and (5) poor quality fruits and seeds.

Potassium. Potassium is rarely present in the soil in sufficient amounts to harm plants. It tends to modify both the fast, soft growth of nitrogen and the early maturity of phosphorus.

The presence of potassium is essential for several reasons.

- It increases the plant's resistance to disease.

- It encourages a strong, healthy root system.

- It is essential for starch formation.

- It is necessary for the development of chlorophyll.

- It is essential for tuber development.

- It encourages the efficient use of carbon dioxide.

Since it is a major element, potash is generally added to soil. The amount is determined by soil tests.

Plants with insufficient potassium have leaves that appear dry and scorched on the edges with irregular *chloratic* (yellow) areas on the surface.

Minor elements of plant food. If plants are grown in a good loamy soil, minor element deficiencies are generally not economically important. Sandy soils may require the application of minor elements.

When using soilless mixes, a fertilizer containing a balance of the minor elements should be added for best results. A number of these fertilizers are available for use, many of which are formulated for special crops.

> Contact a local fertilizer dealer for samples of these fertilizers. Study the labels and record the plant food elements that are listed. Note the small percentages of these elements that are required.

The Importance of Nutrient Balance. The importance of using fertilizers with the proper balance of nutrients cannot be overemphasized. To know how much of any element must be present in a fertilizer, two things must be known: how much of the element is available in the soil or media being used, and how much of each element is required for the crop being grown. Only when these two things are known can the proper fertilizer analysis and amount of fertilizer to be used be determined.

Lime (CaCO₃). Lime acts as a plant food and as a material that affects soil acidity. Soil acidity, in turn, affects the availability of other plant food elements.

Lime furnishes calcium, one of the most important of the minor food elements. Calcium is important in the formation of plant cell walls among other functions.

Soil Acidity (pH)

Most plants grow best in soil with a pH of from 5.6 to 7.0. A pH of 7.0 is neutral; that is, the soil at pH 7 is neither acid nor alkaline (basic). *Alkaline* soil is the opposite of acid soil in pH rating. Hence, on a scale of 1 to 14, values lower than 7.0 indicate acid soils and values higher than 7.0 indicate alkaline soils, figure 3-6. Figure 3-7 gives the pH preferences of common flowers, ornamentals, vegetables and small fruits.

Lime, pH, and Other Plant Food Elements. Lime serves a very important function in changing soil acidity or pH. When a soil test is made and the soil proves too acidic, lime is added to raise the pH.

Whenever it is necessary to lower soil pH for best growth of plants, materials such as sulfur, iron sulfate, aluminum sulfate, or gypsum may be used. This practice may be necessary in alkaline soil areas of the western United States.

Lime also affects the availability of other plant food elements to plants. For example, if a soil is acid with a low pH (5.5-6.5), phosphorus is tied up and not readily available. Adding lime releases phosphorus and makes it available to the plants. As another example, consider the fact that acid soils release iron and aluminum into soil. This poses a problem, since aluminum may be toxic or

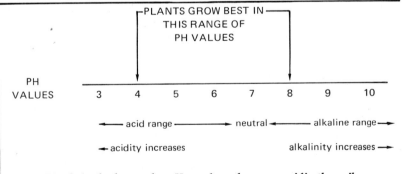

Fig. 3-6 The lower the pH number, the more acidic the soil.

Flowers and Ornamentals

African violet	5.5-7.5	Chrysanthemum	6.0-7.5	Gladiolus	6.0-7.0	Marigold	5.5-7.0
Ageratum, blue	6.0-7.5	Clematis	5.5-7.0	Gloxinia	5.5-6.5	Mock orange	6.0-8.0
Alyssum	6.0-7.5	Clivia	5.5-6.5	Heather	4.5-6.0	Morning glory	6.0-7.5
Amaryllis	5.5-6.5	Coleus	6.0-7.0	Hen and chickens	6.0-8.0	Narcissus	6.0-7.0
Anthurium	5.0-6.0	Columbine	5.5-7.0	Hibiscus, Chinese	6.0-8.0	Nasturtium	5.5-7.5
Arbor vitae	6.0-7.5	Coralbells	6.0-7.0	Holly, English	4.0-5.5	Pansy	5.5-6.5
Arbutus, trailing	4.0-5.0	Cosmos	5.0-8.0	American	5.0-6.0	Peony	6.0-7.5
Aster	6.5-7.0	Coxcomb	6.0-7.5	Hollyhock	6.0-8.0	Petunia	6.0-7.5
Azalea	4.5-6.0	Crab, flowering	5.0-6.5	Honeysuckle bush	6.5-8.0	Philodendron	5.0-6.0
Baby's breath		Crocus	6.0-8.0	Hoya	5.0-6.5	Phlox	5.0-6.0
(gypsophila)	6.0-7.5	Cyclamen	6.0-7.0	Hydrangea, blue	4.0-5.0	Pink	6.0-8.0
Baby's tears	5.0-5.5	Daffodil	6.0-6.5	pink	6.0-7.0	Poinsettia	6.0-7.5
Bachelor's button	6.0-7.5	Dahlia	6.0-7.5	white	6.0-8.0	Poppy	6.0-7.5
Barberry	6.0-7.5	Daisy, shasta	5.0-6.0	Impatiens	5.5-6.5	Portulaca	5.5-7.5
Bayberry	5.0-6.0	Daphne	6.5-7.5	Iris	5.5-7.5	Primrose	5.5-6.5
Begonia	5.5-7.0	Delphinium	6.0-7.5	Iris, Japanese	5.5-6.5	Rhododendron	4.5-6.0
Bird of paradise	6.0-6.5	Deutzia	6.0-7.5	Ivy, Boston	6.0-8.0	Rose	6.0-7.0
Bleeding heart	6.0-7.5	Dieffenbachia	5.0-6.0	English	6.0-8.0	Rose of Sharon	
Bloodleaf	5.5-6.5	Dogwood	5.0-6.0	Grape	5.0-6.5	(althea)	6.0-7.5
Bluebell	6.0-7.5	Eucalyptus	6.0-8.0	Jasmine	5.5-7.0	Rubber plant	5.0-6.0
Boxwood	6.0-7.5	Fern, asparagus	6.0-8.0	Lady's slipper	5.0-6.0	Salvia	6.0-7.5
Bridal wreath	6.0-8.0	bird's nest	5.0-5.5	Larkspur	6.0-7.5	Scilla	6.0-8.0
Burning bush	5.5-7.5	Boston	5.5-6.5	Laurel	4.5-6.0	Shrimp plant	5.5-6.5
Butterfly bush	6.0-7.5	maidenhair	6.0-8.0	Lilac	6.0-7.5	Snapdragon	6.0-7.5
Cacti	4.5-6.0	Forget-me-not	6.0-8.0	Lily, calla	6.0-7.0	Sweet pea	6.0-7.5
Cactus, Christmas	5.0-6.5	Forsythia	6.0-8.0	day	6.0-8.0	Verbena	6.0-8.0
Camellia	4.5-5.5	Foxglove	6.0-7.5	Easter	6.0-7.0	Wisteria	6.0-8.0
Campanula	5.5-6.5	Freesia	6.0-7.5	madonna	6.5-7.5	Yew, Japanese	6.0-7.0
Candytuft	5.5-7.0	Fuchsia	5.5-6.5	tiger	6.0-7.0	Yucca	6.0-8.0
Canna	6.0-8.0	Gardenia	5.0-6.0	Lily of the valley	4.5-6.0	Zinnia	5.5-7.5
Carnation	6.0-7.5	Geranium	6.0-8.0				

Fig. 3-7 pH Preferences of Plants

Vegetables

Artichoke		Cauliflower	5.5-7.5	Lettuce	6.0-7.0	Radishes	6.0-7.0
(Jerusalem)	6.5-7.5	Celery	5.8-7.0	Mushrooms	6.5-7.5	Rhubarb	5.5-7.0
Asparagus	6.0-8.0	Chicory	5.0-6.5	Mustard	6.0-7.5	Rutabaga	5.5-7.0
Beans, lima	6.0-7.0	Chives	6.0-7.0	Okra	6.0-7.5	Sage	5.5-6.5
pole	6.0-7.5	Corn	5.5-7.5	Onions	6.0-7.0	Salsify	6.0-7.5
Beets, sugar	6.5-8.0	Cucumbers	5.5-7.0	Parsley	5.0-7.0	Shallot	5.5-7.0
table	6.0-7.5	Eggplant	5.5-6.5	Parsnip	5.5-7.0	Spinach	6.0-7.5
Broccoli	6.0-7.0	Endive	5.8-7.0	Peas	6.0-7.5	Squash, crookneck	6.0-7.5
Brussels sprouts	6.0-7.5	Garlic	5.5-8.0	Peanuts	5.3-6.6	Swiss chard	6.0-7.5
Cabbage	6.0-7.5	Horseradishes	6.0-7.0	Peppers	5.5-7.0	Tomatoes	5.5-7.5
Cantaloupe		Kale	6.0-7.5	Potatoes	4.8-6.5	Turnips	5.5-6.8
(muskmelon)	6.0-7.5	Kohlrabi	6.0-7.5	Sweet potatoes	5.2-6.0	Watercress	6.0-8.0
Carrots	5.5-7.0	Leek	6.0-8.0	Pumpkins	5.5-7.5	Watermelon	5.5-6.5

Small Fruits

Blackberries	5.0-6.0
Blueberries	4.0-5.5
Grapes	5.5-7.0
Raspberries, black	5.0-6.5
red	5.5-7.0
Strawberries	5.0-6.5

Fig. 3-7 pH Preferences of Plants (continued)

poisonous to some plants. The application of lime makes the aluminum and iron insoluble. On the other hand, liming such crops as blueberries, azaleas, and rhododendron may cause iron deficiency if the pH is raised above 6.0. These crops need large amounts of soluble iron for best growth.

Liming to produce the proper pH also activates soil organisms and encourages the release of plant food. Soil structure is usually improved with the addition of calcium in the form of lime or gypsum.

THE ABOVEGROUND ENVIRONMENT

Just as the entire plant is influenced by the underground environment of the plant roots, the entire plant is also affected by the environment surrounding the top of the plant. The aboveground environment is more changeable and may be more violent in its effect on plants.

The aboveground environment may be explained in terms of the factors affecting plants. These include: (1) temperature,

(2) light, (3) humidity, (4) plant diseases, (5) insects, and (6) gases or particles in the air.

Temperature

The temperature of the air has one of the strongest effects on plant growth. Some plants, such as lettuce, cabbage, and kale, grow best in cool temperatures. Others, such as corn, beans, and tomatoes, prefer hot weather. There are temperatures above which all plant growth stops. At the other extreme, temperatures near and below freezing also stop plant growth and, in fact, kill tender crops. For the best temperatures for raising specific crops, see figure 3-8. Generally the plant growth rate increases as temperature increases up to temperatures of about $90°F$. This varies, but is a good general rule, providing that moisture is available to the plant and wilting does not occur.

TEMPERATURE				SUNLIGHT			
Cool Temperature Plants **$60° - 80°F$**		**Warm Temperature Plants** **$75° - 90°F$**		**Full Sunlight**		**Partial Shade**	
Vegetable Crops		**Vegetable Crops**		**Vegetable Crops**		**Vegetable Crops**	
Beets	Onions	Beans	Pepper	Beans	Melons	Beets	
Broccoli	Parsley	Corn	Squash	Beets	Onions	Broccoli	
Cabbage	Peas	Cucumber	Sweet	Broccoli	Peas	Cabbage	
Carrot	Potatoes	Eggplant	potatoes	Cabbage	Peppers	Carrots	
Chard	Radishes	Melons	Tomatoes	Carrots	Potatoes	Kale	
Endive	Spinach			Corn	Spinach	Lettuce	
Kale	Turnips			Cucumber	Squash	Onions	
Lettuce				Eggplant	Tomatoes	Parsley	
				Kale	Turnip	Radishes	
				Lettuce		Spinach	
						Turnips	
Flowering Plants		**Flowering Plants**		**Flowering Plants**		**Flowering Plants**	
Arbutus		Aster		Rose		Azalea	
Azalea		Marigold		Marigold		Dogwood	
Deciduous azalea		Petunia		(most annual bedding		Lady's slipper	
Primrose		(nearly all summer annual		plants)		Mountain laurel	
Rhododendron		bedding plants)				Rhododendron	

Fig. 3-8 Favorable temperature and light conditions for common vegetable crops and flowering plants

Light

Light must be present before plants can manufacture food. No green plant can exist for very long without light, whether that source is sunlight or light from an artificial source. Plants vary in the amount of light they require for best growth. Some plants prefer full sunlight; others prefer varying degrees of shade.

Light affects plants in other ways. Some plants, such as chrysanthemum, bloom only when the days begin to shorten. (Long nights are necessary for flower buds to form.) This response to different periods of day and night in terms of growth and maturity of the plant is called *photoperiodism.*

Flowering is one way in which plants react to varying periods of light and dark. Plants may be classified in three groups according to this flowering reaction. *Short day* plants, such as chrysanthemum and Christmas cactus, flower when days are short and nights are long. *Long day* plants, such as lettuce and radishes, flower when days are long and nights are short. *Indifferent* plants are plants that do not depend upon certain periods of light or darkness to flower. The African violet and tomato are indifferent plants.

There are other ways in which plants react to length of days. For example, the black raspberry roots from cane tips in five to ten days if the tips are covered with soil in September or October. In contrast, a rooting in midsummer may require up to six weeks. Dahlias develop a fibrous root system during the relatively long days of summer but as days shorten, the roots become thick storage organs.

Plants grow toward their source of light because the plant stem produces more growth hormones on the shady side, causing the stem on that side to grow to a greater length.

Humidity

Most plants are not affected drastically by a minor change in *humidity*, the moisture level of air. Most plants grow best in the 40 to 80 percent relative humidity range. *Relative humidity* is the amount of moisture in the air as compared to the percentage of moisture that the air could hold at the same temperature if it were completely saturated.

Some plants are more sensitive to humidity than others. Provided that the roots are able to replenish moisture lost through plant leaves as fast as it is lost and that the plants do not wilt, low humidity is not a great problem for most horticultural crops. However, when hot, dry winds cause plants to wilt, plant growth is slowed or stopped completely. If wilting is allowed to reach the extreme state of *incipient wilting*, death occurs. When the humidity is very high, (80 to 100 percent relative humidity), other problems may arise. For example, high humidity may cause the spread of fungus diseases.

Plant Diseases and Insects

Any time a plant is suffering from disease or insect damage, production is reduced. The amount of reduction depends on how severe the damage is and what percentage or part of the plant is infected. For example, since leaf damage reduces the area available for producing food, the more leaves that are lost, the more severely total production is reduced. Stem injury may *girdle* (circle) or clog up a stem and kill the entire plant because the supply of water and minerals to the plant top and food to the roots is completely cut off.

Some diseases and insect damage may be prevented by the use of varieties of plants that are resistant to disease and/or insects, or by crop rotation or chemical sprays.

Gases and Air Particles

Carbon dioxide is vital to plants for the production of food. There is rarely a severe enough shortage of carbon dioxide to cause damage to plants. However, greenhouse operators find that by adding carbon dioxide to the air, the growth rate of certain crops may be increased enough to more than pay for the added cost of the carbon dioxide. In field grown crops, there is no economical way to add carbon dioxide. Other growth-restricting factors such as lack of water are usually more important to outside crops and are therefore given more consideration.

Some air pollutants damage plants. Sulfur dioxide from coal furnaces and carbon monoxide from cars are known to reduce plant growth and, in severe cases, to kill plants. It is important to consider other toxic fumes in areas where concentrations are high enough to cause damage to crops.

STUDENT ACTIVITIES

1. Try this experiment to see how sand, silt, and clay particles are easily separated from soil. Take a half-gallon jar, fill it one-half full of soil, and add water to a point within 2 inches of the top. Replace the lid and shake until all the soil is *in suspension* (mixed very well). Set the jar aside and examine it the next day. The heavy sand particles should be on the bottom, silt above them, and clay on top. (Organic matter may be floating on top of the water.) Definite layers should be visible. Classify the soil based on the percent of sand, silt, and clay. Is it a sandy loam, silty loam, or clayey loam? (Use the texture triangle, figure 3-4).

2. Place a small amount of sand in one jar and clay in a second jar. Pour a cup of water into each. Record how long it takes the water to drain down through the media in each jar. For best results, place 2 inches of coarse gravel in the bottom of each jar.

3. Examine the label on a complete plant food. List all the plant food elements and record the percentage of each available. Identify major and minor elements. Notice the small amount of the minor elements available.

4. Experiment with the effect of temperature on the growth rate of plants by placing two plants of the same species in atmospheres with a wide difference in temperature. Note the results.

5. Place a plant in complete darkness and record its growth rate and appearance as compared to a similar plant kept in normal lighting conditions.

6. Take a soil sample. Determine what type of fertilizer should be applied and the rate at which it should be applied for a specific crop.

SELF-EVALUATION

Select the best answer from the choices offered to complete the statement or answer the question.

1. By weight, how much silt and clay do sandy soils contain?
   a. 20 percent
   b. 30 percent
   c. 40 percent
   d. 50 percent

2. A soil having equal parts of sand, silt, and clay is called
   a. an aggregate.
   b. a mixture.
   c. a loam.
   d. none of these

3. The three layers of a normal soil profile are
   a. sand, silt, and clay.
   b. topsoil, subsoil, and root zone.
   c. root zone, topsoil, and subsoil.
   d. topsoil, subsoil, and bedrock or lower subsoil.

4. The moisture-holding capacity of soil is sometimes improved by
   a. adding organic matter.
   b. using tile drains.
   c. irrigation.
   d. mulching.

5. Sandy soils, often called light soils, have
   a. large soil particles and large pore spaces.
   b. good drainage but little moisture-holding capacity.
   c. good aeration.
   d. all of the above

6. The four things that plants receive from soil are
   a. anchorage of roots, food storage, air, and moisture.
   b. anchorage of roots, air, moisture, and food.
   c. nitrogen, phosphorus, potassium, and minor elements.
   d. all of the above

7. Certain plant food elements are called major elements because

    a. plants use them in large amounts.
    b. they are the first elements listed on the fertilizer bag.
    c. they are plentiful in the soil.
    d. none of the above

8. The advantages of a good artificial media are

    a. it is sterile and uniform in content.
    b. it is lighter in weight and therefore easier to handle and ship than natural soil.
    c. it has good drainage and moisture-holding ability.
    d. all of the above

9. Nitrogen causes plants to

    a. produce more flowers and seeds.
    b. resist disease and develop strong roots.
    c. grow much larger than they would otherwise.
    d. grow rapidly and develop a dark green color.

10. Phosphorus causes plants to

    a. produce more flowers and seeds.
    b. resist disease and develop strong roots.
    c. grow much larger than they would otherwise.
    d. grow rapidly and develop a dark green color.

11. Potassium causes plants to

    a. produce more flowers and seeds.
    b. resist disease and develop strong roots.
    c. grow much larger than they would otherwise.
    d. grow rapidly and develop a dark green color.

12. As the outside temperature increases, plant growth normally

    a. increases if moisture is available.
    b. decreases because plants become too hot.
    c. decreases because the plant cannot receive moisture fast enough.
    d. increases because humidity always increases with the temperature.

13. Green plants cannot live without light because

    a. it is necessary for the manufacture of food.
    b. they need light to breathe.
    c. light helps to warm them to the optimum temperature for growth.
    d. none of the above

14. Lime furnishes the plant food element

    a. nitrogen.
    b. phosphorus.
    c. potash.
    d. calcium.

15. Two plants that grow best in an acidic soil (pH 5.0 to 6.0) are

    a. azaleas and blueberries.
    b. corn and onions.
    c. corn and squash.
    d. onions and tomatoes.

16. To raise soil pH and lower soil acidity,

    a. more nitrogen is added to the soil.
    b. lime is added to the soil.
    c. a complete fertilizer is added to the soil.
    d. none of the above

17. To lower soil pH and raise soil acidity,

    a. gypsum or sulfur is added to the soil.
    b. nitrogen is added to the soil.
    c. a complete fertilizer is added to the soil.
    d. lime is added to the soil.

unit 4
Growth Stimulants, Retardants, and Rooting Hormones

OBJECTIVE

To list four types of growth regulators used in the horticulture industry.

COMPETENCY TO BE DEVELOPED

After studying this unit, the student will be able to

- list one example of a substance used to stimulate plant growth.
- explain why chemical retardants are applied to floral crops and name two commonly used retardants.
- explain the use of rooting hormones on cuttings and list several rooting hormones.
- demonstrate the proper application of a rooting hormone to cuttings.
- explain how rootstock is used in dwarfing fruit trees.

Materials List

chrysanthemum and poinsettia plants

rooting hormones (Hormodin #3)

gibberellic acid

B-Nine

American and English holly cuttings (Semi-hard cuttings from different shrubs may be used.)

mist system or plastic enclosed pots or flats

knife for making cuttings

Fig. 4-1 A plant treated with gibberellic acid (right) compared with an untreated plant. Notice how much taller the treated plant has grown. The treated plant also bloomed slightly sooner than the untreated plant; this may or may not have been caused by the gibberellic acid. (USDA photo)

For years, human beings have been developing methods other than pruning to control the growth rate, size, and shape of plants. Some major achievements include the discovery of

- chemical stimulants that cause plants to grow taller.

- chemical retardants that cause plants to cease growing at a certain point.

- hormones that cause cuttings to root faster.

- dwarfing rootstock for tree fruits.

CHEMICAL STIMULANTS

Applications of certain chemicals enable plants to grow taller. The most common chemical of this type is gibberellic acid (GA). An example of the effect this chemical has on plants is shown in figure 4-1. The chrysanthemum on the left which had no chemical treatment, is growing at a normal rate. The plant on the right was treated with gibberellic acid. Notice how much taller the treated plant is.

Gibberellic acid causes the stems of plants to stretch out. The *nodes,* the joints at which buds, leaves, and branches grow out from the stem, are farther apart. Notice how much longer the stems which hold the blossom buds are on the treated plant.

CHEMICAL RETARDANTS

At times, plants may grow too tall and open to be pleasing to customers. In these cases, chemicals are used to retard the growth of the plants, causing them to be shorter and more compact. Not only are these plants more attractive, but the shorter stems are better able to support the flowers, thereby reducing the need for staking or tying. Today, chemical *retardants* (chemicals which *retard*, or slow down growth) are used commercially as a normal part of the growing process of many plants.

Figure 4-2 illustrates the effects of various retardants. All of the chrysanthemums in the photograph are of the same variety and age. Number 1, the test plant, was not treated with a growth-retarding chemical. The other four plants were treated with 8 milligrams of different test compounds. Number 2 was treated with AMO-1618; Number 3 (which shows no dwarfing effect) with CIPC; Number 4 (which shows damage) with a carbamate; and Number 5 with phosphon. Numbers 2 and 5 are saleable plants requiring no staking or support. Another chemical not used in this experiment, B-Nine, is used on a large scale for dwarfing potted plants.

As another example of the effect of a chemical retardant, examine the two holly plants in figure 4-3. The plant on the right has been treated with the growth regulator phosphon. The plant on the left was not treated. The treated plant is not only shorter in growth, but also has produced berries at an earlier age. This early berry production creates a more attractive plant which sells better.

ROOTING HORMONES

When propagating plants from cuttings, it is important that a large percentage of the plants root, and that they root as quickly as possible. Some plants root very easily from cuttings without any chemical treatment. Geranium, azalea, and many soft, succulent houseplants require only a moist, well-aerated medium

Fig. 4-2 Chrysanthemums showing the effects of various retardants. Number 1, the test plant, was not treated. (USDA photo)

Fig. 4-3 The effect of phosphon as a growth retardant is evident in the holly plant on the right. (USDA photo)

and high humidity to root. Many plants must have some assistance if any or very many of the cuttings are to root successfully.

Why is there a difference in ease of rooting? Early researchers discovered that some plants have more natural root-promoting chemicals than others. Indoleacetic acid (IAA) is a natural plant hormone that causes roots to form on plant stems. It is found in various plants in differing amounts.

Plants also root more easily or with more difficulty at different stages of maturity, which affects the hardness of the wood. Some root best when the wood is soft, others when it is almost hard.

The development of chemical rooting hormones made it possible to root certain plant cuttings that were considered impossible to root before. These chemicals also shortened the length of time required to root cuttings.

The chemical most commonly used for rooting cuttings is indolebutyric acid (IBA). Naphthaleneacetic acid (NAA) and alpha-naphthaleneacetic acid are also widely used. Indoleacetic acid, the first naturally occurring plant hormone to be used, is rarely, if ever, used today. It is generally not as effective as indolebutyric acid or naphthaleneacetic acid.

Most rooting hormones are mixed with talc and applied as a powder. The strength of the active chemical varies from 0.1 percent to 0.8 percent, although concentrations as great as 2 percent of IBA have been used on certain evergreens such as camellias and yews. Whatever their concentration, all rooting hormones should contain a fungicide (a chemical that kills fungi) such as captan to prevent rotting of the cutting. Captan also seems to help promote faster rooting.

The top row of holly cuttings in figure 4-4 was treated with indolebutyric acid; the bottom row was left untreated. After twenty-one days, the treated cuttings were rooted. The untreated cuttings had developed no roots.

Fig. 4-4 The top row of holly cuttings, treated with indolebutyric acid, developed roots in twenty-one days; the untreated cuttings did not. (USDA photo)

DWARFING ROOTSTOCKS

As the cost of labor increased, orchardists began looking for smaller fruit trees that could be picked without the use of ladders. Research was started to discover a way to prevent apple trees from growing very tall. Pruning had been used to reduce growth, but the researchers realized that it was a time-consuming, expensive, and only partially effective method. It was then discovered that trees growing from certain types of roots did not grow as large as other trees. These trees also bore fruit at an earlier age. Using this principle, a complete series of rootstock known as the *malling rootstock* was developed in England. (*Rootstock* is a root or piece of root that is used for grafting.) It became possible to control the size and rate of growth of apple trees by selecting the proper rooting stock. Some varieties rooted from the malling rootstock grow close to normal size, while others are only one-quarter to one-half the normal size.

Since the original development of malling stock, many other types have originated, including stock for apple, peach, and pear trees. The dwarf trees themselves can be purchased commercially.

> Check with your local agricultural experiment station to determine which rootstock is best for grafting or budding fruit trees in your area. Consult nursery catalogs for listings of dwarf fruit trees.

THE CHEMICAL BLOSSOM SET

One type of growth hormone is used on tomato blossoms early in the season to cause earlier development of fruit. Normally, the first blossoms on tomato plants often do not *set* (develop) fruit because of cool weather, lack of pollination, or for other reasons. The chemical *blossom set* causes seedless fruit to set on the blossoms and results in tomatoes ready to eat as much as ten days earlier than normal. Figure 4-5 illustrates a cluster of blossoms that was sprayed with the chemical and an untreated

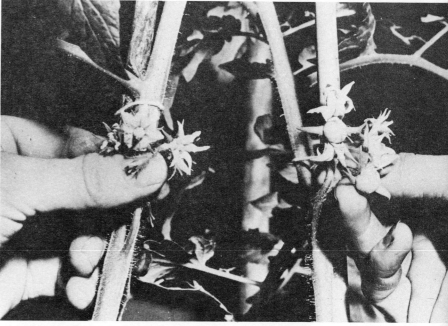

Fig. 4-5 The chemical blossom set applied to the tomato blossoms on the right stimulated early fruit set. (USDA photo)

cluster. Notice that there are small tomatoes on the treated plant and only blossoms on the untreated plant.

> Experiment with some of the commercially used plant growth regulators in figure 4-6, page 52.

STUDENT ACTIVITIES

1. Apply gibberellic acid to an actively growing plant. Keep an identical plant which has not been treated as a check on the results.

2. Apply B-Nine to five poinsettias and five mum plants. Keep five untreated plants of each type as check plants. Record the differences noted in the plants' growing habits. (Be sure to follow directions on the label exactly.)

TRADE NAME	EFFECT	USED ON
A-Rest	shortens stem length	chrysanthemum lily poinsettia
B-Nine	shortens stem length	chrysanthemum hydrangea petunia azalea poinsettia
Cycocel	shortens stem on poinsettia induces heavier flower bud set on azalea	poinsettia azalea geranium
Phosfon	shortens stem length	lily chrysanthemum

Florel	induces basal breaks on
	    roses
	retards growth on
	    poinsettia
	increases number of
	    branches on carnation
	assists in faster rooting of
	    cuttings
	speeds ripening of tomatoes
		rose
		poinsettia
		carnation
		tomato
		cuttings

Growth retardants work best when the humidity is high, but plant foliage is dry. Always follow the directions on the label exactly.

Fig. 4-6 Commercially Used Plant Growth Regulators

3. Start twenty holly cuttings. (If necessary, refer to Unit 10, Semi-hardwood Cuttings for procedural suggestions.) Keep ten untreated plants as checks. Be sure to follow the directions when using the rooting hormone. Record the results.

4. Contact a local orchardist or a state experiment station for details on dwarf fruit trees.

SELF-EVALUATION

1. Plants may be made to grow taller by applying the chemical

   a. naphthaleneacetic acid.
   b. gibberellic acid.
   c. phosphon.
   d. indoleacetic acid.

2. Plants are often treated with chemicals that dwarf or shorten them. This is done to

   a. make them more attractive and to eliminate the need for staking.
   b. make it easier to spray them.
   c. force them to bloom earlier.
   d. all of the above

3. B-Nine is a commercially used chemical that

   a. makes plants taller.
   b. helps cuttings root faster.
   c. causes plants to bloom sooner.
   d. shortens or dwarfs plants.

4. Gibberellic acid causes

   a. plants to grow taller.
   b. plant stems to stretch out.
   c. the development of a greater distance between nodes.
   d. all of the above

5. When applied to holly plants, the growth retardant phosphon

   a. injures the plant.
   b. shortens the plant and causes early berry formation.
   c. causes the plant to grow taller.
   d. none of the above

6. One rooting hormone occurring naturally in plants is

    a. indoleacetic acid.   c. naphthaleneacetic acid.
    b. indolebutyric acid.   d. talc.

7. Holly cuttings treated with indolebutyric acid should be well rooted in about

    a. ten days.   c. fourteen days.
    b. twenty-one days.   d. thirty-one days.

8. The chemical most often used commercially for rooting cuttings is

    a. indoleacetic acid.   c. naphthaleneacetic acid.
    b. indolebutyric acid.   d. talc.

9. The strength of the active chemical in the majority of rooting hormones on the market ranges from

    a. 0.1 percent to 0.8 percent.   c. 5.0 percent to 15.0 percent.
    b. 1.0 percent to 8.0 percent.   d. 0.01 percent to 0.08 percent.

10. A fungicide such as captan added to a rooting hormone

    a. helps prevent cuttings from rotting.
    b. makes the hormone last longer.
    c. causes the hormone to break down chemically.
    d. none of the above

11. Fruit trees may be dwarfed by

    a. pruning.
    b. chemical sprays.
    c. budding or grafting using a special rootstock.
    d. none of the above

Section 3
Holiday
Arrangements

unit 5
Wreaths
and Door Swags

OBJECTIVE

To construct various seasonal decorative pieces, including the evergreen wreath, Delarobia wreath, door swag, and kissing ball.

COMPETENCY TO BE DEVELOPED

After studying this unit, the student will be able to

- list ten different plant materials used in holiday decorations.

- construct an evergreen wreath, Delarobia wreath, door swag, and kissing ball.

- list five different fruit and nut pods which can be used in the construction of a Delarobia wreath.

Materials List

evergreen trees and shrubs used in wreath construction (white pine, Scotch pine, boxwood, noble fir, balsam fir, Norway spruce, blue Colorado spruce, eastern red cedar, holly, rhododendron, and yew)

seed pods of plants, pine cones, spruce cones, oak acorns, walnuts, shellbarks, pecans, filberts

box wreath wire frame measuring 12, 14, 16, or 18 inches

16-gauge and 22-gauge florist wire

straw and sphagnum moss

Holiday decorations, materials for their construction, and holiday plants play an important part in the wintertime business of nurseries, greenhouses, and garden centers. Some of the more popular holiday items are wreaths, door swags, and table center-pieces. These decorations usually consist of a base, some type of moisture-holding material, greenery, and decorative accessories which contrast with the plant material.

EVERGREEN WREATHS

To construct a foundation for an evergreen foliage wreath, a box wreath wire frame is used, figure 5-1. After an appropriate frame has been chosen, it is filled with straw or sphagnum moss. This material is dampened to help hold it in place and to help keep the evergreen foliage fresh. After the straw or moss is packed in the wire frame, the frame is wrapped with a special plastic wreath wrap material, figure 5-2. The wrap holds the straw or moss in place, keeps moisture in, and supports the greenery when it is punched into the frame.

The next step in the construction of a wreath is to select the foliage. The selection of foliage depends upon personal preference, but the final design should be considered. Boxwood has been used in the construction of the wreath shown in this unit, but any type of greenery can be used in basically the same way.

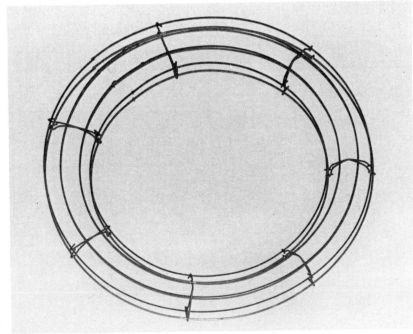

Fig. 5-1 A box wreath wire frame. Frames are available from most florist supply companies in sizes of 10 to 36 inches in diameter. (Richard Kreh, Photographer)

Fig. 5-2 The box wreath wire frame is filled with sphagnum moss and then wrapped tightly with plastic wreath wrap. (Richard Kreh, Photographer)

Fig. 5-3 Sprigs of greenery (in this case, boxwood) are pushed into the wrapped sphagnum moss base until it is completely covered. (Richard Kreh, Photographer)

Sprigs of the greenery in lengths of 3 to 4 inches are then punched individually into the wreath. This operation moves in one direction and is continued around the frame until the entire base is covered, figure 5-3. The material should flow smoothly and evenly for the best finished product. In other words, there should be no open spaces or sprigs which extend farther than the others. The completed wreath is shown in figure 5-4.

DOOR SWAGS

Door swags are a popular decoration for entrance ways to homes during the holiday season. Their construction is less complicated than that of the wreath. To make a door swag, several evergreen branches are simply grouped in an attractive arrangement that can be fastened to a door. The evergreen branches should be 24 to 36 inches long. If more than one branch is used, they may be tied together at the top with 16-gauge florist wire.

Fig. 5-4 The finished boxwood wreath complete with decoration. (Richard Kreh, Photographer)

Fig. 5-5 Door swags are constructed of one bunch of greenery (in this case, noble fir). This florist is adding a cone for interest. (Richard Kreh, Photographer)

As this is wired to the greenery, a small loop is left at the end to serve as a hanger.

Pine cones, lotus pods, or fruit may be fastened to the branches as accent items, figure 5-5. These, also, are attached with florist wire. The most dramatic highlight for a swag is a large bow attached to the base where the hanger is fastened, figure 5-6.

KISSING BALL

Mistletoe, very popular in holiday decorating, is combined with other greenery to create the kissing ball. A kissing ball is made by covering a ball of sphagnum moss with a piece of chicken wire measuring 8 inches by 8 inches. The wire is rolled into a ball,

Fig. 5-6 The addition of a bow to the door swag adds interest and covers the end of the branch that was cut from the tree. (R. Kreh, Photographer)

with the moss held in the center. The final size of the ball should be about the size of a softball. This serves as the foundation or base for the greenery. Other materials may be used, such as a styrofoam ball or a large potato (tuber).

A piece of 16-gauge florist wire is pushed through the center of the ball to serve as a hanger. The ball can then be hung at about eye level or slightly lower. This makes insertion of the greenery much easier. The greenery is cut into pieces about 4 to 5 inches long and inserted into the ball. For best results, start at the side of the ball and work around the bottom in a circular pattern. The sprigs of greenery must be even in length for the best appearance. A small piece of mistletoe is then fastened to the bottom of the ball with an attractive piece of ribbon or a bow.

DELAROBIA WREATH

A Delarobia wreath is a wreath consisting of cones, pods, nuts, and fruits. It is constructed on a box wire wreath frame or a styrofoam ring. A 10-inch ring makes an attractive candle ring display, while the 16- to 18-inch rings form interesting door decorations.

The frame is placed face down on a table. The cones to be used for the wreath are then chosen. The cones which are used for the foundation of the wreath must be equal in size. The cones are inserted into the frame by pushing them in between the wires. Each cone is then wired in place with 22-gauge florist wire, figure 5-7. The wire is attached to the cone and then woven through the frame and secured on the inside or back of the frame, concealing

the wire from view. This process is continued until the outside and inside edges of the frame are lined with cones.

The center of the frame, which has been left open to this point, can be filled with cones placed upside down. This provides a contrast to the other two rows of cones. If a center row is formed with cones, clusters of seeds and nuts can be added to accent the wreath. The entire middle row can be filled with seed pods and nuts if desired, figure 5-8.

Locating Cones, Seed Pods, Nuts, and Fruits

The best source of cones, seed pods, nuts, and fruit is a local florist supply house. However, field trips to the woods can often supply different types of seeds and nuts which are native to a particular area. For example, sweet gum balls, beechnuts, horse chestnuts, acorns, and shellbarks can be collected from trees in

Fig. 5-7 To line the inside and outside of the wreath frame, the pine cones are pushed between the wires of the frame and secured with florist wire. (R. Kreh, Photographer)

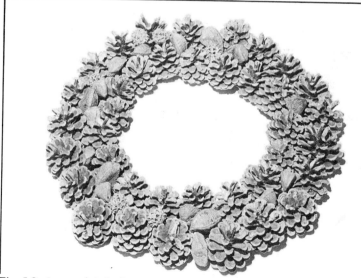

Fig. 5-8 A completed pine cone and nut wreath. This type of arrangement can be stored and used for several years. (Richard Kreh, Photographer)

some areas. Rose hips, hollyhock seeds, and bamboo fruit might be found in your own backyard. A walk in the woods or a vacant lot might produce teasel, lotus, or milkweed pods.

Fastening Seed Pods and Nuts

To fasten seed pods or nuts to a wreath, they must first be drilled. The drilling must be done carefully since nuts often have a tendency to slip out of the driller's hand. Once the nuts or seeds have been drilled, they are fastened to the wreath with 22-gauge florist wire.

The completed wreath can be sprayed with a clear lacquer for a shiny appearance. Clear lacquer is available in aerosol cans at most hardware stores. If a more natural or dull finish is desired, no lacquer is needed.

STUDENT ACTIVITIES

1. Take a field trip to local woods or fields to collect supplies for wreaths. Collect seed pods from trees and shrubs in the area. Cut pine branches of scotch and white pine. (Notice the distinct smell of the white pine.)

2. Visit a local florist supply house and examine materials available for holiday decorations.

3. Ask a local florist to speak with the class about special preparations for the holiday season and the construction of holiday decorations.

4. Visit a library and examine magazines and pamphlets containing ideas for holiday decorations.

5. Construct and sell evergreen and Delarobia wreaths as a special fund-raising project for the class or school.

SELF-EVALUATION

Provide a brief explanation for each of the following.

1. List five types of greenery which can be used in wreath construction.

2. How is the box wreath wire frame which holds the greenery prepared?

3. What lengths of greenery are used in the construction of a wreath, and how are they attached to the frame?

4. What are the two main differences between constructing an evergreen wreath and a door swag?

5. What materials are used to construct the foundation of a kissing ball? What is the finished size of the foundation?

6. What is the function of the moss or straw used in the wreath and kissing ball?

7. How long are the pieces of greenery used in the construction of the kissing ball?

8. What materials are used in a Delarobia wreath?

9. How are cones and nuts attached to the frame of a Delarobia wreath?

10. List five different cones, fruit pods, and nuts that can be used in Delarobia wreaths.

OBJECTIVE

To construct and care for holiday centerpieces.

COMPETENCY TO BE DEVELOPED

After studying this unit, the student will be able to

- construct a holiday centerpiece.
- list ten plant materials which may be used for centerpieces.
- describe the procedures for keeping a centerpiece fresh.

Materials List

pieces of evergreen trees and shrubs

half a block of oasis

one candle

CONSTRUCTION OF CENTERPIECES

In many homes, centerpieces are an important part of the holiday season. They are often found on dining room tables, dry sinks, and hutches. Candles, which have special meaning for many people, are usually part of the centerpiece design.

A variety of evergreens are suitable for use in centerpiece construction. Spruce, white pine, Scotch pine, noble fir, balsam fir, juniper, yew, holly, and boxwood are some of the greens which may be chosen.

Create a pine and holly centerpiece by following the steps on the next page.

unit 6
Creating Holiday Centerpieces

Fig. 6-1 To begin construction of the centerpiece, select half a block of oasis. Soak the oasis in water until no air bubbles form. (Richard Kreh, Photographer)

Step 1. Preparing the Base

Select a half block of *oasis*, a spongelike material, to be used as a base and to hold moisture for the greenery, figure 6-1. Soak it in a pan of water until there are no longer any air bubbles coming out of the oasis. This ensures that it is thoroughly saturated.

Step 2. Inserting the Pine

Place the block on the design table. Take small clippings of white pine 4 to 5 inches long and push them in around the base of the oasis. Be certain that a complete line of greenery is formed and that the pieces of pine are even in length.

> **Caution: Never pull the greenery out of the oasis after it is in place; this causes air pockets which will prevent the greenery from absorbing the available moisture.**

Start the next line of greenery about 1/2 inch above the first line. This allows the design to flow smoothly. Continue until the line is completed.

Add additional rows until the greenery reaches the top of the oasis block, figure 6-2. Care must be taken not to insert the last row so that it is too close to the top of the block, since oasis easily crumbles and breaks apart.

Step 3. Positioning the Candle and Holly

Place a candle in the center of the block. To be sure the candle stays in place, push a toothpick about 1/4 inch into the center of its bottom. This acts as a balance and holds the candle firmly in the oasis. Insert sprigs of variegated English holly

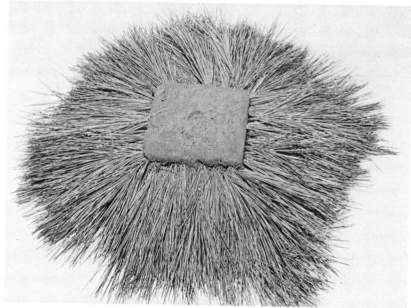

Fig. 6-2 Pieces of pine 4 to 5 inches long are inserted in the sides of the oasis block. (Richard Kreh, Photographer)

4 to 5 inches long into the top portion of the oasis. This particular type of holly has spines on the leaves and is very prickly, so this process must be done carefully. Add enough holly to cover the top of the block and make the centerpiece appear complete, figure 6-3.

Fig. 6-3 Placement of a candle and variegated English holly completes the centerpiece. (Richard Kreh, Photographer)

KEEPING THE ARRANGEMENT FRESH

To keep the centerpiece fresh during the entire holiday season, it must be watered. To water it, submerge the centerpiece in a tub of water deep enough to cover the entire arrangement. Allow it to stand in the water until there are no longer any air bubbles coming out of the oasis. It is necessary to perform this procedure several times during the holidays to keep the greenery looking fresh.

STUDENT ACTIVITIES

1. Take a field trip in a local area to collect greenery for the construction of centerpieces.

2. Design and make a centerpiece.

3. Set up a demonstration on how to water centerpieces after their construction.

SELF-EVALUATION

1. List five types of evergreens that can be used for centerpiece construction.

2. How is oasis used in the construction of centerpieces, and how is it prepared before it is used?

3. How long are the evergreen pieces that are used in the construction of centerpieces?

4. Why should evergreen pieces never be pulled out of the block after they are in place?

5. Describe the way in which pieces of greenery are inserted in the base.

6. What material is used to fill in the top of the centerpiece, and how long are the pieces that are used?

7. Explain how to water a centerpiece.

To create attractive bows for decorative arrangements.

COMPETENCY TO BE DEVELOPED

After studying this unit, the student will be able to

- select ribbon of appropriate size and color for a specific arrangement.
- construct a bow to be used on a wreath or door swag.

Materials List

22-gauge florist wire

wire cutters

floral scissors

one bolt Number 40 red velvet ribbon

---

unit 7
Creating Bows
for Arrangements

Many arrangements, in addition to those used during holiday celebrations, require decorative bows. A well-made bow is an attractive finishing touch which can be made very easily with a little practice.

The bow demonstrated here (known as a *decorator's bow*) is made of velvet ribbon which is 2 3/4 inches wide. A 2 3/4-inch ribbon is also known as Number 40 ribbon to florists. (Florists give each size of ribbon a certain number by which it is known.)

The materials needed for the sample bow are Number 40 velvet ribbon and Number 22 florist wire. From the velvet ribbon,

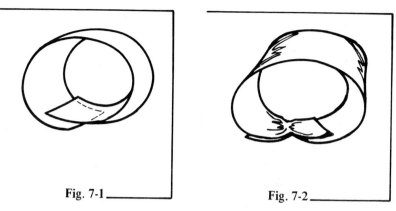

Fig. 7-1

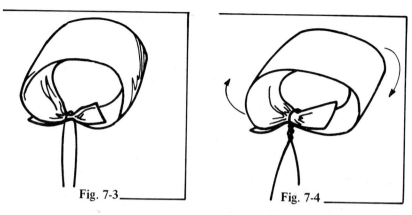

Fig. 7-2

cut one piece 4 inches long (for center of bow), one piece 12 inches, one piece 14 inches, and one piece 20 inches long (for loops). Cut three pieces of florist wire, each 18 inches long.

Step 1

Make a circle with the 4-inch piece of ribbon, allowing the ends to overlap about an inch, figure 7-1. Pinch the ribbon together at the center where the two ends overlap, figure 7-2.

Step 2

Feed a piece of wire through the circle, pulling it tight around the crushed, overlapped side of the loop, figure 7-3.

Step 3

Secure the wire by turning the ribbon two or three times while holding the ends of the wire, figure 7-4. The loop formed acts as the center of the bow and covers all exposed wires when the bow is completed.

Fig. 7-3

Fig. 7-4

Step 4

Make the bottom loop of the bow using the 14-inch piece of velvet. Overlap the ends as shown for the center loop. By pinching the middle of the loop together, form a figure eight with the ribbon, figure 7-5. Take a piece of florist wire and wire the loop in the center, holding the bow tight. Twist the wire by turning the ribbon two or three times, figure 7-6.

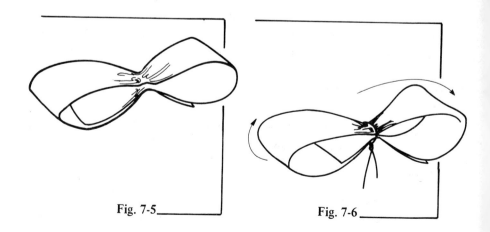

Fig. 7-5

Fig. 7-6

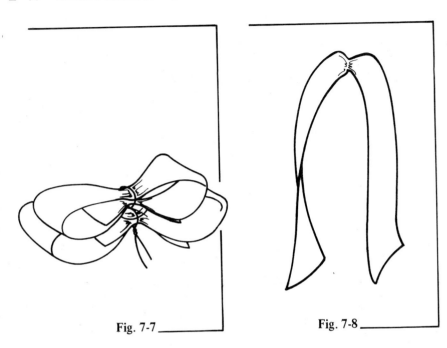

Fig. 7-7 _____   Fig. 7-8 _____

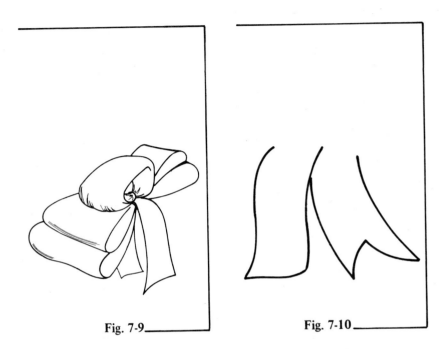

Fig. 7-9 _____   Fig. 7-10 _____

Step 5

Form the top loop of the bow with the 12-inch piece of ribbon. Overlap the ends as before, and then form a figure eight as in Step 4. Crush the loop in the center and wire it with the last piece of wire. Twist the ribbon two or three times to secure the wire.

Step 6

Take the two larger loops and place the smaller of the two on top of the larger one, figure 7-7.

Step 7

Use the 20-inch piece of ribbon to form the tails of the bow. Fold the ribbon in half to determine the middle. Crush the

ribbon together at this point so that it can be easily wired, figure 7-8. Place the tails on top of the two loops from Step 6.

Step 8

With the wire attached to the center loop formed in Step 4, wire the entire bow together. Pull the center tight to form the completed bow, figure 7-9.

Step 9

The tails of the bow should be two different lengths to create interest. The ends of the ribbon may be cut on a diagonal or in a fishtail design for an interesting variation, figure 7-10.

STUDENT ACTIVITIES

1. Visit a local florist and ask to observe their procedure for making bows.

2. Construct decorative bows for use on holiday decorations. Consider selling the bows as a fund-raising project.

SELF-EVALUATION

1. Florist ribbon is sold by number. What is the number given to ribbon that is 2 3/4 inches wide?

2. What gauge (thickness) of florist wire is used in bow construction and how long are the pieces that are required for the bow?

3. What part of the bow does the 4-inch piece of ribbon form? This part covers any exposed wire.

4. What part of the bow does the 14-inch piece of ribbon form?

5. What part of the bow does the 12-inch piece of ribbon form?

Section 4
Plant
Taxonomy

OBJECTIVE

To differentiate between scientific and common plant names and explain the binomial system of naming plants.

COMPETENCY TO BE DEVELOPED

After studying this unit, the student will be able to

- explain why scientific plant names are used.

- explain the difference between genus, species, and variety.

- list five plants by their common and scientific binomial names.

- give four examples of family names (both Latin and common).

Materials List

nursery catalogs

issues of *American Nurseryman Magazine*

Cultural Plants, by L.H. Bailey

---

unit 8
How Plants
are Named

Most plants have more than one common name; some have several. For example, the trout lily is also known as the tiger lily, adder's-tongue, dog's-tooth violet, and yellow snowdrop. The Judas tree and redbud are the same tree, but are known by these different names in different parts of the United States. Common names can be confusing, since two totally different plants may have the same common name. The cowslip is one of these: in New York State, it is a marsh-loving, buttercup-like plant; in England, it is a primrose-like plant found on dry, grassy slopes. Both have yellow flowers; but, apart from their color, they have little else in common. Since common names can be so misleading, it is important that when plants from different areas are spoken of, the same name is used — to use common names when buying or selling could be disastrous.

THE BINOMIAL SYSTEM OF NAMING PLANTS

The early scholars always wrote in Latin or Greek, so naturally, when they described plants or animals, they gave them scientific Latin or Latinized Greek names. However, this way of naming plants also caused problems; the names were often long and difficult. For example, *Nepeta floribus interrupte spicatis pedunculatis* was the name for catnip and *Dianthus floribus solitariis, squamis calycinis subovatis brevissimis, corollis, crenatis* the name for the common carnation.

The famous Swedish botanist, Linnaeus, simplified the matter by developing the binomial (two-name) system for naming plants. This system is still used today. He gave all plants just two Latin names as their scientific name. For example, he renamed the catnip *Nepeta cataria*. The first name is known as the *generic name*; this is the plant's group name. All plants having the same generic name are said to belong to the same *genus*. All plants belonging to the same genus have similar characteristics and are more closely related to each other than they are to the members of any other genus. The second name is the *specific name* or special name. All plants with the same specific name belong to the same *species*. (The Latin word *species* means "kind.") It is difficult to define exactly what a species is, but we can say that plants of the same species have the same characteristics and will consistently produce plants of the same type. Today, species are often subdivided into *varieties*. One variety of a species resembles that of another variety, but there are always one or two differences that are consistent and inherited. For example, the peach tree is

known as *Prunus persica*; the nectarine is *Prunus persica var nectaria*.

The generic name is usually a noun and the species name an adjective. Sometimes, generic names are the names of early botanists; for example *Buddleja* was named in memory of Adam Buddle. Some common generic names include *Acer* (maple), *Chrysanthemum* (mum), *Dianthus* (pink), *Hibiscus* (mallow), *Mimulus* (monkey flower), *Sedum* (stonecrop), *Papaver* (poppy), *Pinus* (pine), and *Pelargonium* (geranium).

The species name, because it is an adjective, often gives important information about the plant. Sometimes, it tells us the color of the plant. For example, *Betula lutea* is the yellow birch (*lutea* means "yellow"); *Betula alba* — the white birch (*alba* means "white"); *Quercus ruba* — the red oak; *Juglans nigra* — the black walnut. Sometimes, the species name tells us if the plant is creeping or erect. For example, *Epigaea repens* is the scientific name for trailing arbutus.

Sometimes, the specific name gives geographical information about where a plant occurs naturally. For example, *Anemone virginiana* is the Virginia anemone; *Taxus canadensis* is the Canadian yew. *Macro* and *micro* are Greek words meaning "large" and "small." Therefore, a plant with the species name *macrocantha* could be expected to have large flowers, such as *Dianthus macrocantha* — the large-flowered pink. Similarly, it would be expected that a plant with the species name microphyllus would have small leaves (*phyllus* means "leaf"); for example, *Philadelphus microphyllus* is the scientific name for the little-leaf mock orange.

When the meanings of the scientific names of plants are understood, the names are interesting and not difficult to learn. The easiest way to become familiar with them is to say them out loud; every time you transplant a seedling, dig a shrub, or sow some seeds, say the Latin name. When scientific names are used, the horticulturist is able to order plants from any part of the world; it is a universal language. The French, Dutch, the Texan, and the New Englander all use the same Latin names. Where common names are often misleading, the scientific name never is. For example, people often confuse the red maple with the Japanese red maple. They are two completely different trees. The red maple (*Acer rubrum*) grows to be over 75 feet tall; it has green leaves in the summer and red foliage in the fall. The Japanese red maple (Acer palmatum), however, has red leaves throughout the year and does not grow above 25 feet tall.

PLANT FAMILIES

Related genera (plural of *genus*) with similar flower structures are grouped together into major units known as *families*. For example, the rose family, known as *Rosaceae*, consists of several genera – *Prunus* (plum), *Fragaria* (strawberry), *Rubus* (bramble), and *Malus* (apple). All the members of each genus of the rose family have relatively simple flowers with separated petals, figure 8-1(A). On the other hand, in the *Solanaceae*, the potato family, the petals are fused or joined to form a corolla tube, figure 8-1(B). The *Solanaceae* includes the genera *Solanum* (potato), *Petunia*, and *Nicotiana* (tobacco). The *Compositae* (daisy family) is the largest of all the plant families. Members of the *Compositae* have two kinds of flowers packed together to form a single head or "flower," figure 8-1(C). The outer flowers (known as ray flowers) may have large or small petals. The inner flowers (disk flowers) always have small petals. The daisy family includes these genera: *Aster, Artemessia, Achillea,* (yarrow), *Helianthus* (sunflower), *Chrysanthemum* (mum), *Senecio* (cineraria), *Calendula* (pot marigold), and many others. Other important families are the *Cruciferae* (cabbage family),

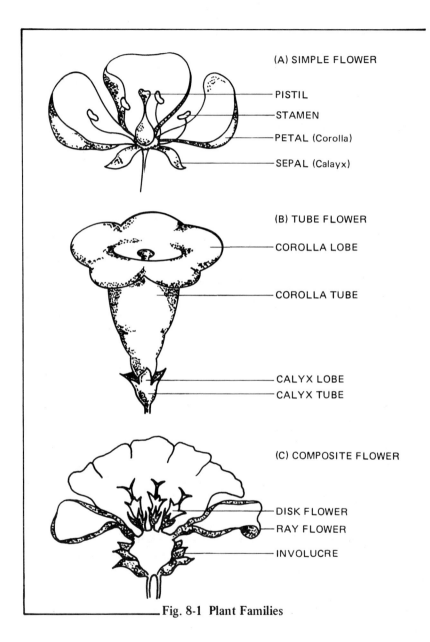

Fig. 8-1 **Plant Families**

Umbelliferae (carrot family or umbellifer family), *Papaveraceae* (poppy family), *Liliaceae* (lily family), and *Graminae* (grass family).

EXPRESSING SCIENTIFIC NAMES

Notice that when the Latin names of plants are printed, they are expressed in italics. This is because when names and phrases are written in a language other than our own, it is conventional to print them in italics or underline them if they are typewritten or hand written. Also, by convention, the generic name is always written first and the species name last. The generic name always begins with a capital letter; the species name with a small letter. Sometimes, when a number of species all belonging to the same genus is the subject, the generic name is abbreviated and the first letter is used. For example, to express several different types of oaks, the generic name for the oak (*Quercus*) is abbreviated to *Q*. The red oak may be expressed as *Q. rubra* and the pin oak as *Q. palustris*.

THE TAXONOMIST

Scientists who identify and classify plants are known as *taxonomists*. An international set of rules has been drawn up to ensure that every different species has a different binomial name and that the scientific name assigned to that plant is the oldest binomial name ever used for that plant. This international set of rules is known as the *International Code of Botanical Nomenclature*.

SELF-EVALUATION

A. Select the best answer from the choices offered to complete the statement.

1. Scientific names are used for plants to

   a. show the chemical makeup of plants.
   b. avoid confusion concerning the names of plants.
   c. increase the number of names of plants.
   d. all of the above

2. Scientific names of plants are expressed in Latin because
   a. it is a dead language.
   b. it is easy for all nationalities to pronounce.
   c. it is an international language and was used by the early scholars to express plant names.
   d. it is an easy language to learn.

3. The name of the person who developed the binomial system for naming plants is

    a. Linnaeus.   c. Socrates.
    b. Plato.   d. Hortus.

4. The generic name of a plant is

    a. placed last and begins with a small letter.
    b. placed first and begins with a small letter.
    c. placed first and begins with a large letter.
    d. placed in the middle of the name in parentheses.

5. A *genus* can be defined as

    a. a group of plants that have more in common with each other than they have with the members of any other genus.
    b. a group of plants that are all alike.
    c. a group of plants that have the same flower structure.
    d. none of the above

6. A *species* can be defined as

    a. a group of plants that are alike in almost every feature and consistently produce like plants.
    b. a group of plants that have more in common with each other than with the plants of any other group.
    c. plants that are all the same size.
    d. plants that all have the same flower color.

7. Latin names of plants are underlined

    a. to make them appear important.
    b. because it is conventional to underline words and phrases that are expressed in a different language.
    c. because it helps to remember them.
    d. because all plant names are always underlined.

8. A person who identifies and classifies plants is known as a

   a. taxidermist.   c. pneumonot.
   b. taxonomist.   d. horticulturist.

B. Match the correct genus with its common name.

   a. geranium   (1) *Nicotania*
   b. tobacco   (2) *Papaver*
   c. oak   (3) *Dianthus*
   d. yew   (4) *Pelargonium*
   e. poppy   (5) *Taxus*
   f. pink   (6) *Quercus*

Section 5
Plant
Propagation

There are many ways of *propagating*, or reproducing, plants. The most common method of reproducing flowering as well as vegetable and cereal crops is through the use of seeds. This is a *sexual* process and requires the union of *pollen* (the male sex cell) with the *egg* (the female sex cell) in the ovary. Male and female cells may be from the same parent (*self-pollination*) or from separate parents (*cross-pollination*).

Seeds are a means of rapidly increasing the number of a certain plant. However, not all plants *come true to seed* (reproduce exact duplicates of the parent plant from seeds). Wheat and barley are examples of plants which do come true from seed. Others, such as rhododendrons, azaleas, and apples, do not come true from seed, causing the offspring to differ from either parent.

Plants that are not produced directly from seeds or do not produce seeds that will grow, such as hybrids, must be propagated by another method to obtain exact duplicates. *Hybrids* are the offspring of two different varieties of one plant, each of which possesses certain characteristics which are desired in the new plants. A female plant of one variety and a male plant of another variety are *crossed* (bred) to produce offspring with the best characteristics of each parent. However, when hybrids themselves are reproduced from seed, their offspring do not have the same characteristics as the parents; rather, they have a variety of combinations of traits possessed by the plants originally used to produce the hybrids. To produce exact duplicates of these plants, asexual reproduction is used. This is not a sexual process, and no seeds are used in this method. Instead, the plant is propagated from one of its parts, such as the leaf, stem, or root.

Propagation may be accomplished by division of roots; by cuttings of leaves, stems, or roots; or by budding and grafting. The strawberry reproduces itself by runners; the lily by tiny bulblets. The reproduction method of the strawberry and lily occur naturally. Cuttings, budding, and grafting, however, have been developed or improved upon by man.

Asexual reproduction is possible because each single cell of a plant contains all the characteristics of the entire plant and can regrow any missing part. Thus, a stem cutting removed from the roots of the plant develops new roots from cells along and at the base of the stem. Root cuttings develop new stem tissue in the same manner.

Plant propagation predates recorded history. Very early human beings planted seeds or divided plants to increase plant numbers and to carry them over the *dormant*, or resting, season. The quality of plants was improved by using seed from the best plants to produce other plants, or by separating naturally propagated plants of superior quality from the parent and planting them.

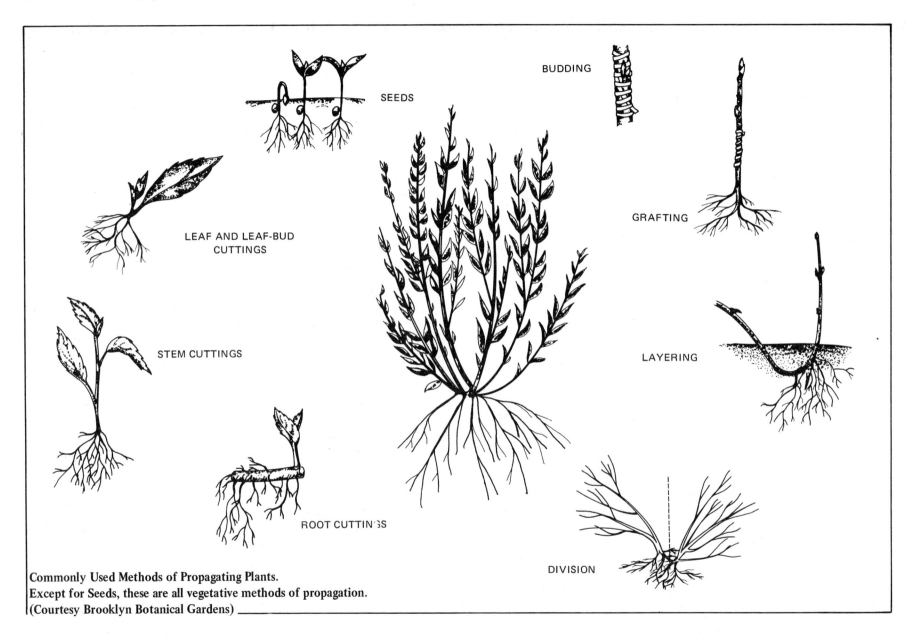

SEEDS

BUDDING

LEAF AND LEAF-BUD
CUTTINGS

GRAFTING

STEM CUTTINGS

LAYERING

ROOT CUTTINGS

DIVISION

Commonly Used Methods of Propagating Plants.
Except for Seeds, these are all vegetative methods of propagation.
(Courtesy Brooklyn Botanical Gardens)

OBJECTIVE

To propagate at least one plant using seeds.

COMPETENCY TO BE DEVELOPED

After studying this unit, the student will be able to

- identify the parts of a seed and the functions of each.
- differentiate between indirect and direct seeding methods.
- prepare a media for seeds, sow seeds, and provide the proper conditions for germination.
- water, fertilize, and harden off seedlings before transplanting.
- transplant seedlings into flats or pots.

Materials List

planting container such as a pot or flat

planting media (sterilized)

seeds of plants commonly grown in the area

labels and marking pen or pencil

leveling board or skew

watering facilities for bottom watering

transplanting containers

watering can

greenhouse or other growing space

---

unit 9
Seeds

PROPAGATION OF PLANTS FROM SEEDS

In the horticultural industry, many plants are started from seed because it is a quick and economical method. Sexual propagation of plants by use of seeds is a process used in the reproduction of many different plants. For successful germination, the proper environmental and cultural conditions must be provided. These conditions include temperature, moisture, light, and media.

Plant	Germination Time Indoors (Days)
Achillea	5 to 7
Alyssum saxatile (basket-of-gold)	5
Anthemis (camomile)	5
Hardy aster	7 to 14
Campanula	10 to 24
Cerastium tomentosum (snow-in-summer)	10
Chrysanthemum	5 to 8
Dianthus (hardy pinks)	4 to 6
Doronicum (leopard's bane)	8
Eupatorium (perennial ageratum)	7
Gaillardia	5
Gloriosa daisy	7
Gypsophila (baby's breath)	7
Heliopsis	4
Hollyhock	7
Iberis (perennial candytuft)	6
Lupine	3 to 4
Myosotis (forget-me-not)	7 to 12
Papaver (Oriental poppy)	6 to 10
Phlox paniculata	21 to 28
Primula (primrose)	21 to 30
Salvia	5 to 21
Shasta daisy	5 to 7
Verbena	10

Fig. 9-1 Perennials easily grown from seed

Seeds of annual plants (plants which complete their life cycle in one year) from named varieties which are not hybrids generally come reasonably true from seed. *Named varieties* are specific individual strains of one type of plant which have been named to indicate their particular traits. The tomato plant has numerous named varieties such as Better Boy and Marglobe, each of which is different in size, days needed to mature, growth habits, and disease resistance. Seed should not be saved from hybrids and planted, since the resulting plants would not come true from seeds. Many perennials are grown from seed, figure 9-1.

Composition of Seeds

The basic parts of a seed are the seed coat, the endosperm (stored plant food), and the embryo, figure 9-2.

Seed Coat. The *seed coat* is the outside covering of the seed which protects the embryonic plant. The seed coat makes it possible for seeds to be transported and stored for long periods of time.

Endosperm (Stored Plant Food). The *endosperm* is the food storage tissue which nourishes the embryonic plant during germination (the first start of growth in a seed).

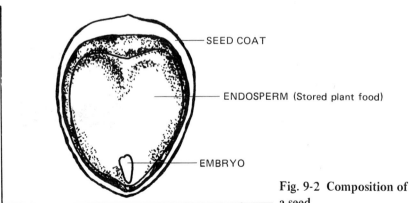

Fig. 9-2 Composition of a seed

Embryo (or Embryonic Plant). The *embryo* is a new plant that is developed as a result of fertilization. During germination, it extends its roots and seed leaves to form a new plant, figure 9-3.

Identification and Selection of Seed

There are several important steps in the selection of seed.

- Identify which seeds are grown locally.

- Select seeds that have been tested for the current year's growing season for germination ability and purity. These tests must comply with state regulations.

- Purchase the seeds from a reliable dealer to assure that the variety is true (pure) and that germination ability is acceptable.

- Choose hybrid varieties for greater vigor, uniformity, and flowering ability.

Germination Media

The best media for germination has a favorable pH level, and an adequate supply of plant nutrients. It is firm, porous, uniform in texture, sterile, and free of weeds, insects, and disease organisms.

A good germinating media contains one or more of the following ingredients.

Soil. The soil should be composed of 45 percent mineral matter, 5 percent organic matter, 25 percent air, and 25 percent water.

Construction Grade Sand. This is the best type of sand to use because it is more porous, thereby allowing for better aeration and drainage. Sand particles have a negative charge; therefore, they do not hold plant nutrients in the media as strongly as particles of soil or peat moss.

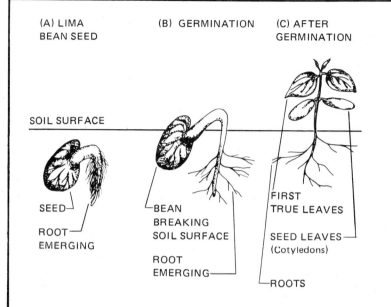

Fig. 9-3 Lima bean germination

Peat Moss. Peat is partially decomposed vegetation which has been preserved under water. The peat is collected from marshes, bogs, or swamps. Peat has a very high capacity for holding water. It contains about 1 percent nitrogen and is low in phosphorous and potassium.

Sphagnum Moss. Sphagnum moss is the dehydrated remains of acid bog plants. It is used in shredded form in seed germination. Sphagnum moss is relatively sterile and lightweight, controls disease well, and has an excellent water-holding capacity.

Horticultural Grade Perlite. Perlite is a gray white material of volcanic origin. It is most commonly used to improve aeration of media. Horticultural grade perlite consists of large particles, thereby providing good drainage and aeration.

Vermiculite. Vermiculite is a very lightweight mineral which expands when it is heated. It is neutral (has a pH of 7) and has a very high water-holding capacity.

Jiffy Mix. Jiffy Mix is composed of equal parts of shredded sphagnum moss, peat, (fine grade Terr-lite vermiculite), and enough nutrients to sustain initial plant growth.

Although germination media is usually made up of one or more of the materials in the list above, good grade, sterilized top-soil provided with the proper drainage is sufficient. Sand or perlite should be added if the soil needs greater drainage and aeration. Peat, sphagnum moss, or vermiculite may be added to improve the moisture-holding capacity of the soil. For general purposes, a good mix consists of one-third soil, one-third sand or perlite, and one-third peat moss, sphagnum moss, or vermiculite. If the soil is heavy (with a high clay content), more sand or perlite may be needed; if it is sandy, less sand and perlite are required.

Soil used in any seed germination media should be sterilized by heating it at 180°F for one-half hour. Sterilized soil may also be purchased at garden stores. It may then be mixed with other materials to form the desired media.

INDIRECT SEEDING

Indirect seeding is a process in which seed is sown in a separate place from where the plants will eventually grow to maturity. The seedlings are transplanted one or more times before reaching the permanent growing area.

Flats

When growing seeds, horticulturists often select a *flat* (a wooden box 14 1/2 inches x 23 inches x 2 3/4 inches or 18 x 18 x 2 3/4 inches with slats in the bottom for drainage), figure 9-4.

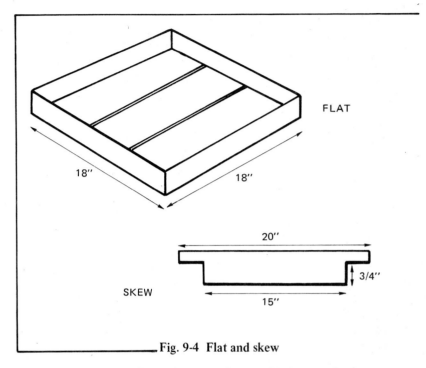

Fig. 9-4 Flat and skew

The bottom of the flat is lined with one thickness of newspaper to prevent the media from falling through the slats. The media selected is placed in the flat and leveled off to about 1/2 to 3/4 inch below the top of the flat. This is done with a tool called a *skew.*

After the flat is filled, rows are made in which to sow the seeds. A row marker may be used, or rows may be made one at a time using a straight board as a guide. It is best to sow seeds in rows because if disease strikes one row of seed, it can be removed without disturbing the others. Also, when several different varieties are in the flat, it is easier to label them by rows and much easier to transplant *seedlings* (young plants that have been germi-nated several days).

When to Seed

As business persons who must operate at a profit to survive, commercial plant growers must be sure that their plants are ready for sale at the correct time for outdoor planting and for holidays. Seeds must be planted on certain dates so that the resulting seedlings are ready for transplanting at the proper time. The chart in figure 9-5 gives planting information for seeds of various plants.

Sowing Seeds

To sow seeds properly, use the following procedure. Take the packet of seeds and shake the seeds to the bottom of the packet. Hold the seed packet with the open end slightly lower than the rest of the packet and gently tap the packet. The seeds will move out slowly and gradually making it easy to sow the seeds properly spaced in rows. Follow

Plant		Date to Sow Seed	Date to Sell (Retail)
Ageratum		March 1 March 15	May 15 early June
Alyssum*		March 15 April 1 April 15	early May late May mid-June
Coleus		March 1 March 15	May 15 early June
Impatiens		March 1	early June
Marigolds*	dwarf tall	to April 1 to April 1	8 weeks 6 to 7 weeks
Petunias		March 1	late May
Salvia		March 1 March 15	May 15 early June
Verbena		February 15 March 1	May 10 late May

Plant	Date to Sow Seed	Date to Sell (Retail)
Zinnia	successive sowings from March 15 to mid-April	6 weeks
Tomatoes	April 1	6 to 7 weeks
Peppers	March 15	8 weeks
Eggplant	March 15	8 weeks
Cucumbers	April	3 to 4 weeks
Melons	April	3 to 4 weeks
Parsley*	March 15	late May (8 to 9 weeks)
Onions*	to March 1	6 to 8 weeks**
Broccoli	April	6 to 8 weeks
Cabbage	April	6 to 8 weeks

*Consider direct seeding. (Soak parsley in water overnight, drain, and dry slightly before sowing.)

**6 weeks with one clipping to a height of 3 to 4 inches

Fig. 9-5 Date to seed for sale of plants on specific dates. (Chart based on experience in West Chicago, Illinois. Courtesy George G. Ball, Inc.)

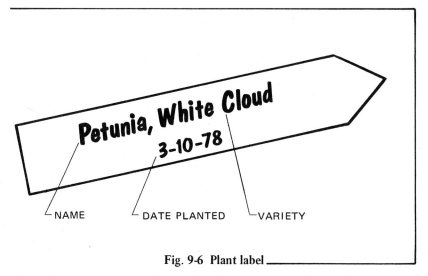

Fig. 9-6 Plant label

directions on the package for determining the distance apart the seeds should be sown.

After the seeds are sown, cover them with shredded sphagnum moss, fine perlite, or fine sand. Cover with a layer of media measuring about twice the thickness of the seed. (The seed package may have directions for depth of planting; if so, follow them for best results.) Very fine seeds, such as the petunia, are not covered when sown.

As soon as the seeds are sown, they are labeled with the name, variety, and date sown, figure 9-6. All labels should be printed clearly with pencil or waterproof marking pen. Ballpoint pen should not be used since it washes out when the seedlings are watered.

Watering the Seed Flat

Water aids in germination by making the seed coat soft so that the embryonic plant can germinate. Water itself is an impor-

tant nutrient and also acts to dissolve other nutrients present in the media, making them available to the growing plants.

To water freshly seeded flats, set the flats in a tub of water. This allows the seeds to be watered by *capillary action* (a process in which the water is drawn up into the spaces between the soil particles to thoroughly moisten the media). After the media is moistened, the flat is removed from the tub. If the flat is watered from the top, care must be taken to avoid washing the seeds out of the flat. Very small seeds should never be watered from the top because they may be washed too deeply into the soil.

Conditions for Germination

Seed flats containing seeds for germination should be located in a semishaded area of the greenhouse and receive a bottom heat of 65° to 70°F (18° to 21°C). (Some cool-season crops germinate well without bottom heat.) This may be accomplished in several ways. Location of the containers above the heat coil or hot water pipes or the use of a special propagating mat which is controlled with a thermostat are two possibilities. After the seeds are sown, the containers are covered with a pane of glass or clear plastic film to retain humidity. The covering should not touch the media while the seedlings are germinating. Some seedlings, such as verbena, dusty miller, pansy, and portulaca, require a three-day period of darkness for germination. For these plants, newspaper is used instead of glass or plastic. The paper is removed as soon as the seeds germinate so that the new seedlings receive light immediately. After germination, the media is kept moist but never wet. The flats are watered with a gentle mist or from the bottom so that the small seedlings are not washed out.

The proper seedling media is low in fertilizer elements. Seedlings should be fed weekly with a water-soluble fertilizer. As the seedlings approach transplanting size, a cooler temperature (55° to

Fig. 9-7 A flat of Super Seedlings grown by a specialist and sold to a plant grower. (Courtesy George J. Ball, Inc.)

60°F) (13° to 16°C) is provided to prepare the seedlings for the shock of transplanting. This is known as the *hardening-off* process. The process may also include a modest withholding of water to slow active growth.

Some commercial companies sell hardy seedlings in styrofoam flats. These seedlings are shipped ready for transplanting, figure 9-7.

Seed may be sown in pots, pans, or trays in a manner similar to that used for flats. The horticulturist should remember to provide drainage and uniform moisture. The containers should be watered from the bottom if possible. Plastic, newspaper, or glass is placed over the container to retain uniform moisture.

Transplanting Seedlings

After seeds germinate, they develop seed leaves or *cotyledons*, the first leaves to appear on the plant, figure 9-8. The plant should be allowed to grow until the first true leaves are present before it is transplanted.

When handling seedlings, hold them by their true leaves using the thumb and forefinger. Do not hold by the stem; if the stem is badly bruised, the seedling could die. A bruised leaf is not nearly

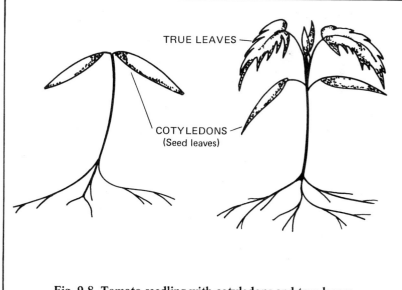

Fig. 9-8 Tomato seedling with cotyledons and true leaves

as serious an injury. While carefully holding the seedling, use a pot label to reach under the roots and lift, pushing the seedling out of the germinating media. This exposes the roots. Do not shake the media off the roots — exposure to air causes them to dry out. Plant the seedlings one by one about 2 inches apart in a flat. Using a *dibble* (a tool used to make the hole for transplanting seedlings), insert the seedlings to a depth slightly deeper than that at which they grew in the seedling flat. Gently press the media around the roots. Water the seedlings at the soil surface with a gentle stream of water to settle the soil around the roots. The new seedlings are now ready to be grown to saleable size. (An alternate method involves using a trowel or fork when digging seedlings for transplanting.)

It is sometimes convenient to transplant seedlings into peat pots or market packs rather than flats. When a peat pot is used, one plant is placed in a pot and later transplanted directly to the garden with the plant remaining in the pot. Market packs generally hold six to twelve plants and are sold to the customer as such. The plants are then separated from the pack and planted individually in a permanent location.

Another indirect seeding technique is to plant a single seed in a Jiffy 7 peat moss pellet. These pellets, when soaked in water, expand to seven times their original size. They contain all the necessary nutrients to feed the small seedling until it is planted in a permanent site. This eliminates the seedling transplanting stage prior to permanent planting and allows the young seedlings to be transplanted to the garden with less transplant shock. The Jiffy 7 pellets are placed in plastic trays or flats for ease of handling. Manufacturer's directions should be followed when using these pellets.

Single seeds of squash, cucumber, and melon may also be planted in 2-inch peat pots in preparation for direct planting to the garden at a later time. These crops transplant with difficulty and the peat pot permits transplanting without disturbing the root system.

TRANSPLANTING TECHNIQUE
(for seedlings or rooted cuttings)

Materials List

seedlings or cuttings for transplanting

pots in which to grow plants

broken crock, gravel, or coarse sphagnum moss to cover drain hole

potting media

labels

watering can or hose

Procedure

1. Work from a large table. Place the pots in a row on the worker's right side.

2. Place all other material to the worker's left side.

3. Cover drain hole with pottery or coarse sphagnum. (If using peat pots, this is not necessary.)

4. Hold the plant to be transplanted in the pot with the left hand.

5. Fill the pot to the rim with soil or soilless mix, gently holding the plant at the proper planting depth with the left hand.

6. Gently firm the planting media around the plant by pushing down along the edge of the pot with the thumbs. Do not push hard enough to tear or bruise plant roots.

7. Label the pots or growing area according to variety, name, and date.

8. Move plants to growing area and water until water runs through the pot and out the drain hole.

Fig. 9-9(A) Transplanting to Pots

TRANSPLANTING TECHNIQUE

Procedure

1. Work from a large table. Spread fingers of the left hand around the plant stem and over the top of the soil surface.
2. Invert the pot and gently tap the pot edge on the edge of the table.
3. Remove the pot from the root ball as it loosens.
4. If roots are pot bound, gently unwind them.

——————— **Fig. 9-9(B) Removing Plants from Pots** ———————

Seedlings of nursery stock plants such as trees and shrubs are generally planted in a flat and transplanted directly to the nursery row outdoors. Seeds of these plants may also be seeded directly in the nursery row.

DIRECT SEEDING

Many seeds are planted directly in the permanent growing area. This is referred to as *direct seeding*. It is the most economical method of seeding. Plants such as corn, melons, beans, beets, peas, lettuce, carrots, and most other vegetable crops are grown by this process. Some vegetables, such as tomatoes, peppers, cabbage, and eggplant, are generally transplanted as plants to the garden. Some trees and shrubs are also grown by direct seeding.

In direct seeding, the planting media is the soil. The soil is prepared by removing all large clods or lumps of earth so that the seeds are uniformly covered. In some cases, manure, grass clippings, or compost may be added to the soil to improve its structure. In direct seeding, it is important to plant the right variety at the right time in the right soil. Weather conditions largely determine germination and initial growth.

The same conditions are necessary for germination in direct seeding as when seeding in a flat. There must be sufficient moisture and aeration; the seedbed must allow firm contact with the seed; and the temperature must be high enough to support the germination process. Requirements for germination of specific seeds are given in the directions on the label of the seed packet.

SPECIAL TREATMENT OF SEEDS FOR GERMINATION

Some seeds have a hard seed coat which must be soaked or scratched before the seeds are able to germinate. Examples of this type of seed are the red bud, Judas tree, and sweet clover. Sandpaper, scratching, or an acid bath may be used to weaken the seed coat.

Other seeds require a moist, cold rest period (dormant stage) at temperatures below 37°F (3°C) for eight weeks or longer. Examples of seeds which require a rest period are the apple, peach, pear, tree peony, maple, and yew. These may be buried in moist sand and kept cool.

Still other seeds must go through alternate wetting and drying. Some must have light to germinate; others must have darkness. These specific requirements must be met when these seeds are planted. Always follow instructions on seed packets or enclosed planting directions.

STUDENT ACTIVITIES

1. Plant seeds in a pot or flat. Care for the seeds through the transplanting stage. Keep a record of the date the seeds were planted, when they germinate, and when they are transplanted.

2. Collect various seed packets and compile a list of information given on the packets. Discuss the importance of such information.

3. Soak a bean seed in water for a few days. Compare the soaked seed with a dry seed. What difference do you notice in the seed coat?

4. Separate a soaked bean seed and see if you can identify the parts of the seed. Sketch and label the parts.

SELF-EVALUATION

Select the best answer from the choices offered to complete the statement or answer the question.

1. Producing plants from seed is a type of _____ propagation.

   a. asexual
   b. bisexual
   c. sexual
   d. unsexual

2. Seeds are composed of the

   a. seed coat, endosperm, and embryonic plant.
   b. seed coat, root, and stem.
   c. eye, starch, and seed coat.
   d. root system, starch coat, and seed coat.

3. When two separate parent plants are involved in the pollination process, it is known as

   a. self-pollination.
   b. cross-pollination.
   c. bisexual pollination.
   d. asexual pollination.

4. Many horticultural crops are started from seed because

   a. there are very few people who know how to propagate them any other way.
   b. it is not possible to propagate them in any other way.
   c. it is quick and economical.
   d. both b. and c.

5. A good growing media in which to plant seeds must

   a. drain well.   c. be sterile.
   b. hold moisture.   d. all of these

6. The first part of the new plant to emerge from the seed is the

   a. stem.   c. root.
   b. leaves.   d. endosperm.

7. Hybrid seeds are developed because they

   a. provide greater vigor.   c. are more uniform.
   b. produce better crops.   d. all of these

8. The best temperature for germinating most seeds is

   a. 50° to 60°F.   c. 80° to 90°F.
   b. 75° to 80°F.   d. 65° to 70°F.

9. Pansy and portulaca seeds require three days of darkness for germination to occur. This may be accomplished by

   a. covering the flat or pot with newspaper.
   b. burying the flat or pot in the ground.
   c. planting the seeds at an unusually great depth.
   d. none of the above

10. Which two of the following materials are mixed in the seeding media to give good drainage and aeration?

   a. peat moss and vermiculite   c. sand and perlite
   b. sand and sphagnum moss   d. perlite and peat moss

11. Soil sterilization requires heating the soil at

   a. 120°F for two hours.   c. 160°F for two hours.
   b. 180°F for half an hour.   d. none of these

12. The date to start seeds is very important because

   a. there must be greenhouse space made available for them.
   b. seed houses have seeds available only at certain times.
   c. the seeding media must be ready.
   d. the plants must often be ready for sale at a certain time.

13. The label in a flat of seeds should include the

   a. name or type of plant, variety, and date seeded.
   b. date seeded and the selling date.
   c. percent of germination listed on seed pack.
   d. all of the above

14. Which of the following are perennials listed in the text as easily grown from seed?

   a. hardy aster, hollyhock, crab apple
   b. verbena, lupine, dianthus
   c. rhododendron, marigold, petunia
   d. petunia, tomato, sweet corn

15. The growth of seedlings is slowed down by withholding water and lowering the ground temperature. This process, called hardening off, is done to

   a. keep the seedlings from growing too quickly.
   b. prepare the seedlings for transplanting shock.
   c. hold the seedlings until they can be sold.
   d. none of the above

16. Seedlings are held by the first true leaves instead of the stem when transplanting because

   a. there are no cotyledons to grasp.
   b. the stems are too slippery and seedlings are dropped and lost.
   c. the stem may be bruised in handling and cause death of the plant.
   d. this bruises the leaves, which causes better growth.

17. The germinating media which sticks to the seedling roots should not be knocked off because

    a. shaking the plant may bruise it.
    b. the roots will dry out more rapidly.
    c. it assures that the seedling will be planted at the right depth.
    d. it could spread disease to other seedlings.

18. Direct seeded plants are planted

    a. in the soil where they grow to saleable size.
    b. directly in flats in the greenhouse.
    c. in greenhouses only during winter months.
    d. all of the above

19. Which of the following would make a good seed germination media?

    a. one-third sterilized soil, one-third sand, and one-third perlite
    b. one-third sterilized soil, one-third peat moss, and one-third sphagnum moss
    c. one-third sterilized soil, one-third perlite, and one-third peat moss
    d. one-third sterilized sand, one-third perlite, and one-third vermiculite

OBJECTIVE

To collect softwood and semi-hardwood cutting wood, prepare cuttings, and place them in the propagating structure for rooting.

COMPETENCY TO BE DEVELOPED

After studying this unit, the student will be able to

- select plants suitable for propagating by use of cuttings and determine if the maturity of wood is correct for optimum rooting.
- propagate at least ten softwood or semi-hardwood plants from cuttings following procedures outlined in the unit, including
  - collecting wood from the parent plant,
  - making the actual cutting,
  - treating the cutting with the proper rooting hormone when necessary,
  - preparing the rooting media and placing the cutting, and
  - caring for the cutting after it is placed in the media.

A minimum of nine of the cuttings should root for development of competency.

unit 10
Softwood and Semi-hardwood Cuttings

Materials List

parent plant material from which to make cuttings
plastic bags for collection
labels and label marking pen
propagating knife
propagating container (flat, pot, etc.)
rooting media (sphagnum moss, perlite, or sand)
rooting hormone
watering can
polyethylene plastic cover or sprinkling system

The most commonly used and commercially important method of asexual plant reproduction is propagation by the use of cuttings. *Cuttings* are leaves or pieces of stems or roots used for propagating plants. Most of the economically important perennials and many annuals are reproduced in this way. (*Perennials*, such as chrysanthemums, are plants that grow year after year without replanting. *Annuals*, such as petunias, normally complete their growth cycle in one year.)

There are various kinds of cuttings which are taken from different parts of the plant and at different stages of plant maturity (age). Cuttings may be taken from the leaf, bud, stem, or root, figure 10-1. They may be taken when the plant tissue is soft and tender (*softwood cuttings*), or when the plant tissue is hardened and more woody (*hardwood cuttings*). Some plants root more easily from softwood, others more easily from hardwood.

To root, cuttings require essentially the same conditions that seeds need to germinate: moisture, oxygen, and warmth (40°F [4°C] or above). In addition, cuttings need light — something that seeds do not need to germinate. Light is essential in allowing the cuttings to produce food through the process of photosynthesis. Cuttings may begin to grow roots without light, especially the first stages of *callus* formation (growth of tissue over a wounded plant stem), but generally, light is necessary for food production. Food production supplies energy to the rapidly developing roots. Stored hardwood cuttings are an exception since they are able to develop roots without light.

Root formation on cuttings is stimulated because of the interruption of the downward movement of carbohydrates, hormones, and other materials from the leaves and growing tips. Since these materials cannot travel beyond the point at which the stem has been cut, they accumulate and stimulate the healing and rooting process.

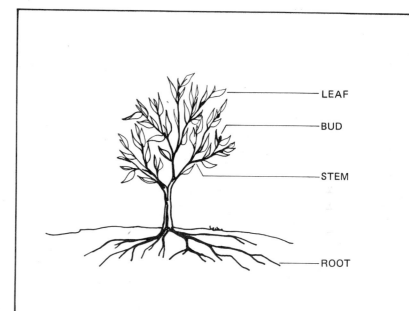

LEAF

BUD

STEM

ROOT

Fig. 10-1 Cuttings may be taken from the bud, stem, leaf, or root.

Rooting hormones are sometimes used to aid in root formation. Some plants root more easily than others because they produce a higher level of natural rooting hormones. These plants need less synthetic or man-made rooting hormones to root satisfactorily.

SOFTWOOD AND SEMI-HARDWOOD STEM CUTTINGS

One method of asexual reproduction by cuttings is softwood and semi-hardwood cuttings of the stem. These cuttings are taken after the current or present season's growth has at least partially matured or hardened. The wood for semi-hardwood cuttings should be able to bend, but generally snaps if bent at an angle greater than 70 to 80 degrees. The softwood cutting is taken while there is still active *terminal* (tip) growth. This type of wood is very soft and breaks with a snap. The horticulturist best learns about maturity of wood through experience; it is not necessary to be exact with most plants.

Nine Steps in Taking Softwood and Semi-Hardwood Cuttings

1. Collect the cutting wood of proper maturity from a selected parent plant.

2. Label the cutting wood.

3. Fill the propagating container with rooting media.

4. Make the cutting.

5. Treat the cutting with rooting hormone.

6. Insert the cutting in the rooting media.

7. Water to settle the media around the cutting.

8. Insert the label in the rooting container.

9. Control the atmosphere around the cuttings by covering with plastic or placing under a mist system.

Collecting Cutting Wood

The first step in making the cutting is to find a parent plant that has made at least 2 to 6 inches of new growth during the current year. This current season growth may be from the tip of a branch, or a shoot that is growing from the base of the plant, figure 10-2. Check to be sure the wood is of proper maturity or hardness and that the plant is not wilted or dried out. If the plant appears to be suffering from lack of moisture, water it well and wait until the shoots and leaves are *turgid* (swollen with moisture). This may require waiting overnight.

Early morning is the best time to take cuttings from the parent plant. The shoots and leaves contain more moisture at this time than later when the sun has had time to dry them out. Remem-

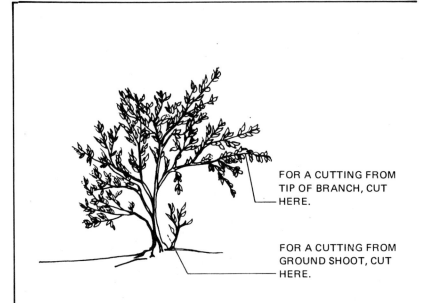

FOR A CUTTING FROM TIP OF BRANCH, CUT HERE.

FOR A CUTTING FROM GROUND SHOOT, CUT HERE.

Fig. 10-2 Softwood and semi-hardwood cuttings may be taken from the tip of a branch, or from a shoot growing at the base of the plant.

ber, the new cutting will have no roots with which to gather moisture. It should be started with as much moisture as possible.

Caution: Knives used in propagation are sharp. Never cut toward yourself or another person.

Immediately after taking the cutting from the parent plant, place it in a bucket of water or a plastic bag containing water. This prevents drying of the wood after it is cut from the plant and before the actual cutting is made from it. At this time prepare a label with the variety name and the date on it. Place the label in the container with the cutting wood. Be sure the label is marked with waterproof pencil or waterproof felt tip pen.

Softwood and semi-hardwood cuttings are made when plants have leaves on them. There are various degrees of tissue hardness or maturity, depending on the plant being propagated. Figure 10-3 gives the best rooting date and desired maturity of various plants.

Plant Name	Rooting Hormone	Month to Take Cutting in Mid-Atlantic States	Cutting Wood Maturity	Remarks
Arborvitae* (*Thuja occidentalis*) (*Thuja orientalis*)	Hormodin #2	Oct.–Dec.	hardwood	Wounding cutting helps.
Azalea (evergreen)	Rootone or Hormodin #1	early July	semi-hardwood	Use only terminal growth.
Barberry (*Berberis deciduous*)	Hormodin #2	July–Sept.	semi-hardwood	
Barberry (evergreen)	Hormodin #2	Nov.–Dec.	hardwood	
Boxwood* (*Buxus*)	Hormodin #2	July or Nov.–Dec.	semi-hardwood or hardwood	
Cotoneaster	Hormodin #2	July	semi-hardwood	
Dogwood (*Cornus*)	Hormodin #3	June–July	softwood	new growth after flowering
Deutzia	Rootone or Hormodin #1	June–July Nov.–Dec.	softwood hardwood	
English ivy*	Rootone or Hormodin #1	Nov.–Dec.	hardwood	
Euonymus*	Rootone or Hormodin #1	June–July–Aug. Dec.–March	evergreen-semi-hardwood deciduous-hardwood	
Firethorn* (Pyracantha)	Rootone or Hormodin #1	July–Oct.	semi-hardwood	
Forsythia	Rootone or Hormodin #1	June–July Dec.–March	softwood in container hardwood, outside	

Fig. 10-3 Rooting Date and Desired Maturity of Various Plants

Plant Name	Rooting Hormone	Month to Take Cutting in Mid-Atlantic States	Cutting Wood Maturity	Remarks
Holly (*Ilex*)* (evergreen)	Hormodin #3	July–Aug.	semi-hardwood	terminal growth
Hydrangea	Rootone or Hormodin #1	June–July Dec.–March	softwood hardwood	
Juniper (*Juniperus*)*	Hormodin #3	Nov.–Dec.	hardwood	
Lilac (*Syringa*)	Hormodin #2	directly after flowering	very soft wood	Use fungicide.
Mock orange (*Philadelphus*)	Rootone or Hormodin #1	July Dec.–March	softwood hardwood	
Pachysandra*	Rootone or Hormodin #1	Nov.–Dec.	hardwood	
Privet (*Liqustrum*)	Rootone or Hormodin #1	July Dec.–March	softwood hardwood	
Rhododendron	Hormodin #3	July and Sept.–Oct.	semi-hardwood	Wound cuttings; use only terminal growth.
Spirea (*Spiraea*)	Rootone or Hormodin #1	July Dec.–March	softwood hardwood	species: *vanhouttei*
Viburnum	Rootone or Hormodin #1	May–June July–Aug.	softwood semi-hardwood	fragrant types all others
Weigela	Rootone or Hormodin #1	July–Sept. Dec.–March	softwood hardwood	
Wisteria	Rootone or Hormodin #1	July	softwood	
Yew (*Taxus*)	Hormodin #3	Dec.–March	hardwood	

*May be taken in fall and midwinter.

Fig. 10-3 Rooting Date and Desired Maturity of Various Plants (Continued)

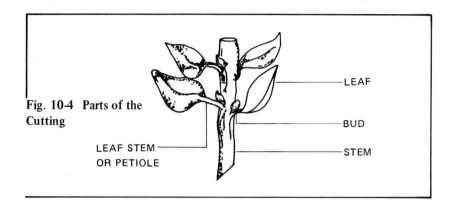

Fig. 10-4 Parts of the Cutting

LEAF
BUD
STEM
LEAF STEM OR PETIOLE

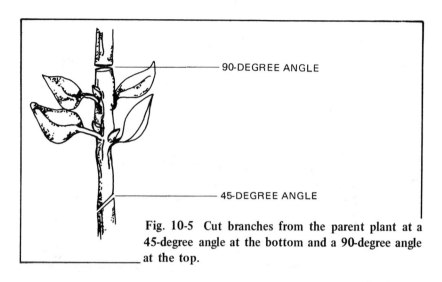

90-DEGREE ANGLE

45-DEGREE ANGLE

Fig. 10-5 Cut branches from the parent plant at a 45-degree angle at the bottom and a 90-degree angle at the top.

Leaves help the propagator keep the cuttings right side up — they point upward. The buds in the leaf axil are always on top on the leaf stem and toward the tip, figure 10-4. It is necessary to know the top of the cutting because the stem must be placed in the rooting media with the bottom end down. Stem tissue is highly polarized and roots form much easier on the bottom of the cutting. To avoid confusion, cut the branches from the parent plant at a 45-degree angle on the bottom end and straight across at a 90-degree angle on the top, figure 10-5. This is important only if the tip of the branch is cut off, a process which is usually necessary if the wood is too soft on the branch tip.

The cutting wood is now brought back to the bench or table where cuttings are to be made.

> Make the cuttings as soon as possible;
> Never allow the wood to dry out.

Preparing the Container and Rooting Media

Before making the cuttings, a container for holding the media must be prepared. A flat, pot, or any container at least 4 inches deep may be used, figure 10-6. The size of the container (other

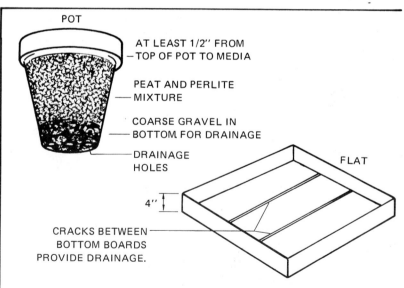

POT

AT LEAST 1/2" FROM TOP OF POT TO MEDIA

PEAT AND PERLITE MIXTURE

COARSE GRAVEL IN BOTTOM FOR DRAINAGE

DRAINAGE HOLES

FLAT

4"

CRACKS BETWEEN BOTTOM BOARDS PROVIDE DRAINAGE.

Fig. 10-6 Any container at least 4 inches deep, such as a plant pot or flat, may be used to hold cuttings.

than its depth) is determined by how many cuttings are to be made. The container must have holes in the bottom to allow for drainage. It is best to place coarse sand, gravel, or some other coarse material 3/4 inches deep in the bottom to aid in drainage. The rooting media, at least 3 inches deep, is placed on top of this coarse material.

> It is essential that the media for rooting be loose to allow circulation of air, be able to hold moisture, and be sterile.

Remember that roots need oxygen and moisture to live. The cut end of the cutting makes it easy for organisms which cause rotting to enter the plant, so the media must be sterile, or free from any living organisms. A mixture of one-half horticultural perlite to help loosen the structure of the rooting media and one-half coarse sphagnum moss provides a good media. Washed construction grade sand may be used in place of the perlite and other types of peat moss may be used, however.

The media may depend on the type of watering system which is to be used. Excellent success can be obtained with the peat-perlite mixture in containers that are to be covered with plastic until the cuttings root. When mist is used (a process in which fine droplets or water are sprayed on plants through nozzles), pure perlite or pure sand is satisfactory. If pure perlite or sand is not used, the percentage of peat moss should be reduced to 25 percent or less. The high percentage of peat moss recommended for pots and flats sealed with plastic would hold too much moisture under mist, thereby withholding sufficient oxygen and possibly causing the cuttings to rot. The rooting mixture should be mixed before it is wet so that a uniform mixture results.

Filling the Container

After the rooting media is thoroughly mixed, it is placed in the container, leveled, and firmed. There should be at least 1/4 inch between the top of the container and the level of the media surface so that water does not run off the surface as the media is watered. The surface is leveled with a flat board, skew, or brick. The leveling is important so that water spreads evenly over the surface and all cuttings are watered equally. The media is watered until water runs through the bottom of the container. If the peat moss used is dry, the moss may require an overnight soaking in water, since it resists wetting when completely dry. After soaking the peat moss in water, squeeze out the excess water before mixing and placing in the container. Hot water is recommended to thoroughly moisten peat moss and artificial planting soils.

Making the Cuttings

Now that the cutting wood is selected and the container is prepared, the cuttings are made. Inspect the cutting wood for proper maturity. If the tips of branches or shoots are too soft for the particular plant being propagated, cut them off and use more mature wood nearer the base, figure 10-7. This does not apply to plants such as the rhododendron, a plant in which the terminal end must be used as the cutting. This is because buds only appear near the terminal end on the rhododendron.

Make as many 3-inch or 4-inch cuttings as possible from each stem or shoot. A very sharp knife or pruning shears must be used so that a clean cut is made. Remember to cut the bottom on a slant and the top of the cutting straight across to indicate which end should be placed in the rooting media. The bottom of cuttings are cut slanted for another reason. The new cutting has no roots, and the slanted bottom allows more surface area to contact

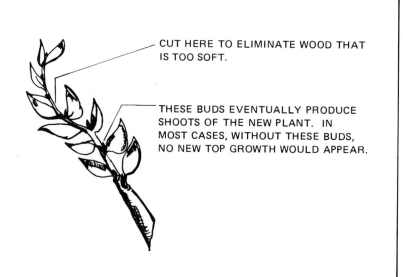

Fig. 10-7 If the tip of the branch or shoot is too soft, the more mature wood closer to the base should be used.

the rooting media for moisture uptake or absorption. It is not important to cut at or near a node unless the plant stem is hollow. On hollow stems, cut just below the node on the bottom and just above a node at the top, figure 10-8.

It is important to include two or three buds on each cutting. New shoots can grow only from leaf buds which appear in the axil of the leaf petiole or leaf stem.

Treating With Hormone and Inserting in Media

After the cutting is made, it is treated with the proper concentration of rooting hormone. Follow directions on the label, and be sure the substance is recommended for the plants being rooted. Rooting hormones are chemicals which help cuttings

grow roots more quickly and grow a larger number of roots. Hormones used for various plants are shown in figure 10-3, page 100. If the hormone does not contain a fungicide, the cutting may be sprayed, dipped, or watered with a fungicide such as captan to help prevent rotting. Captan has been reported to stimulate rooting in addition to controlling fungus infection.

The cutting is immediately placed in the rooting media to one-half its length but never more than 2 inches deep. Cuttings should be spaced so that no leaves overlap. Large leaves may be trimmed. Do not press the media around the cutting very firmly. Add water to settle the mixture around the cutting.

Labeling

The cuttings must now be labeled, using either the same label that was prepared when the wood was cut from the parent plant, or by preparing new labels. The name of the rooting hormone should be added to the information already on the label (variety name and date, including the day, month, and year). Remember

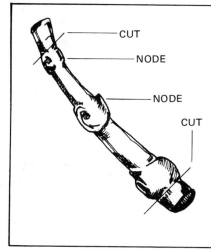

Fig. 10-8 Stem cutting on hollow stem. Cutting is made below a node at the bottom and above a node at the top. (Leaves have been omitted from this illustration to allow clear view of the nodes.)

Fig. 10-9(A) Rhododendron cutting ready to be inserted in plastic food bag. The bag holds moisture and assists in rooting. (Richard Kreh, Photographer)

Fig. 10-9(B) Same cutting in covered pot, sealed for rooting. It is now placed in a spot where it will receive light, but no direct sunlight. (Richard Kreh, Photographer)

Fig. 10-10(A) Flat with rooting media being enclosed in a water-tight wire framework. (Richard Kreh, Photographer)

Fig. 10-10(B) Flat with plastic in place. The plastic is tucked under the bottom of the flat to seal in moisture. (Richard Kreh, Photographer)

to use a waterproof pencil or pen. The label is placed in the rooting container.

Controlling the Atmosphere

The new cutting has leaves, but no roots. Because of this, these leaves lose moisture through *transpiration* (loss of water through the leaves or stems of plants) more quickly than it can be drawn up through the slanted cut end. Therefore, it is essential to keep the relative humidity around the cutting close to 100 percent to reduce moisture loss.

If a greenhouse is available, one way in which high humidity can be maintained is to place the cuttings in a misting propagation bench. This system uses a greenhouse bench equipped with misting nozzles which spray a fine mist on the cuttings from a piped water system. The system may be operated manually, or set up to water the cuttings automatically at specific intervals. An alternative method for use without a greenhouse is to enclose the containers with a film of polyethylene plastic (a clear plastic which holds in moisture but lets in light and a small amount of air.) Very little water is lost from the container, and there is no need to water the cuttings until they have rooted.

When a pot is used as a container, a plastic food bag may be slipped over it and tied in place or tucked under the pot as shown in figure 10-9. A flat may be covered by drilling three holes in each side and placing stiff wire over the flat. This structure is covered with polyethylene film which is tied or tucked under the bottom of the flat, figure 10-10. Containers covered with plastic must be placed out of direct sunlight by locating them against a north wall, in the shade of a tree, or under a bench. Light aids formation of carbohydrates — and thus rooting — and should be available. However, the cuttings should not receive direct sunlight since this raises the temperature inside the container too much, thereby killing the cuttings. If they must be placed in the sun, the plastic may be covered with cheesecloth or burlap to give shade and hold down the temperature. When the misting system is used and the cuttings are not in a tight container, direct sunlight is beneficial.

Rooting

As time nears for the cuttings to be rooted, open the plastic covered container or go to the misting propagation bench and check the cuttings by holding each cutting and tugging gently. If the cutting does not easily slip out, roots are developing. If there are no roots, merely close the container or leave the cuttings under mist, depending upon the system used, and wait another week to ten days. As long as the cutting has not rotted or turned dark and is holding its leaves, it will probably root if it is given more time. Growth on the tip or sides of the cutting is normal, but does not indicate that the cutting has rooted.

When the cuttings have a root ball measuring 2 or 3 inches across (see (C) of figure 10-11), it is time to harden them off in preparation for transplanting (removing and planting in a new place). The hardening off must be done very slowly over a period of seven to fourteen days so that the cuttings are not exposed to a drying atmosphere too suddenly. They are accustomed to high humidity and must adapt to new, drier conditions. Open the plastic a little more each day over the fourteen-day period until it is totally removed. Once the plastic is completely opened, care must be taken not to allow the rooting media to dry out. The cuttings must now be watered to keep the media moist but not wet. If a misting propagation bench is used, the misting cycle is changed so that the misting time becomes shorter and shorter until no mist is used and the cuttings are watered normally.

After the hardening-off process, the cuttings are ready to be *lined out* (planted outside in the nursery row under shade or sprinklers) or transplanted to other containers for further growth.

HERBACEOUS CUTTINGS (SOFTWOOD)

Herbaceous cuttings are made from succulent greenhouse plants such as the geranium, chrysanthemum, coleus, and carnation. Most florist crops are propagated in this way. Cuttings are made 3 to 6 inches long with leaves left on the upper, or terminal, end. Several cuttings may be cut from a plant stem. This allows the horticulturist to obtain many more cuttings than if only leaf or stem tips are used. Leaves are trimmed from the basal, or lower, end of the cutting, with three or four leaves left remaining on the top end. A fungicide should be used on all cuttings to prevent rotting. A sterile rooting media is also essential.

Herbaceous cuttings are rooted under the same high humidity conditions required for softwood cuttings. Bottom heat helps speed rooting. Under proper conditions, rooting is rapid. For this reason, this type of cutting is highly recommended for the beginner. A mild concentration of rooting hormone such as Rootone #1 may be used, but is not necessary. A high degree of success may be expected, with more than 90 percent of cuttings rooting.

> Using a flat, flowerpot, or mist system in the greenhouse, try propagating at least one of the herbaceous plants listed. Care for them in the same manner as noted for the softwood cuttings.

Additional plants easily propagated as herbaceous cuttings include the lantana, Swedish ivy, wandering Jew, and begonia.

STUDENT ACTIVITIES

1. Practice making softwood and semi-hardwood cuttings by completing each step of this exercise.

   a. Collect cutting wood for at least ten softwood or semi-hardwood cuttings. Label the cuttings and keep moist until placed in the rooting mix. Check with instructor on the proper maturity of cutting wood.

   b. Select a container and fill it with the proper rooting mix. (Provide for drainage.) Water properly as determined by the instructor.

   c. Using a sharp knife, make ten cuttings and insert in the rooting media. Treat with rooting hormone, label, and space properly. (The text may be used for reference.)

   d. Place the cuttings under a mist system or cover with plastic. Ninety percent of the cuttings should root. Check to be sure all of the previous steps in the exercise have been completed correctly. If nine out of ten cuttings root, you are a competent propagator. If less than 90 percent root, repeat the project if time allows. If not, review the steps in the procedure and trace your errors.

Suggestion. Spray cuttings with a systemic fungicide or dip them in a fungicide when they are made. This is usually sufficient treatment to prevent rotting if a sterile media is used.

2. Plants which are more difficult to root, such as evergreens, may be tried as an additional activity. A good choice might be the rhododendron. The following are extra steps necessary for this plant, which can be difficult to root.

   a. Take a cutting with three or four leaves from terminal current season's growth in July or from September 15 to October 30. (Trim leaves if necessary.)

   b. Cut to 2 1/2 inches long.

   c. Trim a slice off each side of the cutting 1 inch from the base and extending to the base just deeply enough to cut into the wood.

   d. Treat with Hormodin #3, 0.8 percent indolebutyric acid, or another rooting hormone of equivalent strength.

   e. Complete by following the same steps given for other cuttings.

Fig. 10-11 (A) shows rhododendron cutting as taken from parent plant. (B) shows cutting after trimming, with stem and leaves trimmed and the stem sliced on each side at the base. (C) illustrates a well-rooted rhododendron cutting ready to be planted in the cold frame. (Richard Kreh, Photographer)

3. The information mentioned in this unit for taking stem cuttings also applies to other types of cuttings, such as leaf and root. The procedure involved in taking leaf and root cuttings is only slightly different. The differences are noted here for the student who wishes to experiment with these methods of propagation.

Fig. 10-12 (A) African violet cutting as taken from the parent plant.

(B) Rooted cutting showing the small plant growing from the base of the leaf stem and roots. It is ready to be planted as a new plant. The old leaf is pruned off at this stage or soon after.

(C) Adult African violet plant in bloom. (Richard Kreh, Photographer)

Leaf Cuttings

The African violet is a good example of a plant propagated by leaf cuttings. Examine figure 10-12 for details and try this easy-to-root plant. See figure 10-13 for details on making mallet cuttings.

It will be necessary to advance to another unit while the cuttings are rooting. A careful check should be made each day, however, to see if the cuttings are progressing and to be sure that no damage has been done to the containers or the cuttings.

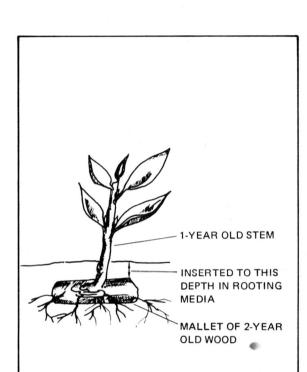

1-YEAR OLD STEM

INSERTED TO THIS
DEPTH IN ROOTING
MEDIA

MALLET OF 2-YEAR
OLD WOOD

Fig. 10-13 Mallet cutting

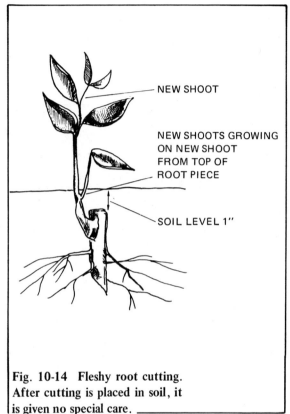

NEW SHOOT

NEW SHOOTS GROWING
ON NEW SHOOT
FROM TOP OF
ROOT PIECE

SOIL LEVEL 1"

Fig. 10-14 Fleshy root cutting.
After cutting is placed in soil, it
is given no special care.

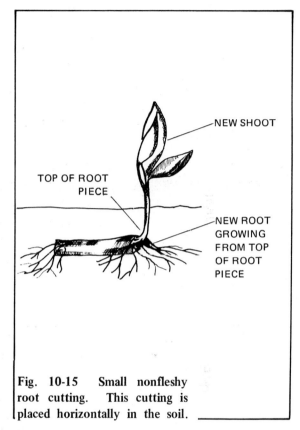

TOP OF ROOT
PIECE

NEW SHOOT

NEW ROOT
GROWING
FROM TOP
OF ROOT
PIECE

Fig. 10-15 Small nonfleshy
root cutting. This cutting is
placed horizontally in the soil.

Root Cuttings

Root cuttings may be made from any plant that will sprout or sucker from the root. Roots from 1/8 to 1/2 inch in diameter cut into pieces 1 to 4 inches long are placed in sand in a flat or other container and watered well. Glass or clear plastic over the flat helps hold moisture. When new shoots sprout from the flat, the new plants are moved to the nursery row. Plants commonly propagated this way include the Oriental poppy, wisteria, spirea, red raspberry, and trumpet creeper. Figures 10-14 and 10-15 illustrate the propagation of plants by use of root cuttings.

SELF-EVALUATION

A. Complete each statement on the left regarding the propagation of plants from cuttings with the proper term on the right. Include both the letter and the word in your answer.

1. Cutting wood must be of the proper _____, or hardness.

2. A rooting media that lets in _____and holds _____ must be used.

3. The media is placed in a container which measures at least _____in depth.

4. The cutting wood must be kept _____and identified with a(an) _____.

5. To aid in the rooting process, the cutting is treated with a(an) _____.

6. The cutting is inserted in the media no deeper than

    _____.

7. The rooting media is not pressed around the cutting to settle the media; rather, it is _____.

8. The container holding the cutting is labeled according to _____, _____, and _____.

9. If the cuttings are placed under plastic, they should be given plenty of _____, but no direct sunlight.

10. If the cuttings are placed under _____, direct sunlight is preferred.

11. As rooting time nears, the horticulturist checks for roots by gently _____ on the cuttings.

12. The cuttings are hardened off when they have sufficiently _____.

13. After hardening off, the cuttings are ready to be

    _____.

a. moisture
b. 6 inches
c. light
d. pulling
e. rooting hormone
f. maturity
g. sterile
h. oxygen
i. 4 inches
j. variety
k. fungicide
l. moist
m. 2 inches
n. mist
o. rooted
p. label
q. watered
r. misting
s. date
t. hormone
u. transplanted

B. Provide a brief answer for each of the following.

  1. What is asexual plant propagation?

  2. What type of plants are usually propagated by asexual propagation?

  3. Why is it necessary to reproduce plants asexually?

  4. List the nine key steps in taking softwood or semi-hardwood cuttings.

OBJECTIVE

To perform the six steps in taking hardwood cuttings so that success in rooting 60 to 80 percent of the cuttings is achieved.

COMPETENCY TO BE DEVELOPED

After studying this unit, the student will be able to

- list five plants commonly propagated by hardwood cuttings.
- select, collect, and label mature cutting wood from one of the five identified parent plants.
- make and store ten hardwood cuttings.
- line out cuttings in the field at the proper planting time.
- keep accurate records of the percentage of cuttings that root.

Materials List

cutting wood

string with which to tie bundles

storage area

propagating knife

labels and waterproof pen or pencil with which to mark labels

---

unit 11
Hardwood Cuttings

Propagation by hardwood cuttings is one of the easiest and least expensive methods of vegetative or asexual propagation. Cuttings are prepared in the dormant winter season when time is usually more available to the propagator. Hardwood cuttings may be shipped long distances or stored for long periods of time.

Expensive misting equipment or rooting benches are not needed and the necessary cold storage may be done out of doors in low-cost facilities if temperatures are 50°F (10°C) or lower. Many deciduous woody plants are easily propagated by hardwood cuttings. Figure 11-1 lists deciduous and narrowleaf evergreens propagated by hardwood cuttings.

Some Deciduous Plants Generally Propagated by Hardwood Cuttings

blueberry	mock orange
currant (ribes)	mulberry
deutzia	Persian lilac
dogwood	plum
euonymous	pomegranate
fig	privet
forsythia	rosa multiflora
gooseberry	spirea
grape	weigela
honeysuckle	wisteria
kerria	

Some Narrowleaf Evergreens Propagated by Hardwood Cuttings

chamaecyparis	spruce
hemlock	thuja
juniper	yew
pine	

Fig. 11-1 Deciduous and narrowleaf evergreens propagated by hardwood cuttings.

Propagation by hardwood cuttings differs from softwood and semi-hardwood cuttings in several ways:

- the time of year in which the cuttings are taken,
- the hardness, or maturity, of wood used for the cuttings,
- the usual absence of leaves on the cuttings, and
- the storage of cuttings in place of immediate planting.

Six Steps in Taking Hardwood Cuttings

1. Identify the plant to be propagated and select proper cutting wood.
2. Collect wood for cuttings.
3. Make the cuttings.
4. Store the cuttings.
5. Line out or plant the cuttings.
6. Determine rooting percentage of the cuttings.

SELECTING THE CUTTING WOOD

Wood for hardwood cuttings is generally taken from current season's growth, as is done with softwood cuttings. The wood is cut from the ends of branches or from long shoots that grow from the base of the plant, figure 11-2, page 114. This material may be gathered soon after the plants become dormant. Plants that are *dormant* have lost their leaves and are preparing for the winter rest cycle. Cuttings may be taken throughout the winter months. The wood taken is current year's growth, but it is now mature, or hardwood. Plants that are healthy, vigorous, and grown in full sunlight yield cutting wood with more stored carbohydrates and more vigor in rooting.

Fig. 11-2   Three strong young shoots growing from the base of a blueberry plant. These shoots (located to the right of the parent plant) make excellent hardwood cuttings. Notice that the leaves have all dropped off the plant; this is because the plant is dormant. (Richard Kreh, Photographer)

COLLECTING THE CUTTING WOOD

Using a sharp knife, cut the selected wood from the parent plants. Since the wood has no leaves, drying out is not as much of a problem as it is with softwood cuttings. Label the cuttings according to variety name and date collected. The wood may be made into 6- to 8-inch cuttings for immediate use, or stored in a cool moist place (below 50°F or 10°C) for later use, figure 11-3. If stored, the branches should never be allowed to dry out or become too wet. A covering of moist (but not wet) sawdust, sand,

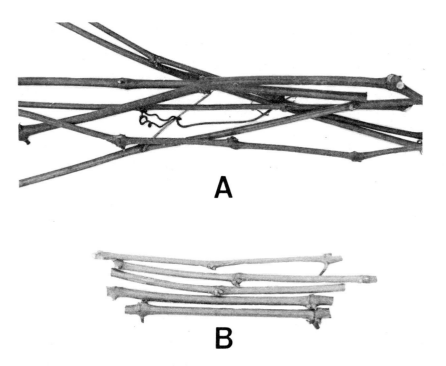

Fig. 11-3   (A) illustrates grape hardwood cutting wood as collected from the plant. (Stems are longer than shown in the photograph.) (B) shows hardwood cuttings cut to proper length and ready to be tied in bundles and stored. (Richard Kreh, Photographer)

or peat moss works well to maintain moisture. There should be little variation in temperature during storage since temperature variation tends to hasten drying.

TAKING CUTTINGS

Cuttings, usually 6 to 8 inches long, are made with a sharp knife or pruning shears. For hollow-stemmed plants, such as the grape, the bottom cut is made just below a node and the top cut about 1 inch above a node or bud (see (B) of figure 11-3). The diameter of cutting wood is not extremely important. Medium-sized wood, however, tends to survive storage and root better than wood that is very thin.

Cuttings should be made early enough so that they may be stored 6 to 8 weeks prior to planting.

STORAGE OF CUTTINGS

The cuttings are treated with a rooting hormone for better growth. The finished cuttings are then tied in bundles for storage. They should be placed so that all the tops are on the same end of the bundle and buried in moist sawdust or similar material. The bundles should not be tightly sealed; the cuttings are alive and need oxygen to remain healthy. A tight seal could also cause excess moisture to collect, resulting in rotting.

The storage period allows the cut ends of the cuttings to callus and the rooting process to begin. Callus formation which occurs during storage is an advantage when cuttings are placed outdoors for rooting under drying conditions, since calloused cuttings root more quickly. New roots begin to grow during storage and may be visible at this point.

The cuttings may be buried outside in sand-filled containers in a well-drained area. Outside temperatures must be cold enough to prevent growth beginning at the tops of the cuttings during the storage period. During the first four weeks of storage, the temperature should be 50° to 55°F (10° to 13°C). This relatively high temperature is favorable for callus formation. After callus has formed, the temperature may be lowered to below 40°F (4°C), but never below 32°F (0°C). This lower temperature prevents growth on the tops before the cuttings are lined out (planted).

LINING OUT CUTTINGS

As soon as the soil is ready in the spring, the cuttings are planted outside. This process is known to nursery workers as *lining out*. Before lining out the cuttings, the soil must be prepared. A well-drained, sunny site with some wind protection is best.

Rows are dug deeply enough so that the cutting can be covered with soil with only the top bud remaining above the soil level. If they are to be left only for one year, the cuttings are placed 6 inches apart in rows 1 foot apart, figure 11-4. When the cuttings are to be left more than one year, or when especially fast growing plants are involved, the cuttings are spaced 9 to 10 inches apart and in rows 2 to 3 feet apart.

Fig. 11-4 Hardwood cuttings lined out in the field. Only the top bud is left above the soil level. (USDA Photo)

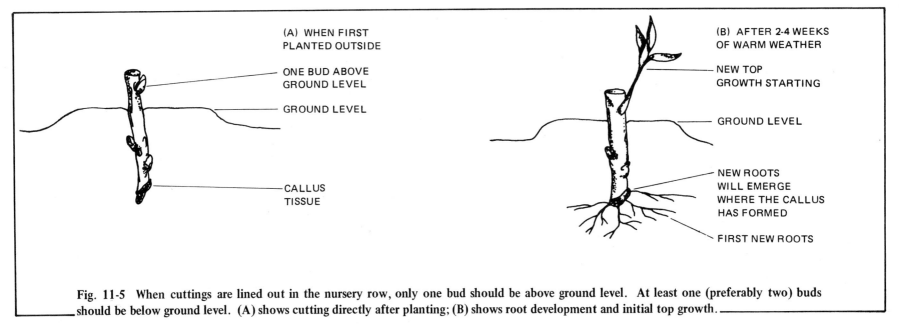

(A) WHEN FIRST PLANTED OUTSIDE

ONE BUD ABOVE GROUND LEVEL

GROUND LEVEL

CALLUS TISSUE

(B) AFTER 2-4 WEEKS OF WARM WEATHER

NEW TOP GROWTH STARTING

GROUND LEVEL

NEW ROOTS WILL EMERGE WHERE THE CALLUS HAS FORMED

FIRST NEW ROOTS

Fig. 11-5 When cuttings are lined out in the nursery row, only one bud should be above ground level. At least one (preferably two) buds should be below ground level. (A) shows cutting directly after planting; (B) shows root development and initial top growth.

A small shovel or other tool is used to break the soil. The cuttings are placed in the soil at the depth described above. The cutting should not be pushed into the soil under pressure; this could damage the tissue or break the cutting. After the cutting is placed in the row, the soil is pressed firmly around the cutting. The soil surface may then be mulched to help hold moisture and control weeds or cultivated for weed control. The soil should never be allowed to dry out around the cuttings. Provision should be made for watering as needed.

The number of plants which will eventually root depends upon the variety of the plant. However, between 60 and 80 percent of the cuttings can be expected to root and grow.

As the weather becomes warm, the cuttings develop leaves and shoots although they may not yet be rooted. However, if growth continues into the summer months, it is almost certain that the cuttings have rooted. The care required beyond this stage varies with the plant being propagated.

Narrowleaf evergreens propagated by hardwood cuttings are often very difficult and slow to root. In general, the slow growing *Juniperus* species root easily, the yews fairly easily, and the upright junipers, spruces, and hemlocks with difficulty. Pines are extremely difficult to root. Rooting hormones with high concentrations of indolebutyric acid (0.8 to 1.0 percent actual chemical solution or powder) are usually beneficial. These cuttings are taken from late fall to late winter and are best rooted in a greenhouse under high light intensity and in high humidity. This is achieved by placing the cuttings under plastic or using a misting system as specified for softwood and semi-hardwood cuttings. Cuttings should also be dipped in a fungicide such as captan to prevent rotting. Rooting media and other environmental conditions are essentially the same as those for softwood and semi-hardwood cuttings.

Hardwood cuttings of deciduous plants do not require the extensive care given to softwood and semi-hardwood cuttings. Since no leaves are present in the initial rooting stages, the demand for moisture is not high, and these cuttings need not be placed under plastic or in a mist system. The evergreen hardwood cutting, which does have needles at the time of rooting, must be handled as softwood cuttings are.

Cuttings of deciduous plants are usually dug for transplanting when they are dormant and without leaves. They may be dug bare root. Evergreens, however, which have needles, should be dug with a ball of soil attached to the roots. Care must be taken to keep the root ball intact.

STUDENT ACTIVITIES

1. Visit a local nursery to observe a demonstration of hardwood cutting techniques.

2. A. Select at least one plant commonly propagated by hardwood cuttings. Perform the six basic steps to produce at least ten cuttings.

    B. Set up a calendar to determine

        1. the storage temperature and dates to change temperature.

        2. the approximate date for lining out certain cuttings.

        3. inspection dates so that such things as moisture conditions, temperature, and rooting date can be noted.

    C. Line the cuttings out in a nursery row in the spring and follow up to determine rooting success. Keep a record of observations including dates when shoots and roots first develop.

3. Start a small nursery on the school grounds or at home in which to propagate and grow plants for future use. This is an inexpensive way to keep a supply of laboratory materials on hand.

SELF-EVALUATION

A. List three differences between hardwood and softwood cuttings.

B.   Complete each statement on the left with the proper term on the right. Include both the letter and the word in your answer.

1. Ideally, hardwood cuttings are _____ to _____ inches long.

2. Hardwood cuttings are stored for the first four weeks at a temperature of _____.

3. Wood of _____ thickness makes the best cuttings.

4. Cuttings should not be sealed airtight since _____ would be excluded causing rotting or even death.

5. After the first four weeks, cuttings are stored at a temperature of between _____ and _____.

6. A storage temperature of between 32° and 40°F prevents cuttings from growing _____.

7. A soft wound-healing growth called _____ forms on the basal end of the hardwood cutting.

8. It is best if cuttings are stored a minimum of _____ weeks prior to planting or lining out.

9. Cuttings are planted so that only the top _____ is above the soil level.

10. In any group of cuttings which are properly cared for, between _____ and _____ percent of the cuttings should root.

a. oxygen
b. 6 to 8 inches
c. medium
d. 2
e. 32° – 40°F
f. bud
g. 6
h. 60 – 80
i. tops
j. 50° –55°F
k. callus
l. 4 to 8 inches
m. moisture

To propagate one plant by separation and one by division.

COMPETENCY TO BE DEVELOPED

After studying this unit, the student will be able to

- describe the processes of separation and division and explain the major difference between the two.
- identify five specialized plant structures used in propagation and explain how each is used.
- list five plants propagated by separation or division and the type of specialized structure used in each case.
- write a brief definition of *bulb, corm, tuber, tuberous root,* and *rhizome.*

Materials List

    bulbs, corms, tubers, tuberous roots, rhizomes

    sharp knife

    storage container and packing material

---

unit 12
Separation
and Division

Many plants are propagated by one of two processes usually associated with one another — the process of separation, and the process of division. In both cases, plants are propagated by use of specialized underground plant parts. The primary function of these plant parts is to act as an organ for food storage. These plant parts are found on herbaceous perennials having shoots which die back at the end of the growing season; the underground part lives through the winter (dormant season) and sends forth a new top the following growing season. The second function of these plant parts is that of vegetative or asexual reproduction.

Separation is a method of propagation in which natural structures produced by certain plants are removed from the parent plant to become new plants. Bulbs and corms are examples of removable structures. These parts are usually removed while the parent plant is in a dormant or resting stage. Some bulb-producing plants propagated by separation are the tulip, amaryllis, lily, narcissus, daffodil, hyacinth, and grape hyacinth. Examples of corm-producing plants are the crocus, gladiolus, and colchicum.

Division is a method of propagation in which parts of the plant are cut into sections, each capable of developing a new plant. Division is possible with plants which grow rhizomes, stem tubers, and tuberous roots. Examples of plants with rhizomes are the iris, canna, and calla lily. Those with stem tubers include the caladium, peony, and Irish potato. Tuberous roots are found on the dahlia, gloxinia, and bleeding heart.

PROPAGATION BY SEPARATION

As mentioned earlier, plants which are propagated by separation are those which produce bulbs or corms as specialized underground plant parts which are responsible for food storage and propagation of the plant. Separation is a natural process.

Bulbs

Parts of the Bulb. A *bulb*, such as the tulip bulb, is a plant structure containing many parts but primarily composed of leaf scales, figure 12-1. Immediately next to and outside of the foliage leaves are bulb scales which produce small *bulblets* (tiny bulbs) at their base. These tiny bulblets grow larger and become small bulbs which may be separated and planted as individual plants. After

several growing seasons, the small bulbs become large enough to bloom. This produces *offset bulbs*. The process of developing offset bulbs is a simple commercial method for propagating the tulip, daffodil, bulbous iris, and hyacinth.

Offset daffodil bulbs are called *splits* or *slabs* when first separated from the mother bulb. After one year of growth, the slab or split is called a *round bulb*, and is capable of flowering the following season. Within the second year, a second flower bud

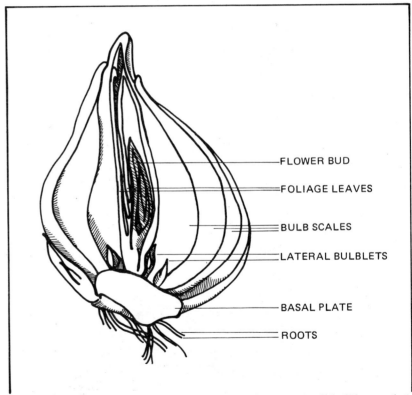

Fig. 12-1 **Parts of the tulip bulb. The bulb scales grow small bulblets at their base. The bulblets eventually grow larger and may be removed and grown to flowering size.**

FLOWER BUD

FOLIAGE LEAVES

BULB SCALES

LATERAL BULBLETS

BASAL PLATE

ROOTS

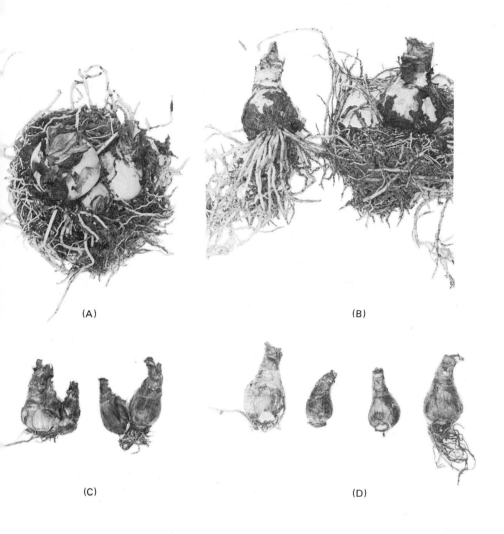

forms. At this time, the bulb is known as a *double nose* and is capable of producing two flower stalks. The round and double nose sizes are sold commercially. Large mother bulbs are kept to produce more splits. Splits or slabs are planted to grow for another year to reach flowering size.

Types of Bulbs. The tulip bulb is a fairly solid structure and therefore does not require special care in handling. The lily, however, is a loosely scaled bulb which cannot withstand rough handling or drying. The tulip is a *laminate* or *tunicate* bulb; these types of bulbs have dry membraneous outer scales that protect them against drying and injury. The lily represents a *nontunicate* or *scaly* bulb, which has no tough outer cover.

Propagation Procedure. Bulbs may be dug and separated after the foliage of the plant dies back and the plant is in a dormant state. The new bulbs are then stored at 65° to 68°F (18° to 20°C) and planted at the proper time. For tulips, planting time is in the fall. The minimum size for flowering of tulip bulbs is 10 centimeters. Separated bulbs which are smaller than this must be grown for 1 to 3 years before they reach flowering size.

Dry bulbs are stored at 55° to 60°F (13° to 16°C). Bulbs should be washed clean of all soil before storage and treated for rot.

The tops of bulb-producing plants should not be cut off until they have turned brown. Early removal of tops greatly reduces the amount of food produced and stored, and thus reduces bulb size. This could result in the bulb being too small to flower the following season.

Potted plants, such as the amaryllis, should be allowed to continue to grow for 6 to 8 months after blooming before water is withheld to force them into dormancy. This allows time for the

Fig. 12-2 **(A) An amaryllis bulb with two smaller offset bulbs attached. (B) The same bulbs, after the offsets have been separated from the *mother*, or parent, bulb. (C) Two daffodil bulbs with smaller offset bulbs attached. (D) The same bulbs, after the offset bulbs have been separated from the parent plant. The two offset daffodils may require another year to grow to blooming size. (Richard Kreh, Photographer)**

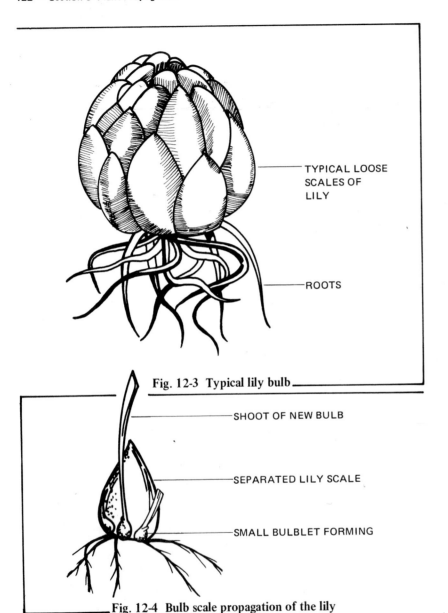

TYPICAL LOOSE
SCALES OF
LILY

ROOTS

Fig. 12-3 Typical lily bulb

SHOOT OF NEW BULB

SEPARATED LILY SCALE

SMALL BULBLET FORMING

Fig. 12-4 Bulb scale propagation of the lily

bulb to grow and to store sufficient food reserves to bloom during the next growing period.

The other plants mentioned earlier such as narcissus, hyacinth, and grape hyacinth are propagated in a similar manner.

Lily Propagation. The lily, figure 12-3, is also propagated by bulbs, but at a much slower rate. Mother bulbs are split at the base to force production of new bulblets. Some lilies naturally grow small bulblets on the stems and even in leaf axils. Commercially grown lily bulbs are separated and individual bulb scales are planted to produce new bulblets, figure 12-4. This is a more rapid method of increasing the number of bulbs since the lily bulb contains many loose, easily separated bulb scales.

The Easter lily is propagated commercially with underground stem bulblets, figure 12-5. The flowering stems are pulled from the ground in late August through mid-September and the small bulblets and stems kept moist by sprinkling them. About mid-October, the small bulblets are placed 4 inches deep and planted about 1 inch apart to grow the first season. They are moved again in September of the following season and planted about 6 inches deep, spaced 6 inches apart. By the end of the second year, they are dug and sold as flowering bulbs. At this time, bulb size should be 7 inches in circumference or larger. When dug, bulbs must be handled carefully to prevent bruising, and must not be allowed to dry out. The best storage is in moist, but not wet, sphagnum moss. A temperature of just below freezing holds lily bulbs in storage in an inactive state.

Corms

Another natural plant structure which can be used in propagation is the corm. A *corm* is a very solid, compact stem with nodes and internodes. While the bulb is composed of leaf scales,

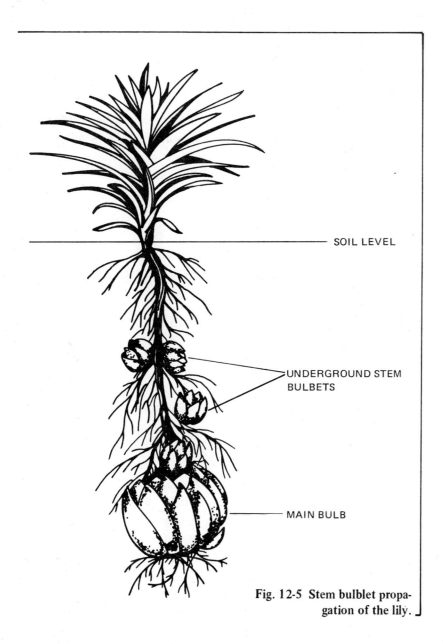

Fig. 12-5 Stem bulblet propagation of the lily.

- SOIL LEVEL
- UNDERGROUND STEM BULBETS
- MAIN BULB

the corm consists of a very short specialized stem for food storage. Corms have a dry covering which protects them from injury and drying, figure 12-6. In addition to being an organ for food storage, the corm is used as a reproductive structure. Development of small *cormels* is the principle means of reproducing (by separation) such plants as the gladiolus.

Propagation Procedure. Cormels form naturally. If they are of flowering size, corms are planted more shallow than normal, about 2 to 3 inches deep. More cormels will form at this depth than if they are planted deeper. When the plant top dies back, the plant may be dug and the small cormels separated and grown to larger size. The plant top should be allowed to die back normally by the effects of frost, or grown at least three months after blooming

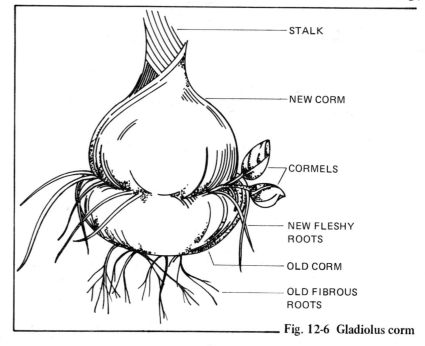

- STALK
- NEW CORM
- CORMELS
- NEW FLESHY ROOTS
- OLD CORM
- OLD FIBROUS ROOTS

Fig. 12-6 Gladiolus corm

so that the food supply manufactured in the top and stored in the specialized stem (the corm) is sufficient to develop good-sized cormels.

After frost or pulling of plants, the corms and cormels are separated from the rest of the plant and dried for storage. They should be treated with a fungicide and stored at 40°F (5°C) in a well-ventilated area with 80 percent humidity to prevent too much drying.

A hot-water treatment for disease control may also be necessary for cormels. Contact a commercial grower for information concerning the best treatment.

In the field, new cormels are planted in rows to grow larger, much as large seeds are. Two years are usually required for them to reach blooming size.

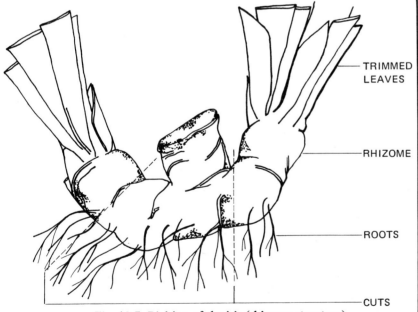

TRIMMED LEAVES

RHIZOME

ROOTS

CUTS

Fig. 12-7 Division of the iris (rhizome structure)

PROPAGATION BY DIVISION

Division differs from separation in that it is not a natural process; the parts used for propagation do not separate naturally from the mother plant. Parts of the plant which are to become new plants must be cut from the mother plant with pruners or a knife. Rhizomes, stem tubers, and tuberous roots are used in propagation by division.

Rhizomes

Rhizomes are underground stems which grow horizontally and produce roots on the bottom and stems on the top. A rhizome may be thought of as plant lying on its side with the stem covered with soil. An example of a plant which reproduces in this way is the iris.

Propagation Procedure. When plants become crowded, they may be divided for the purpose of reproducing or increasing numbers of plants.

Rhizomes generally grow very near the surface of the soil. They are removed from the soil by digging underneath the plant with a garden (spading) fork or shovel and lifting it out of the ground. All soil is washed from the plant so that the parts are clearly visible for division. Division is done by cutting the rhizome into sections. Since a rhizome is an underground stem, the same care must be taken as with a stem cutting to make sure that each section has at least one bud (eye) and preferably several.

The iris is used here to illustrate division of a rhizome. Iris are divided in late summer after bloom. Each clump is washed clean and cut apart, figure 12-7. The tops should be cut back to about one-third of their original height to balance the root loss.

After being cut apart, the new rhizomes are dusted with a fungicide, especially on the cut surfaces. They may then be planted in a new location.

Tubers

A *tuber* is a swollen end of an underground side shoot or stem. Tubers are distinguished by their eyes. Each eye produces a separate plant as it sprouts, developing a shoot with roots at the base of the new shoot. The tuber contains stored food on which the new plant feeds until new leaves take over the job of food production. One common tuber is the Irish potato.

Propagation Procedure. Propagation of plants by tubers is done by cutting the tubers into small pieces. Each piece must contain at least one bud or eye, figure 12-8. The pieces are planted in the same manner as larger seeds are. No storage is required; the cut surfaces are allowed to dry and the pieces are planted immediately.

Tuberous Roots

Tuberous roots are thickened roots which contain large amounts of stored food. They differ from tubers in that they are roots and have buds only at the stem end. Roots are produced at the opposite end, figure 12-9.

Propagation Procedure. Tuberous root crops are propagated by dividing the crown, or cluster of roots, when the plant is dormant. The plant is dug in the fall after frost has killed the top or it has died back for other reasons. The soil is washed away from the roots and they are allowed to dry. The roots are stored in dry sawdust, peat, or other materials at a temperature of 40° to 50°F (4° to 10°C) to prevent shriveling or complete drying out. Just before planting in early spring, the clumps or crowns are cut apart

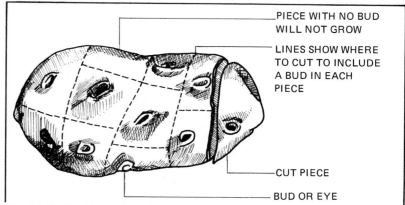

PIECE WITH NO BUD WILL NOT GROW

LINES SHOW WHERE TO CUT TO INCLUDE A BUD IN EACH PIECE

CUT PIECE

BUD OR EYE

Fig. 12-8 Division of Irish potato (propagated by production of tubers)

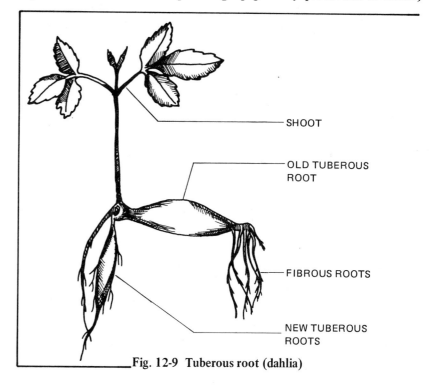

SHOOT

OLD TUBEROUS ROOT

FIBROUS ROOTS

NEW TUBEROUS ROOTS

Fig. 12-9 Tuberous root (dahlia)

so that each piece has a bud. If sprouts are present, it is easier to see if a bud is part of the cut piece. The new pieces are planted in the garden or nursery row after the danger of frost has passed.

The sweet potato, a tuberous root, sends out many shoots from adventitious buds. (*Adventitious* buds are those which occur at sporadic and unexpected places on the vegetative structure.) The shoots which grow from these buds root very quickly if the tuber is placed in a moist rooting media. The shoots are pulled off and planted as rooted shoots without any part of the old tuberous root attached, figure 12-10.

> If time permits, choose other plants which are propagated by separation and division and experiment with them.

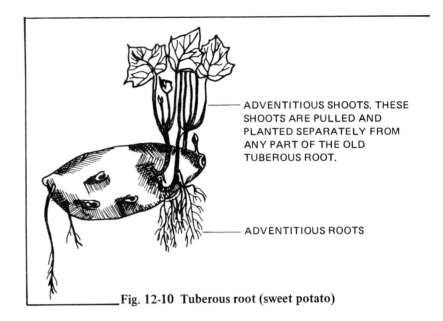

ADVENTITIOUS SHOOTS. THESE SHOOTS ARE PULLED AND PLANTED SEPARATELY FROM ANY PART OF THE OLD TUBEROUS ROOT.

ADVENTITIOUS ROOTS

Fig. 12-10 Tuberous root (sweet potato)

STUDENT ACTIVITIES

1. Propagate at least one plant by separation and one by division. Plant on the school grounds or at home if possible.

2. Start a sweet potato in a glass of water or in moist sand. Observe the roots growing from one end of the tuberous root and the shoots from the other. This is known as *polarity*. The commercial grower increases the number of shoots for propagation by giving the sweet potato root a hot-water treatment to eliminate the polarity.

3. Cut a sweet potato root in half and heat the pieces at 110°F for 26 hours. Place in moist sand or water. Observe the greater number of adventitious shoots which grow when polarity has been eliminated. Cover the base of the shoots with moist sand as they grow and roots will develop.

SELF-EVALUATION

A.   Select the best answer from the choices offered to complete each statement.

1. The main difference between separation and division is that

   a. separation is a natural method of propagation; in division, plants must be cut apart.
   b. division is a natural method of propagation; in separation, plants must be cut apart.
   c. There is no real difference.
   d. separation is done in the spring of the year; division in the fall.

2. The specialized plant parts used in division and separation have two functions. One is for propagation; the second is

   a. to divide or multiply the plant.
   b. to furnish food for human beings and wildlife.
   c. to store food.
   d. none of the above.

3. A corm is a specialized

   a. root.   c. leaf.
   b. stem.   d. plant.

4. Daffodil bulbs are stored at

   a. 55° to 60°F.   c. 80°F.
   b. 60° to 70°F.   d. 40°F.

5. The tops of bulb-producing plants should be allowed to continue growing until they die naturally because

   a. they add beauty to the flower garden.
   b. they continue to manufacture food for storage in the bulb.
   c. they grow larger and provide good mulch for the bulbs.
   d. they shade the bulbs and keep them cool.

6. Potted plants such as a potted amaryllis can be forced into dormancy by withholding

    a. fertilizer.   c. warmth.
    b. light.   d. water.

7. Lily bulbs are loosely sealed bulbs and must be stored in moist sand, peat, or sawdust to prevent

    a. drying out.   c. root growth.
    b. forcing new tops.   d. none of these

8. The Easter lily is propagated commercially with underground

    a. roots.   c. tubers.
    b. stem bulblets.   d. corms.

9. The gladiolus is propagated through formation of small

    a. seeds.   c. tubers.
    b. roots.   d. cormels.

10. In propagation by division, the section of dahlia root which is cut from the mother clump must have a

    a. bud.   c. number of small rootlets.
    b. separation scar.   d. all of these

11. A rhizome is an underground

    a. leaf.   c. stem.
    b. root.   d. tuber.

12. An example of a plant propagated with rhizomes is the

    a. tulip.   c. gladiolus.
    b. potato.   d. iris.

13. Tuberous roots are thickened

    a. roots.   c. corms.
    b. stems.   d. bulbs.

14. Bulbs and corms reproduce by a process known as

    a. division.
    b. cuttage.
    c. offshoots.
    d. separation.

B.   Match each plant in the right-hand column with the structure used in its propagation in the left-hand column.

1.	bulb	A.	iris
2.	corm	B.	tulip
3.	tuber	C.	dahlia
4.	tuberous root	D.	gladiolus
5.	rhizome	E.	Irish potato

OBJECTIVE

To perform three methods of grafting so that success in growing 60 percent of the grafted plants is achieved.

COMPETENCY TO BE DEVELOPED

After studying this unit, the student will be able to

- identify three types of grafts from a sketch.
- list at least one plant that is propagated by each of the three grafting methods.
- list the three reasons for grafting.

Grafting of plants is an ancient technique. It was practiced as early as 1,000 B.C. Centuries later, it was employed widely by the Romans. *Grafting* is a process by which two different plants are united so that they grow as one. The *scion* is the newly installed shoot or top of the plant. The *rootstock* (or *stock*) is the seedling or plant used as the bottom half of the graft. The rootstock becomes the root system of the newly grafted plant. The union between scion and rootstock is a physical union (growing together of the tissue) which allows free movement of plant sap across the graft and from the rootstock to the new top and back again.

Many times, grafting is used as a method of rapidly increasing the number of a desirable plant. Grafting is also used to give plants stronger, more disease-resistant roots. In grafting, it is important that both the rootstock and scion be disease free. The two must also be compatible so that they can grow together to form a strong union.

unit 13
Grafting

If rootstock is carefully selected, it will have little effect on the newly attached top of the plant. The top will grow much the same as it grew on the parent plant before the graft and will generally duplicate all the characteristics of the parent plant. An exception to this is when rootstock is used to dwarf plants. In these cases, rootstock reduces vigor and the overall size to which the new plant will grow. Dwarf fruit trees are a good example of this.

Grafting is not as widely used as budding in the commercial propagation of plants but is often used

- to topwork a large tree. This is done by grafting a different variety or varieties to many of the limbs of the tree.
- to insert a different variety on part of the limbs of a tree for cross-pollination.
- to propagate plants that may be difficult to bud.

REQUIREMENTS FOR SUCCESSFUL GRAFTING

The following conditions must be met to assure a successful graft.

Compatibility. The two plants must be related to one another closely enough so that the stock and scion are able to grow together. As a general rule, it is best to graft a scion to a rootstock of the same type of plant. For example, apple is grafted to apple; pear is grafted to pear. Certain varieties are exceptions, however. Some Japanese plums, almonds, and apricots, for example, experience best growth on peach roots. The almond and apricot are both able to grow on peach roots, but cannot be grafted to one another.

It is extremely important to know which families and varieties of plants grow together best when grafted. When selecting rootstock for commercial production, also consider the climate and conditions of the local growing area.

Scion Wood. The scion wood should be one year old and of vigorous growth.

Timing. Grafting is generally done when the stock and scion are dormant or have no leaves. The rootstock may be actively growing, but the scion should not be. Of course, when evergreens are grafted, there are needles present on the shoot.

Matching of Tissue. The cambium layer of the two matched plant parts, the scion and the rootstock, must come in close contact and be held tightly together. The cambium must not be allowed to dry out.

Waterproofing. Immediately after the graft is made, all cut surfaces must be covered with a waterproofing material such as grafting wax, plastic ties, or rubber ties.

A discussion of three methods of grafting follow: the whip (tongue) graft, the side veneer graft, and the cleft graft.

WHIP, OR TONGUE, GRAFT

Materials List
rootstock and scion wood
propagating knife
tying material (plastic propagation tape or rubber ties)
label and label marking pen or pencil

The whip (or tongue) graft, accomplished during winter months, is used when small (1/4- to 1/2-inch) material is being grafted. Fruit trees are examples of plants propagated by the whip

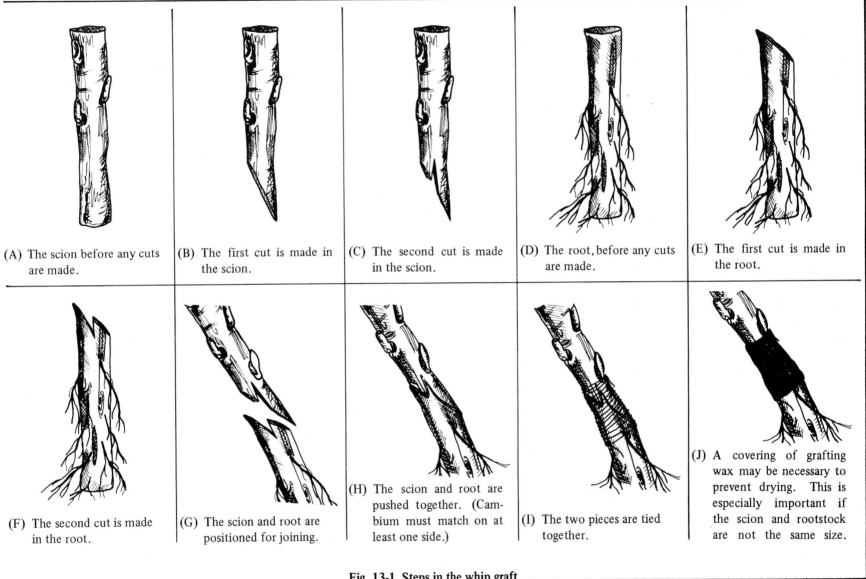

(A) The scion before any cuts are made.

(B) The first cut is made in the scion.

(C) The second cut is made in the scion.

(D) The root, before any cuts are made.

(E) The first cut is made in the root.

(F) The second cut is made in the root.

(G) The scion and root are positioned for joining.

(H) The scion and root are pushed together. (Cambium must match on at least one side.)

(I) The two pieces are tied together.

(J) A covering of grafting wax may be necessary to prevent drying. This is especially important if the scion and rootstock are not the same size.

Fig. 13-1 Steps in the whip graft

graft. If possible, the scion and root should be of the same diameter. The scion should contain three buds. The root piece should be at least 4 to 8 inches long and contain small fibrous roots. Figure 13-1 illustrates the steps in the whip graft.

The grafting cut is made below a bud on the stock. It should slant at such an angle that a smooth surface about 1 to 1 1/2 inches long is produced. A very sharp knife allows the cut to be made with a single pass and usually gives a straight, smooth surface. This smooth surface is essential so that a close contact between the root and scion is made all along the cut surface. The cut made on the root piece should be exactly the same length and slope as that on the scion. This allows the two parts to fit together evenly. The cambium, the thin tissue just under the bark of the scion or stem, must match the growing area at the edge of the root piece on at least one side. If the two pieces are the same diameter, it will match on both sides.

Before the two pieces are placed together, a second cut is made on the first cut surface in the reverse direction, starting about one-third of the distance from the tip and cutting parallel to the first cut. This cut is made about half as long as the first cut. The two pieces are then slipped together with the tongues interlocking. Great care must be taken to match the cambium on at least one side.

The two pieces are then tied tightly together with plastic propagation tape or rubber bud ties and stored in moist sand or peat moss to heal. Storage for the first 3 to 4 weeks is at 50° to 55°F (10° to 13°C). After that, storage is at 32° to 40°F (0° to 4°C) to prevent top growth. The grafted plant may also be planted directly in the nursery row. In these cases, the graft union must be below ground level.

The new plant should send out shoots and begin to grow with warm spring weather.

SIDE VENEER GRAFT

> **Materials List**
>
>   rootstock and scion wood
>
>   propagation knife
>
>   tying material
>
>   mulch
>
>   polyethylene tent or other wilt-proofing material
>
>   aluminum foil
>
>   labels and label marking pen or pencil

The side veneer grafting method is a very effective way to graft evergreens such as the blue Colorado spruce. To obtain precisely that shade of blue which these trees possess, a small piece of stem from a selected blue spruce must be grafted to a small spruce seedling used as rootstock. The selection of the stem or scion is very important, since these seedlings vary in color and very few have the particular shade of blue that is so desirable.

Grafting Procedure

The procedure for this graft is a simple one. Small seedlings may be grown in pots or in a nursery row. They graft more easily if they are about lead pencil size or a little larger. The grafts are made in early spring just before growth begins.

A shallow cut about 1 1/2 inches long is made into one side of the seedling rootstock just above the soil level, figure 13-2(A), page 134. The cut is not complete at the bottom, but stops so that a second short downward cut leaves a ledge when the cut

(A) A long, shallow cut is made on one side of the rootstock.

(B) A second cut is made to remove the piece of wood.

(C) Two cuts are made on the scion to shape it so that it fits the cut made in the rootstock. The dark area is that piece which is cut out.

(D) The scion is inserted into the rootstock. The cambium layers must match on at least one side.

(E) The scion is tied tightly in place with string or rubber bud ties. A plastic tent may be used at this point. Mulch may be applied.

(F) The completed graft. A stake may be used to tie the new graft to prevent breakage.

Fig. 13-2 Steps in the side veneer graft

piece is removed, figure 13-2(B). The 4-inch scion, or new top piece, is cut with a similar long cut on one side and a short cut on the opposite side at the bottom, making an end that fits on the ledge left on the seedling rootstock, figure 13-2(C). All cuts should be smooth and even so that the scion and rootstock surfaces fit together along the entire cut surface, figure 13-2(D).

The cambium layer of the scion and seedling rootstock must match on at least one side. The two are placed with cut surfaces matching and are tied securely together with rubber bud ties or similar material, figure 13-2(E).

Placing mulching peat moss or other moist material over the graft union is very helpful in promoting healing and helps to insure a higher percentage of successful unions.

When grafts are exposed to sun and drying winds, a polyethylene tent may be tied to a stake and drawn over the seedling and the graft until the graft has healed and the new growth has started on the top. Aluminum foil is secured on the south side of the tent to keep the temperature inside the tent down. When the tent is removed, it should be opened gradually over a period of a week to harden off the new graft.

Spraying the stem or shoot grafted on the rootstock with one of the new plastic materials such as Wilt-pruf is also helpful in holding down moisture loss and drying of the new graft until the union heals.

As new growth starts, the top of the seedling rootstock is gradually cut away to force buds on the newly attached scion into active growth. The entire top of the seedling should be removed as soon as new growth is well established on the scion. The grafting process is now complete, but care must be taken in handling the new grafts so that the top is not broken loose from the rootstock.

CLEFT GRAFT

Materials List

> rootstock and scion wood
>
> heavy propagation knife
>
> light hammer for driving knife
>
> cleft grafting tool or other wedge
>
> grafting wax
>
> labels and label marking pen or pencil

The cleft graft is used most often in topworking trees or grafting to a rootstock that is considerably larger than the scion. It is done in early spring just before the buds start to swell. The scion is cut from one-year-old wood and should include about three buds.

Grafting Procedure

The selected rootstock or tree to be topworked must first be cut back to the point at which the graft is to occur. The limb or trunk is sawed off at a right angle, figure 13-3(A), page 136. A heavy knife is used to split the rootstock or understock through the center until a crack opens wide enough in which to place the scion. A metal or wooden wedge is then placed in the center of the stem to hold the split open, figure 13-3(C). A piece of scion wood is selected and two scions are cut, each containing three buds. The scion is cut with a long, smooth tapered cut on each side making a wedge-shaped stick. The cut should be about 1 inch long and such that the wedge is slightly thicker on one side than the other, figure 13-3(D).

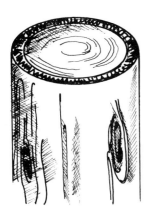

(A) The rootstock is sawed off at a right angle to the main branch.

(B) The rootstock is split with a heavy knife and hammer.

(C) The split is held open with a wedge.

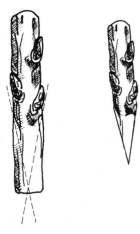

(D) The scion is cut in a long, smooth wedge shape.

(E) The scions are placed in the rootstock. The scions must make close contact with the rootstock for their entire distance. The scions must be cut at the same slant as the split in the rootstock.

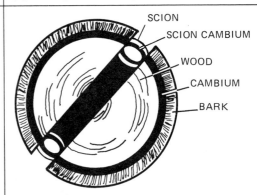

(F) Bird's-eye view of scion in rootstock. The scion sets in from the outside of the bark because its bark is thinner. The cambium of the scion and rootstock must match.

Fig. 13-3 Steps in the cleft graft

The thicker side is cut so that there is a bud directly above the top of the wedge. This thicker side is then placed against the outside of the rootstock where it is pressed firmly in contact with the cambium of the understock, figure 13-3(E). The scion is pushed down into the split in the rootstock until the entire cut area is inside the split. Do not push the scion hard enough to cause tearing of the tissue. If necessary, the wedge is driven in further to widen the split in the rootstock – this allows the scion to slip in more easily. Since the understock is much larger and has thicker bark than the scion, care must be taken to match the cambium layer of the scion with the cambium layer of the rootstock, not the bark of the rootstock. If properly placed, the scion will not be smooth with the outside of the understock, but will be set in, figure 13-3(F). The cambium layers must match for a successful graft.

After the two scions are in place, the wedge used to hold the split open is carefully removed. The grafter should check to be sure that the cambium layers are in contact for the full length of the cut area on the scion. The pressure of the split closing is usually sufficient to hold the scions in place. No nailing or tying is necessary. The entire cut surface is then covered with grafting wax to prevent drying.

If both scions grow, one should be removed after the first season so that there is not a sharp crotch on the tree or limb at the graft union.

STUDENT ACTIVITIES

1. Collect wood for propagation and accomplish at least one of the three grafts presented in this unit.

> **Caution:** **Grafting knives are sharp. Always cut away from yourself or other people in the area.**

Ask the instructor to check the graft for accuracy and for suggestions leading to a better graft.

2. Working either individually or with another student, prepare at least one graft, plant the resulting grafted plant, and allow it to grow. (Necessary supplies are the scion wood, rootstock, a sharp knife, and waterproof material to tie the graft union.) After the plant is grafted, label it according to variety. If possible, invite an experienced propagator to assist in the actual grafting activities.

SELF-EVALUATION

A. Provide a brief answer for each of the following.

1. Define the term *grafting*.

2. List the three reasons for grafting.

B. Select the best answer from the choices offered to complete the statement or answer the question.

1. Whip, or tongue, grafting is the best method to use when the rootstock and the scion are

a. the same size in diameter.　　　　c. of the same variety.

b. the same age.　　　　d. of different varieties.

2. The goal in grafting is to unite two plants

a. so that they grow as one.

b. in a physical union so that they grow together.

c. that are difficult to bud.

d. all of the above

3. In grafting, great care must be used to match which layer of the scion and rootstock?

a. woody　　　　c. bark

b. cambium　　　　d. outside

4. The ideal time of year to cleft graft is

a. late summer, when buds are mature.

b. early spring, just as the buds start to swell.

c. in winter, when the plants are dormant.

d. in spring, when new growth is young and tender.

5. Whip, or tongue, grafting is accomplished in

a. winter, when plants are dormant.

b. early spring.

c. late summer, after buds mature.

d. early summer, while there is still active growth.

6. The side veneer graft is the most effective way to propagate the

a. apple.　　　　c. tree peony.

b. cherry.　　　　d. blue Colorado spruce.

7. The top part of a grafted Colorado spruce should be protected very carefully from drying because

   a. spruce trees are naturally sensitive to drying.
   b. the graft has no roots to replenish moisture.
   c. it is an evergreen and dries quickly.
   d. none of the above

8. In cleft grafting, one side of the scion is cut thicker than the other

   a. so that there is more pressure on that side resulting in close contact with the cambium.
   b. so that the scion is handled more easily.
   c. so that the scion makes more rapid growth.
   d. all of the above

9. All cut areas that are left exposed in the grafting process should be covered with waterproof material

   a. to prevent bleeding of sap.
   b. to prevent drying of the plant tissue.
   c. to prevent disease organisms from entering the graft.
   d. to prevent insects from entering the graft.

OBJECTIVE

To propagate at least one plant by the budding process.

COMPETENCY TO BE DEVELOPED

After studying this unit, the student will be able to

- name at least three plants commercially propagated by budding.
- list the seven steps in the budding process.
- identify on a diagram a piece of budwood, rootstock, a budding knife, and rubber bud ties.
- differentiate between a fruit bud and a leaf or vegetable bud by cutting each from a bud stick and labeling.
- propagate a plant by the T-budding process.

The process of budding is actually a form of grafting; in fact, budding is sometimes called *bud grafting*. Budding differs from grafting in several ways, however. In budding, a single bud is used instead of a scion. Because of this, many more plants can be reproduced from the same amount of parent wood. Budding is also accomplished more quickly. Another difference is the time at which the operation takes place — budding is accomplished when the rootstock is in active growth rather than in the dormant (resting) stage. Budding is done either in spring (March or April), in June, or in late summer or fall (July through September).

Budding is generally done on small, one-year-old seedlings. Small branches in the tops of trees are also budded. Whichever type of rootstock is chosen, it is important that it be actively growing, disease resistant, and able to give the desired growth.

unit 14 Budding

If there is no sign of disease and the bark separates easily from the wood, the operation should be successful. It is also important that mature vegetative or leaf buds of the desired variety are available. As in other forms of grafting, the scion and rootstock must be compatible.

Some plants commonly propagated by budding include apples, pears, peaches, plums, cherries, roses, and the citrus group. Late summer is the best time to propagate fruit trees by budding. The rootstock generally grows from seed planted in fall to the size of a lead pencil by July or August of the following year. Leaf buds on the stock or parent plant are generally mature enough for use at that time.

METHODS OF BUDDING

Methods of budding include T-budding, patch budding, inverted T-budding, flute budding, I-budding, and chip budding, among others. The most commonly used method is the T-budding process, also known as shield budding. The T-budding operation is especially popular in the propagation of tree fruits and roses. It is a much quicker process than grafting; some rose budders insert up to several thousand buds a day if there are other people present to help do the tying.

STEPS IN THE BUDDING PROCESS

1. Plant seeds for seedling rootstock (the previous fall).
2. Select the variety of budwood to be propagated.
3. Determine the correct date to bud, as indicated by both the bud maturity of the desired plant and the active growth of the rootstock.
4. Cut the budwood, label it, and protect it in such a way that it does not dry out.

5. Perform the budding process.
6. Check to see if buds have taken.
7. Cut off the rootstock above the bud (the following spring).

COLLECTING BUDWOOD

Bud sticks, small shoots of the current season's growth, are collected on the same day the buds are to be inserted, figure 14-1(A). The budwood is kept wrapped in waterproof paper so

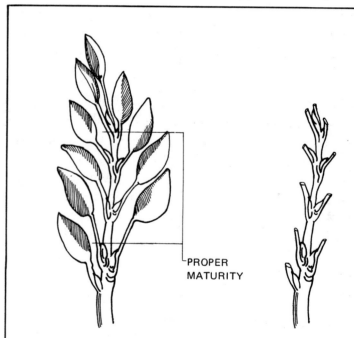

(A) Budwood with leaves still attached. Buds in the bracketed area are probably best for budding because of maturity.

(B) Budwood with top and bottom trimmed and leaves removed. The leaf petiole remains as a handle for holding the buds.

Fig. 14-1 Preparing the budwood

that it does not dry out. The bundle of bud sticks is then labeled according to variety and date cut. Shoots that have a vigorous growth pattern usually have the best and most usable buds on them.

Remember that vegetative buds are necessary for propagation — fruit buds will not grow into new plants. Vegetative and fruit buds can be easily distinguished. The vegetative or leaf bud is slimmer and more pointed compared with the fruit bud. Fruit buds generally appear on the base or bottom end of the shoot.

Collect budwood of the proper variety. Carefully cut all the leaves from the stick, leaving a short piece of the leaf or petiole to act as a handle to hold the bud after it is cut from the stem, figure 14-1(B). Wrap in waterproof paper or place in a plastic bag.

ROOTSTOCK

The rootstock, which at this point has been developing from seed for one year, should now be about the size of a lead pencil for proper budding. Remember that the rootstock and budwood must be of related species, such as apple with apple, and cherry with cherry.

The seedling rootstock must be in active growth, indicated by soft, rapidly growing branch tips. The final test for determining if the seedling is receptive to the bud is to make a T-shaped cut and separate the bark from the wood of the stem. The bark should slip loose easily and the wood underneath appear moist and smooth, with no tearing or stringing of tissue. When this is established, the process may be continued.

CUTTING AND INSERTING THE BUD

Materials List	
budding knife	root stock
rubber bud ties	bud wood
plastic bag to collect budwood	
label and waterproof pen or pencil	

The next step is to prepare the seedlings being used for rootstock. The budding process is easier if a team of two people work together. One person makes the cut with a sharp budding knife and inserts the bud while the second person ties the bud securely in place.

> **Caution:** The person making the cut should exercise care when handling the knife.

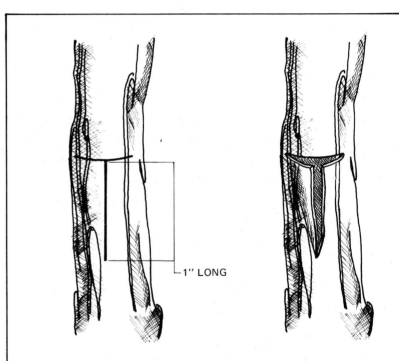

(A) A T-shaped cut is made in the rootstock. The horizontal cut is about one-third the diameter of the rootstock.

(B) The corners of the bark are lifted so that the bud can be inserted easily.

Fig. 14-2 Preparing the rootstock

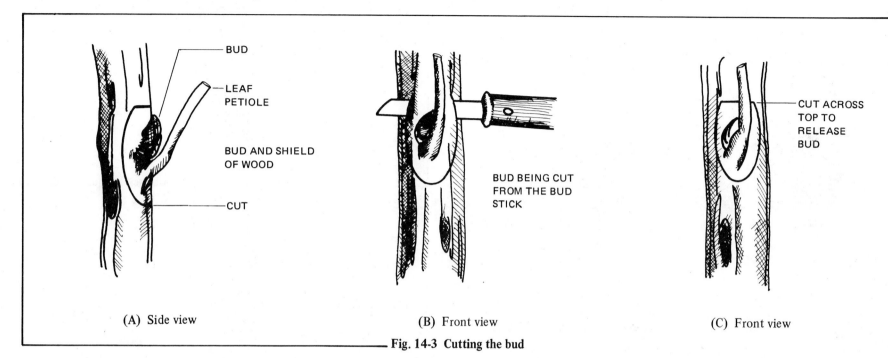

(A) Side view   (B) Front view   (C) Front view

Fig. 14-3 Cutting the bud

The person making the cut kneels beside the row of seedlings, bending them toward him or her. At a spot about 1 to 2 inches above ground level and where the stem is smooth, a 1-inch vertical cut is made through the bark. The cut should be made on the north side of the stem if possible. This protects the bud from the sun both in summer and winter. A second horizontal cut is made across the tip of the vertical cut to form the T shape for which this process is named. The knife is given a twist so that the bark and wood separate, figure 14-2(A). If necessary, the two sides of the vertical cut should be lifted at the top with the edge of the knife blade, opening the slit in the bark, figure 14-2(B). As mentioned earlier, if the bark separates easily, the rootstock is growing rapidly enough to bud.

The next step is to cut the leaf bud from the bud stick. Generally, those buds in the middle portion of the bud stick are used for budding. Buds on the terminal end may not be mature enough, and buds on the basal end may be flower buds. The buds which are used may vary with the maturity of the bud stick. Whichever is used, the bud in the axil of the leaf stem must be visible and well developed.

The bud is cut with a small shield of bark and a sliver of wood attached, figure 14-3. The cut is made starting about 1/2 inch below the bud and cutting just deeply enough to include a small sliver of wood. The cut is continued under the bud and past it to about 1/4 inch above the bud. A second cut made at a right angle to the bud stick at the end of the first cut releases the bud

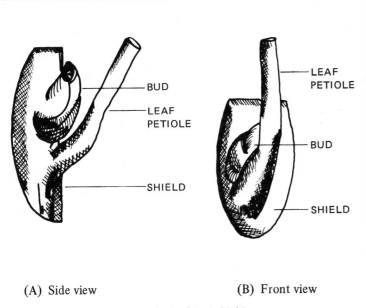

(A) Side view   (B) Front view

Fig. 14-4 The leaf bud shield

from the bud stick. The bud shield should appear as the one in figure 14-4.

The bud is immediately inserted into the T-shaped cut made in the rootstock and pushed down into the open slit in the bark until the top of the bud shield is even with the top of the T cut in the rootstock. This allows the bud shield to be completely inserted underneath the bark of the rootstock and to fit snugly against the wood, figure 14-5.

The cut area is now tied with a rubber bud tie to hold the new bud firmly in place. The entire cut area is covered with the bud tie; only the bud itself is exposed. The bud tie is tucked under itself at each end to hold it in place. These ties, made of a special material, are designed to disintegrate or break in about three weeks. They do not require removal.

(A) The bud is inserted in the T cut on the rootstock by pushing the bud shield down and underneath the flaps of bark.

(B) The bud is now in place underneath the bark. The top of the bud shield is even with the horizontal cut on the rootstock.

(C) The bud is tied firmly in place with a rubber bud tie. Nothing more is done until the following spring.

Fig. 14-5 Insertion and tying of bud

Fig. 14-6 Budder inserting the shield of bark with an apple bud in the T cut. A small tree is being used as the rootstock. (Courtesy New York State Agriculture Experiment Station, Paris Trail, Photographer)

In about three weeks, an inspection is made to see if the bud has *taken* (has begun to grow and is still alive). If the leaf petiole has dropped off and the bud appears to be plump and of normal color, the bud has probably grown to the rootstock. Nothing else is done to the bud or the rootstock until early the following spring. At that time, the top of the seedling rootstock is cut off directly above the bud as shown in figure 14-7(A) and (B). This forces the new bud into active growth, which then develops into the new top of the plant, figure 14-7(C). Any suckers or shoots that sprout out below the bud from the rootstock are pinched off. This process must be continued until no more new sprouts emerge. The budding process is then completed. After one or two years' growth of the new bud, it is ready to be transplanted to a permanent site.

(A) Front view. The rootstock is cut directly above the bud in the spring before growth begins.

(B) Side view. The cut is made at a slant starting about 1/4" above the bud.

(C) New shoot growing and healing of the wound.

Fig. 14-7 Continuation of the budding process the following spring

STUDENT ACTIVITIES

1. Take a field trip to a nursery to watch a budding operation.

2. Plant seeds to grow rootstock.

3. Practice the budding process on plant parts provided.

4. Bud plants in a nursery row.

5. Write a description of the physical condition of a bud that has grown quickly and of one that is dead. Pinpoint the differences in appearance.

SELF-EVALUATION

Select the best answer from the choices offered to complete the statement or answer the question.

1. When budding,
   a. the bud stick must have mature buds.
   b. a single bud is used instead of a scion.
   c. only vegetative buds are used.
   d. all of these

2. Budding is done when the rootstock
   a. is in active growth and about the size of a lead pencil.
   b. has matured for the season.
   c. is available.
   d. is about the same size as the scion.

3. Vegetative or leaf buds may be differentiated from flower buds because they
   a. are plumper.
   b. are more slender or pointed.
   c. are generally located at the base of the bud stick.
   d. are not yet mature.

4. Bud sticks are cut from
   a. actively growing wood.
   b. healthy wood.
   c. the current season's growth.
   d. all of these

5. Rootstock for budding is

   a. grown from seed planted the previous year.   c. of a closely related plant species.
   b. generally about the size of a lead pencil.   d. all of these

6. The bud is inserted into the rootstock

   a. on the south side of the rootstock 2 inches above the ground.
   b. on the north side of the rootstock 2 inches above the ground.
   c. on the north side of the rootstock 6 inches above the ground.
   d. none of the above

7. From which portion of the bud stick are the best quality buds obtained?

   a. end   c. middle
   b. base   d. none of these

8. Three plants commercially propagated by budding are the

   a. apple, peach, and rose.   c. azalea, rhododendron, and viburnum.
   b. apple, rhododendron, and cherry.   d. rose, peach, and chestnut.

9. The bud is cut with a small shield of

   a. bark.   c. bark and wood.
   b. wood.   d. none of these

10. The seedling is in the proper state of growth for budding if upon opening the T-shaped cut in the seedling, the bark separates from the wood

    a. with difficulty.
    b. with a small piece of wood attached.
    c. enough so that the bud can be inserted.
    d. easily and cleanly.

11. The leaf stem or petiole is left on the bud because

    a. it acts as a handle for holding the bud.
    b. it helps the bud to heal.
    c. it simplifies tying the bud in place.
    d. the bud would be damaged if it were removed.

12. The bud is tied tightly in place with a bud tie so that

   a. it is held firmly in place against the rootstock.
   b. animals are prevented from loosening it.
   c. the flow of sap is stopped at that point.
   d. the rootstock is gradually girdled.

13. An inspection to see if the bud has grown and is alive should be made in about

   a. 6 weeks.
   b. 3 weeks.
   c. 1 year.
   d. 6 months.

14. The spring after the bud is attached, the seedling rootstock top is cut off

   a. just below the bud.
   b. even with the bud.
   c. directly above the bud.
   d. none of these

15. In the development of a new bud, what must be pinched off as they appear on the rootstock below the bud?

   a. roots
   b. suckers
   c. scars
   d. knots

16. The new plant may be transplanted to a permanent site in

   a. 2 or 3 years.
   b. 6 months.
   c. 1 or 2 years.
   d. 3 or 4 years.

17. Which of the following occurs in budding but not in grafting?

   a. A single bud is used and the rootstock is in active growth.
   b. A single bud is used and the rootstock is larger than that used in grafting.
   c. A vegetative leaf bud is used and it is placed closer to the ground on the rootstock than in grafting.
   d. A single bud is used and the rootstock is dormant.

OBJECTIVE

To propagate one plant by one of the layering processes described in the unit.

COMPETENCY TO BE DEVELOPED

After studying this unit, the student will be able to

- describe two situations in which layering is used to propagate plants.
- list five plants commonly propagated by layering and two plants which propagate naturally by layering.
- list the eight steps in the layering process.
- perform simple layering as described in the text.

Materials List

parent plant

shovel

peat moss or sawdust

two stakes, one with hooks

string

brick or stone

propagating knife

hammer

---

unit 15
Layering

Layering is a method of asexual propagation in which roots are formed on a stem or root while it is still attached to the parent plant. The stem or root which is rooted is called a *layer*. The

Fig. 15-1 Natural layering of spirea. In (A), the plant has been dug from the ground. The circled area shows the point at which one layer is to be cut from the parent plant. The three stems shown at the right of the pruner's hand in (A) are shown cut free in (B). (Richard Kreh, Photographer)

(A)

(B)

layer is cut free from the parent plant only after rooting has taken place. This is the major difference between layering and other methods of asexual reproduction.

As a method of propagation, layering has advantages and disadvantages. It is a relatively simple process, but requires more time than many other methods of asexual propagation, since it requires a great deal of work by hand. Also, fewer plants can be started from each parent plant than with the use of cuttings. The major advantage of layering is the degree of success with which some plants will root when layered as compared with results achieved with other methods. (Some plants which are difficult to root by other methods may be rooted by layering.) In fact, many plants root naturally by layering, figure 15-1.

Cane fruits such as the black raspberry and the boysenberry root easily and economically by *tip layering*, one form of natural layering. In this method of natural layering, a shoot with current season's growth bends to the ground and then turns sharply upward. Roots develop from the point at which the shoot meets

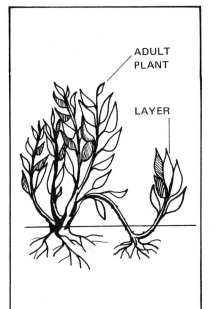

ADULT PLANT

LAYER

Fig. 15-2 Tip layering

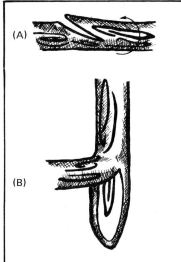

(A)

(B)

Fig. 15-3 The stem is cut halfway through where it is to be inserted in the soil (A) and then twisted 90 degrees (B).

with the soil, figure 15-2. Other plants which root naturally by layering are strawberries (by runners), red raspberries and spirea (by root suckers), and African violets (by crown division).

Some plants do not reproduce naturally by layering, but can be propagated in this manner with assistance from the horticulturist. Two methods used for this purpose are simple layering and air layering.

SIMPLE LAYERING

In *simple layering*, a branch from a parent plant is bent to the ground where it is partially covered at one point with soil. The terminal end remains exposed. (See figure 15-3.) This process is usually accomplished in early spring.

Steps in Simple Layering

1. Select a stem with one-year-old wood that can be bent to soil level.

2. Make a cut about halfway through the top of the stem at the point at which the stem is to be inserted in the ground, figure 15-3. Use a sharp knife for the best cut.

> **Caution: Do not cut toward yourself or another person.**

3. Twist the stem 90 degrees and raise the stem to an upright position.

4. Dig a hole or trench 3 to 6 inches deep at the point at which the twist in the stem touches the ground.

5. Dust the stem wound with rooting hormone and push the stem into the opening in the ground.

6. Drive a peg with hooks on it next to the stem to hold it upright, figure 15-4. Another peg holds the layer in the ground.

7. Fill the trench with soil, and mulch the top to help hold in moisture. A brick or stone may be placed on top to keep the layer in position. Keep the soil moist (by watering, if necessary). The layer should be rooted by the next spring (about one year).

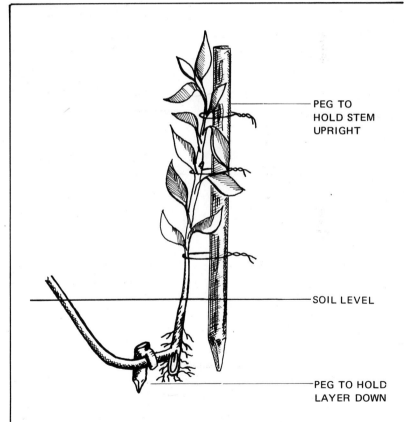

PEG TO HOLD STEM UPRIGHT

SOIL LEVEL

PEG TO HOLD LAYER DOWN

Fig. 15-4 Pegs hold the layer upright and in the ground.

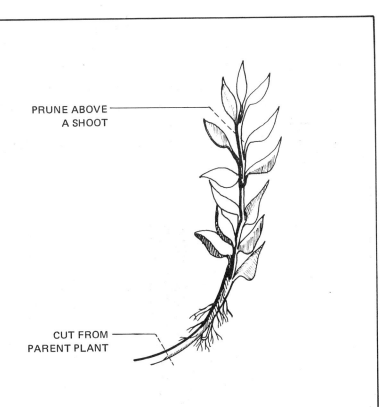

PRUNE ABOVE
A SHOOT

CUT FROM
PARENT PLANT

Fig. 15-5   Rooted layer cut from parent plant. The top may be pruned to form a better balance between the plant and its root system.

8. Cut the layer free of the parent plant when it is well rooted and transplant to a new location, figure 15-5. If the top of the plant is largely developed and only a small root system is present, cut and remove part of the top to balance the plant's moisture requirements with its root system.

Since the root system of the new plant is so limited, it is necessary to water it liberally. Some shade may also be necessary until the root system is well established.

AIR LAYERING

The ancient Chinese practiced a technique called *air layering*, a process which eliminates burying part of the parent plant in the soil. Instead, a part of the plant stem is slit, or girdled, and then surrounded by a moist growing medium in some type of enclosure. Roots form where the plant has been wounded. The materials available at the time made the task much more time consuming and difficult for the Chinese than it is today. The major problem in the past was keeping the layered area moist. The development of polyethylene film has made air layering a highly successful procedure. It is used commercially to propagate certain tropical and subtropical trees and shrubs which are difficult to root by other procedures. The Persian lime (*Citrus aurantifolia*) is commercially propagated in this way. The magnolia and the rubber plant (*Fiscus elastica*) may also be propagated by air layering. In fact, most plants may be propagated using this method, but usually it is not economically feasible.

Air layers are generally made in the spring on wood of the previous year's growth or in late summer on partially hardened growth of the current season. Older wood is much more difficult to root.

Steps in Air Layering

1. Girdle the selected stem in a band 1/2 to 1 inch wide. (To *girdle* is to completely remove the bark and cambium around a plant.) Girdle at a spot that is 6 to 12 inches from the top of the branch so that 6 to 10 leaves are left between the girdled spot and the branch tip. These leaves are necessary to produce carbohydrates and hormones necessary for good root development. Since the stem is girdled, these materials, produced in the leaf, accumulate at the girdled point on the

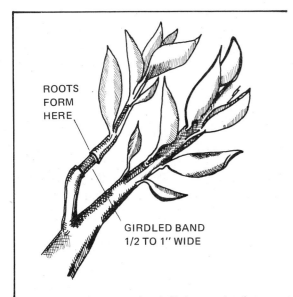

ROOTS
FORM
HERE

GIRDLED BAND
1/2 TO 1″ WIDE

Fig. 15-6   The stem is girdled to stimulate root formation.

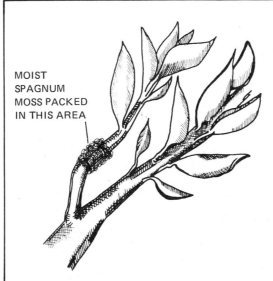

MOIST
SPAGNUM
MOSS PACKED
IN THIS AREA

Fig. 15-7   Sphagnum moss is packed around the girdled area.

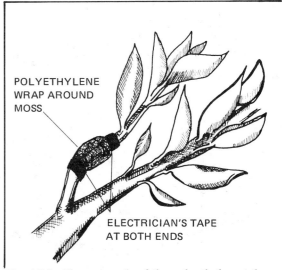

POLYETHYLENE
WRAP AROUND
MOSS

ELECTRICIAN'S TAPE
AT BOTH ENDS

Fig. 15-8   The two ends of the polyethylene tube are twisted around the branch and fastened securely with electrician's tape.

stem of the plant, greatly stimulating root formation. Scrape the girdled area to ensure that all bark and cambium are removed, thus slowing the healing process, figure 15-6.

2. Treat the girdled area with rooting hormone (Hormodin #3, Rootone #10, or a concentration of 0.5 to 0.8 percent indolebutyric acid in water). Treat the bark edges on the branch tip side of the girdled area especially well.

3. Pack about two handfuls of moist (but not wet) sphagnum moss around the girdled area, figure 15-7. Squeeze all the moisture out of the moss — the cut area will rot if the moss is wet.

4. Wrap a 10-inch square piece of polyethylene plastic around the sphagnum moss so that the moss is completely covered. No moss should protrude from underneath the polyethylene wrap. Fold together both ends of the polyethylene film at the same time. Wrap the film until a snug fitting tube is formed. Twist the two ends of the tube around the branch and fasten securely with plastic electrician's tape, figure 15-8. Extend the tape far enough onto the plant stem so that there is a watertight seal. The polyethylene allows the passage of air with no moisture loss. Additional watering is not necessary.

5. Rooting may be observed through the polyethylene plastic. When strong roots have developed, the layer may be removed for transplanting. If the plant being layered is deciduous, wait until the leaves fall from the plant and it is dormant. If the plant is an evergreen, remove the layer when no new active growth is occurring. Simply cut the layer free below the rooted area.

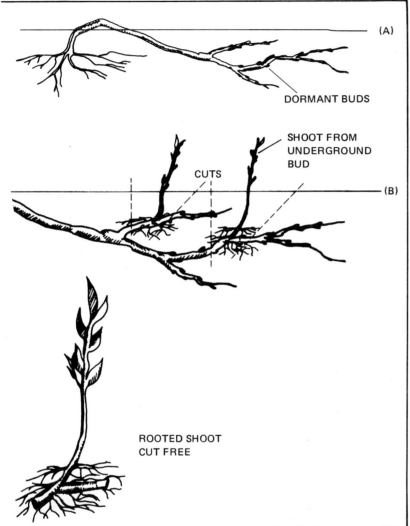

(A)

DORMANT BUDS

SHOOT FROM
UNDERGROUND
BUD

CUTS

(B)

ROOTED SHOOT
CUT FREE

Fig. 15-9 Trench layering. (A) shows the mother plant in the trench exhibiting dormant buds. Soil is added to the trench periodically to cover the developing shoots. At the end of the growing season, the layers are separated from the mother plant, as shown in (B).

If the top of the new plant has a great deal of growth in relation to the root system, plant it in a pot or flat and place it in an area of high humidity for a short time. After additional root development has occurred, it may be gradually hardened off by exposure to a drier atmosphere in preparation for the permanent planting site.

OTHER METHODS OF LAYERING

Three other methods by which plants are layered by the horticulturist are trench layering, stool layering, and compound layering.

Trench Layering

In *trench layering*, the mother plant is bent to the ground and buried in a trench, figure 15-9(A). As shoots arise from the buried buds, roots form on the covered portion of the plant. The shoots may then be separated from the mother plant, figure 15-9(B).

Stool Layering

Stool (or *mound*) *layering* begins with the planting of a rooted layer in soil. After one season's growth, the parent plant is cut back to soil level, figure 15-10(A). The stem is then covered with a mound of soil, figure 15-10(B). Soil is added to the mound periodically as the shoots grow, figure 15-10(C). At the end of the season, the new shoots are rooted and dormant, figure 15-10(D). The shoots are cut free and planted in early spring. One year has passed from the beginning of the layering process to the planting of the new shoots.

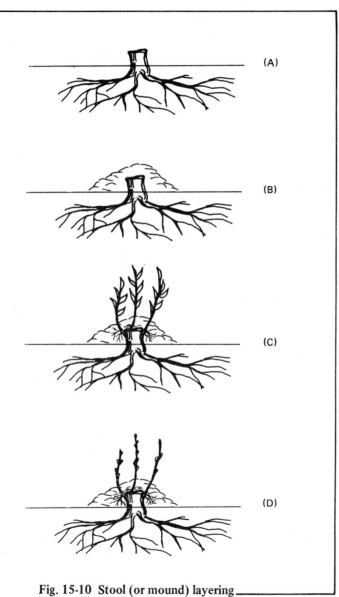

Fig. 15-10 Stool (or mound) layering

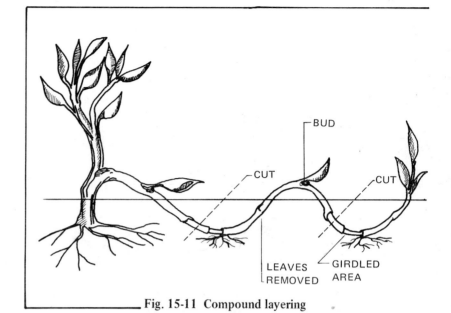

Fig. 15-11 Compound layering

Compound Layering

Compound layering, accomplished in springtime, is very similar to simple layering except that a stem is covered by soil at two or more points along its length. Long stemmed, vinelike plants such as wisteria, clematis, and grape, work well with this system, figure 15-11.

The stem is girdled at a point below ground near where new roots are expected to form. The portion of the stem above soil level must have at least one bud to form a new shoot which will grow into the top of the new plant.

After rooting takes place, usually by the end of one growing season, the stem is cut to include roots and a new shoot on each new plant. Several plants may be produced from a single stem by compound layering.

STUDENT ACTIVITIES

1. Visit a horticulturist and observe a layering process. Record the steps in the process in a notebook.

2. With a group of students, perform a layering process on a parent plant. Keep records on the experiment, including the date on which the plant is layered and the date on which it roots.

SELF-EVALUATION

1. Layering is used to propagate plants which are

    a. best grown from seed.
    b. difficult to root or root naturally by layering.
    c. easily cut apart.
    d. too thick to propagate by other methods.

2. The main difference in propagating by layering opposed to cuttings is that in layering, the newly propagated plant part is

    a. cut free from the parent plant.   c. wounded.
    b. older.   d. still attached to the parent plant.

3. Which of the following plants are commonly propagated by layering?

    a. apple, pear, peach, cherry, plum
    b. black raspberry, boysenberry, strawberry, red raspberry, African violet
    c. bean, tomato, pea, corn, squash
    d. rhododendron, azalea, viburnum, holly, lilac

4. When a rooted layer being transplanted has a small root system and a large top, the plant's moisture requirements are balanced by

    a. cutting back the top to a smaller size.
    b. watering the plant more often.
    c. using a mulch to conserve moisture.
    d. none of the above

5. The age of the wood affects the speed at which layers root. Younger (one-year-old) wood generally roots

a. more slowly than older wood.
b. at the same speed as older wood.
c. more rapidly than older wood.
d. more heavily than older wood.

6. In air layering, cutting the bark off the stem, known as girdling,

a. stimulates root formation just above the point of girdling.
b. helps in breaking off the stem when it is rooted.
c. gradually starves the stem.
d. all of the above

7. The cut area on the layered stem is dusted with a rooting agent to

a. prevent fungus diseases from killing the stem.
b. cause roots to form faster.
c. kill any insects that may be present.
d. aid in attracting moisture.

8. The main reason for covering the sphagnum moss which is packed around air layers with polyethylene plastic is that

a. it holds moisture in and allows the roots to breathe.
b. the roots can be observed growing inside.
c. it is the least expensive material available.
d. it prevents circulation of air.

Section 6
Greenhouse Crops

unit 16
Poinsettias

OBJECTIVE

To produce a saleable poinsettia crop.

COMPETENCY TO BE DEVELOPED

After studying this unit, the student will be able to

- prepare a soil mixture for growing poinsettias.
- transplant ten rooted poinsettia cuttings.
- identify one growth regulator and use it correctly.
- outline a growing schedule for poinsettias.
- identify four insects that affect poinsettias and the control measures for each.
- list the common names of three diseases that affect poinsettias and the control measures for each.

Materials List

2 1/4-inch rooted poinsettias

loamy soil

sand

horticultural grade perlite

Pro-Mix B

4-, 5-, and 6-inch pots

Cycocel or some other growth regulator

Benlate fungicide

assorted insecticides (Cygon 2E, orthene, diazinon, Sulfotepp, Sevin)

Fig. 16-1 Poinsettia varieties grown for the winter holiday season

The poinsettia has long been a traditional flower used during the winter holiday season. Poinsettias were first introduced into the United States from Mexico by J.R. Poinsett, the first ambassador to Mexico. Poinsettias were used as exotic plants in conservatories and botanical gardens until 1900. Through the work of Albert Ecke, poinsettias were then grown in southern California as cut flowers. Today, the growing of poinsettias is a big part of many horticultural operations.

THE POINSETTIA FLOWER

The poinsettia flower is a small, yellow flower that grows at the terminal end of the plant. Just below the flower are the *bracts*, the leaves that give poinsettias their color. The most common bract color is red, but through years of research, other colors have been developed. Some of these colors are white, pink, white variegated, and red and white variegated.

Poinsettias require a short day for the production of the flower, which in turn is necessary to produce the color in the bracts. Poinsettias should be grown in full sunlight. During months with long days, the plants develop leaves and increase stem length. The plant forms the flower buds naturally during late September and early October. During November and December, color develops in the bracts.

GROWING POINSETTIAS

Planting and Spacing

Wholesale growers can supply 2 1/4-inch rooted poinsettia cuttings for planting. Before purchasing the cuttings, check to be sure they are healthy. When planting the cuttings, several things must be considered. The size of the pot is important, since it partially determines the size of the full grown plant. The size of the plant also depends upon the number of plants per pot and greenhouse space allowed for each potted plant. Each of the above factors also determines the price of the saleable plant.

Figure 16-2 gives the correct pot size and number of rooted cuttings per pot, treatment, and spacing for each pot in the greenhouse.

Pot Size	Plants per pot	Treatment	2 Flowers per Square Foot	
			Spacing inches	Square ft/pot
4 inches	1	none	9 x 9 inches	0.56
4 inches	1	pinched	15 x 15 inches	1.56
5 inches	3	none	15 x 15 inches	1.56
5 inches	1	pinched	15 x 15 inches	1.56
6 inches	3	none	15 x 15 inches	1.56
6 inches	1	pinched	15 x 15 inches	1.56

Fig. 16-2 (from *Ball Red Book*, 13th edition, 1975, courtesy George J. Ball Company)

Pinching

Pinching poinsettias is a process in which the grower removes the terminal end of the plant. Pinching is accomplished by taking the top of the plant between the thumb and index finger and carefully breaking the top of the poinsettia completely off. Four or five nodes should remain above the soil level. Nodes occur at each point at which a leaf is attached to the stem. Therefore, four or five leaves should remain on the plant after it is pinched. After pinching, new shoots will develop from the bud in the axis of the node. As this shoot grows, it will produce a flower. After the flower develops, color will form on the bracts of the poinsettia. The major reason for pinching poinsettias is to increase the number of bracts per plant. It also causes the plant to appear fuller.

Soil Mixes

Poinsettias may be grown in many different medias. Whatever type of media is used, it must be well drained. This allows for maximum root development. Poinsettias also require a generous supply of fertilizer and moisture.

One suitable soil mixture consists of one-quarter soil (loam), one-quarter construction grade sand, one-quarter perlite (horticultural grade) and one-quarter peat moss. Another mix that works well is Pro-Mix B (in bags containing 6 cubic feet) and horticultural grade perlite (in bags containing 2 cubic feet). Perlite aids in drainage of the mixture. The Pro-Mix B and perlite should be mixed to evenly distribute the perlite.

It is important that all soil mixes be free of any disease organisms. If soil is used, it should be sterilized at 180°F for thirty minutes.

Using Growth Regulators

Growth regulators were developed to produce a better quality plant. Growth regulators have two main purposes: to control the height of the plant, therefore causing the plant to

Date	Temperature (°F)		Greenhouse Procedures
	Night	Day	
September 10	72	80-85	Arrival of 2 1/4-inch rooted cuttings from a reliable greenhouse grower. Transplant three rooted cuttings into a 6-inch azalea pot. Treat with a fungicide to control disease. Mix with water: Dexon — 35% wettable powder (WP), 1 1/4 teaspoon per 5 gallons water and Benlate — 50% wettable powder (WP), 1 1/4 teaspoon per 5 gallons water.
September 11	72	80-85	Start slow release fertilizer program. Apply 1 level tablespoon osmocote 18-9-9 per 6-inch azalea pot. Start a constant liquid feeding program.
September 23	65	80-85	Space each 6-inch pot in an area of 15 inches by 15 inches. Spray with Cycocel 3000 ppm (3 1/2 oz. per 1 gallon water). Lower temperature to 65°F. It is during this time that the flower buds form. Formation of flower buds is necessary for the production of color in the bracts of the poinsettia.
October 10	68	80-85	Increase night temperature to force color formation in bracts.
October 15	68	80-85	Drench soil with Dexon — 35% WP, 1 1/4 teaspoon per 5 gallons water.
November 15	68	75	Color should be showing in upper bracts. Use a fan to provide continuous air movement in the greenhouse. Drench soil with Dexon — 35% (WP), 1 1/4 teaspoon per 5 gallons of water.
December 5	62	75	Lower night temperature to develop a bright color in the bracts. By this time the bracts should have achieved maximum size.
December 10	60	75	Poinsettias are ready for sale.

WP — Wettable Powder
ppm — parts per million

Fig. 16-3 Growing schedule for poinsettias

be saleable, and to improve the quality of color in the poinsettia bracts by darkening them, figure 16-4.

Some poinsettia growers use a soil drench growth regulator. This type is mixed with water and applied to the soil mixture. It is then absorbed by the poinsettia plant, which in turn controls the height of the plant. One example of this type of growth regulator is A-Rest.

Another type of growth regulator is applied as a foliar spray. The poinsettia foliage is sprayed until droplets form on the leaves. The leaves absorb the regulator, thus controlling their height. One

Fig. 16-4 The effect of growth retardant on poinsettias. Left to right: (1) no treatment (control plant); (2) 0.125 milligrams/pot; (3) 0.25 milligrams/pot; and (4) 0.50 milligrams/pot. (Courtesy George J. Ball Company)

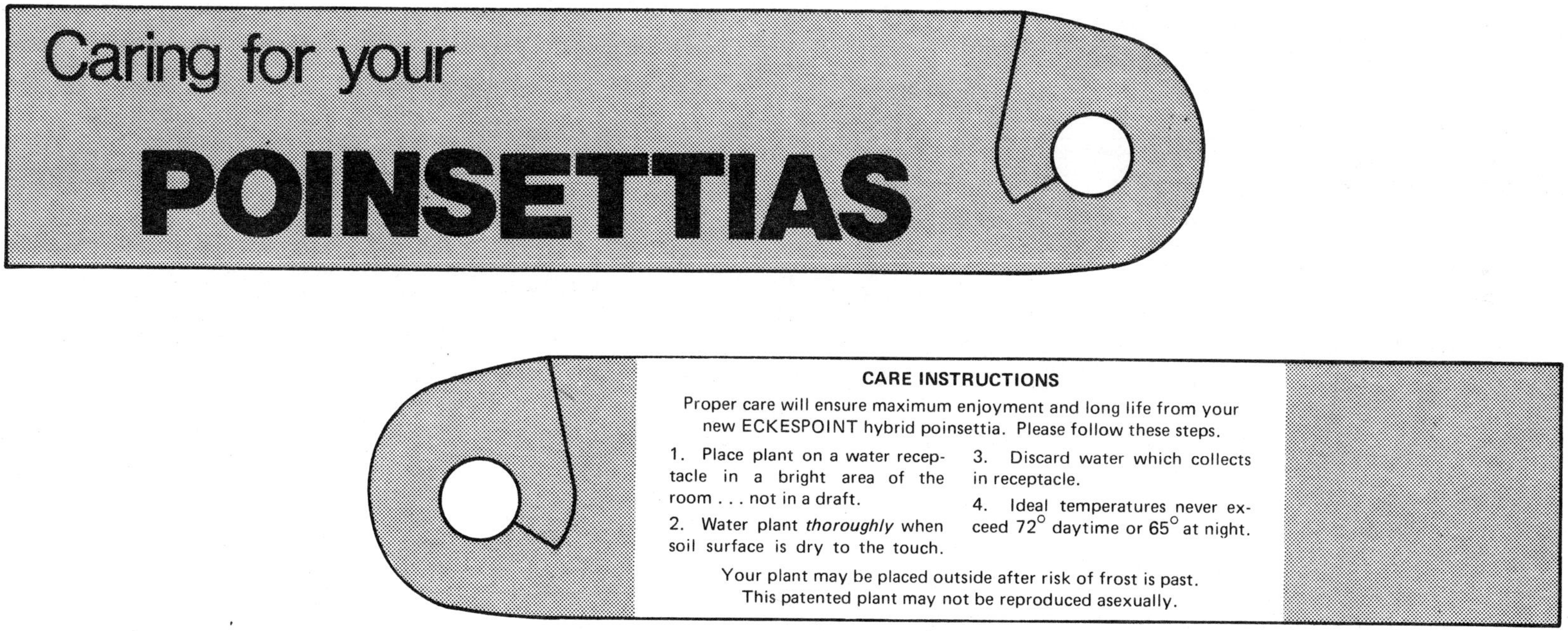

Caring for your

POINSETTIAS

CARE INSTRUCTIONS

Proper care will ensure maximum enjoyment and long life from your new ECKESPOINT hybrid poinsettia. Please follow these steps.

1. Place plant on a water receptacle in a bright area of the room . . . not in a draft.

2. Water plant *thoroughly* when soil surface is dry to the touch.

3. Discard water which collects in receptacle.

4. Ideal temperatures never exceed 72° daytime or 65° at night.

Your plant may be placed outside after risk of frost is past.
This patented plant may not be reproduced asexually.

Fig. 16-5 Instructions for care of the poinsettia

Pest	Materials	Formulation	Dosage per 100 gal.	General
Aphids	dimethoate (Cygon)	2 EC	1 pint	1 spray or 1 drench
	Meta-Systox	25.2% EC	1 pint	1 spray or 1 drench
Ants	diazinon	50% WP	1 pound	Apply on bench legs and wall bases, walls, and soil.
Mealybugs	dithio (Sulfotepp)	aerosol or smoke	Follow label directions.	4 applications, 3 days apart, repeat every 3 weeks until controlled
	diazinon	50% WP	1 pound	3 spray applications, 10 days apart
	dimethoate (Cygon)	2 EC	1 pint	2 spray applications, 14 days apart
Scales	dithio (Sulfotepp)	aerosol or smoke	Follow label directions.	For crawler stage only. 4 applications 3 days apart
	diazinon	50% WP	1 pound	2 spray applications, 14 days apart
	dimethoate (Cygon)	2 EC	1 pint	2 spray applications, 14 days apart
	Sevin	50% WP	2 pounds	2 spray applications, 14 days apart
Slugs and Snails	metaldehyde	20% liquid	1 quart	1 spray application to soil
	metaldehyde	dust or granules		Follow label directions.
Spider Mites (Red Spiders)	Kethane	18 1/2% WP	1 pound	3 spray applications, 10 days apart
	Chlorobenzilate	25% WP	1 pound	3 spray applications, 10 days apart
	dithio (Sulfotepp)	aerosol or smoke	Follow label directions.	4 applications, 2 days apart
	Pentac	50% WP	1/2 pound	2 spray applications, 10 days apart
Whiteflies	dithio (Sulfotepp)	aerosol or smoke	Follow label directions.	4 applications, 3 days apart; repeat every 3 weeks until controlled
	Thiodan	50% WP	1 pound	2 spray applications, 10 days apart
	Cygon	2 EC	1 pint	1 spray application
	diazinon	50% WP	1 1/2 pounds	2 spray applications, 10 days apart

Important terms in Chart

drench — to apply the material to the soil by watering

spray — to apply material in a mist form, giving coverage from top to bottom of foliage area

WP — wettable powder

EC — emulsifiable concentrate

SPECIAL NOTE: It is also advisable to consult local university extension services for special recommendations.

Fig. 16-6 Pest control schedule. (from *Ball Red Book,* 13th edition, 1975. Courtesy George J. Ball Company)

Name	Plant Symptoms	Organism Characteristics	Control
Rhizoctonia (Stem and Root Rot)	Plant has a brown rot on the stem at soil line; roots have brown lesions; plant appears stunted with yellowing leaves from the bottom. In severe cases, plant will fall over.	Fungus carried in the soil and by infected plants; spreads very easily by water. Grows best under conditions of high humidity and high temperatures.	Remove infected plants from the greenhouse; drench with fungicides. Terraclor — 75% WP (1 1/4 tsp./5 gal. water) Benlate (1 1/4 tsp./5 gal. water) Keep soil somewhat dry.
Botrytis (Gray Mold)	Rotting of young and immature tissue on leaf edges	Fungus with airborne spores. Occurs in very humid areas with moderately low temperatures.	Keep a clean greenhouse; provide good air circulation; remove all dead plant material. Fungicides to use: Termil — fumigation captan — 50% WP (2 tsp./5 gal. water) Benlate — 50% WP (1 1/4 tsp./5 gal. water)
Pythium (Water Mold, Root Rot)	The root tips are rotted; rot moves up the stem; plants are stunted; lower leaves turn yellow and drop; soil remains wet.	Fungus is carried by infected plants and soil.	Reduce amount of moisture. Drench with Dexon — 35% WP (1 1/4 tsp./5 gal. water)

Fig. 16-7 Diseases of poinsettias

example of this type of regulator is Cycocel. The rate of application is determined by the manufacturer's directions.

MAINTAINING POINSETTIAS

Transporting the Plant

When poinsettias are purchased from the local florist and transported to a new location, they should be handled very carefully. The plants should be protected from cold temperatures, and special care should be taken that the branches are not broken. This can be accomplished by carrying the plant in a sleeve.

Sleeves are made for the purpose of protecting plants. They are constructed in several materials such as paper or plastic. Sleeves are tapered in a funnel shape. The base of the sleeve is the same size as the pot, allowing the plant to be completely covered by the sleeve.

When removing the sleeve from the plant, cut it off with scissors. Lift the sleeve upward and off the plant. Be very careful not to break any of the plant's branches or bracts.

Figure 16-5, page 163, illustrates a sample care tag accompanying poinsettias which are purchased from florists. By following these instructions, poinsettias can be maintained throughout the entire holiday season.

Controlling Insects

For optimum growth, poinsettias must be protected from insects. For best protection, be sure the greenhouse is kept clean. Figure 16-6, page 164, lists pests that attack poinsettias and control methods for each.

Controlling Disease

Just as poinsettias must be protected from attack by insects, so must they also be protected from attacks by disease organisms. One of the more common problems of poinsettias growers is root rot. To help prevent this problem, purchase rooted poinsettias from wholesalers who supply disease-free stock. When poinsettias are transplanted, treat them with a fungicide for further control.

Although it may seem that poinsettias have more than their share of attacks by disease organisms, keep in mind that fungicides provide good control. Figure 16-7 lists common poinsettia diseases and methods of control.

STUDENT ACTIVITIES

1. Demonstrate the proper techniques involved in mixing soil for growing poinsettias.
2. Propagate poinsettias from cuttings. Compare the full-grown plants with those propagated from 2 1/4-inch rooted cuttings.
3. At the proper stage of development, pinch the poinsettias grown in activity 1 to increase the number of flowers per plant.
4. Contact a local university and inquire about field days on poinsettia production. If possible, become involved in some of the activities.

SELF-EVALUATION

Select the best answer from the choices offered to complete the statement or answer the question.

1. A poinsettia flower is the
   a. large blue flower at the terminal end of the plant.
   b. large yellow flower at the terminal end of the plant.
   c. small red flower at the terminal end of the plant.
   d. small yellow flower at the terminal end of the plant.

2. Poinsettias produce their color in the leaves below the flower called the

    a. branches.   c. canes.
    b. bracts.   d. none of these

3. What do poinsettias require for flower production?

    a. a long day   c. no light at all
    b. a short day   d. none of these

4. Poinsettias form the flower buds naturally during

    a. late June and early July.
    b. late January and early February.
    c. late September and early October.
    d. late May and early June.

5. Pinching poinsettias is a process of removing

    a. flower buds.   c. terminal ends.
    b. bracts.   d. none of these

6. The major reason for pinching poinsettias is to

    a. increase the number of flowers per stem.
    b. decrease the number of flowers per stem.
    c. separate cuttings for propagation.
    d. none of the above

7. Which of the following are growth regulators that can be used on poinsettias?

    a. 2, 4-D   c. Dexon
    b. Cycocel   d. B-Nine

8. Poinsettias are affected by

   a. aphids.
   b. scales.

   c. mites.
   d. all of these

9. Pesticides applied to soil in which poinsettias are grown are

   a. sprayed.
   b. drenched.

   c. wettable powders.
   d. none of these

10. One of the most common diseases suffered by poinsettias is

    a. flower drop.
    b. root rot.

    c. blight.
    d. none of these

OBJECTIVE

To produce a commercially acceptable potted chrysanthemum crop, beginning with rooted cuttings.

COMPETENCY TO BE DEVELOPED

After studying this unit, the student will be able to

- outline in writing a growing schedule for potted mums for any preselected market date.
- pot rooted cuttings.
- regulate formation of chrysanthemum flower buds by control of night length.
- determine the proper cultural requirements, such as fertilizing and watering, for production of a marketable chrysanthemum crop.

Materials List

rooted chrysanthemum cuttings

growing media and 5-inch or 6-inch pots

lights or shade cloth as need to control length of night

chemical growth retardants such as B-Nine or A-Rest

fertilizer

insecticide and fungicide if needed

---

unit 17
Chrysanthemums

According to sales figures, the chrysanthemum (mum) is the most popular cut flower sold in the United States. It also competes very well in sales as a potted plant. Because it is now possible to artificially control the length of day and night in the greenhouse, the mum can be grown throughout the entire year. The most popular markets for mums are Mother's Day and Easter. Memorial Day also provides a good market.

The mum is considered an easy plant to grow. It offers a wide range of colors and keeps well in the home after being sold. In this unit, the mum is considered a winter crop, flowering for sale at Easter time. It will be prepared for sale as a potted plant.

DETERMINING MARKET AND VARIETY

If sale of mums at Easter is desired, a schedule that assumes that the plants will be blooming before that date must be established. This necessitates that the plants be started several weeks before the Easter date.

The growing schedule is largely dependent upon the variety grown. Some varieties bloom in eight weeks after the start of short days; others bloom in ten, twelve, and fourteen weeks. This is known as the *response time* (response to short days). Varieties of chrysanthemums are grouped in terms of response time. For example, a catalog might specify an *8 week variety* or *10 week variety.*

REGULATING LIGHT

Mums are considered a short season crop. That is, the plant sets flower buds and blooms only when nights are long and days are short. To grow mums satisfactorily, the length of day and night must be regulated at two times.

1. When rooted cuttings are first potted (for one to three weeks), days must be kept long and nights short. (Plants should be exposed to no more than seven hours of continuous darkness.) This allows the plant to establish good vegetative growth before the formation of flower buds.

2. When summer flowering is desired, days must be kept short and nights long to encourage the proper timing of flower bud formation and flowering.

Mums naturally set flower buds in the fall. This means that unless lighting is regulated, the newly started cuttings will set blossom buds in the fall and winter, when nights are long. This causes the plants to blossom when they have not attained enough growth to be attractive.

To prevent flower buds from forming too soon, artificial lighting must be used to lengthen the day or more accurately, to shorten the night. To delay flower bud formation, the plants must not be exposed to an uninterrupted period of darkness over seven hours in length. The best way to accomplish this is to add a period of light in the middle of the night.

The intensity of light to which the plants should be exposed is about 7 foot candles. Bulbs with reflectors are usually necessary for best results. Two 4-foot beds may be lighted with one row of 100-watt bulbs. Space the bulbs 6 feet apart and 5 feet above the bed's soil level.

Example. If sunset occurs at 5:00 P.M. and the sun rises at 7:20 A.M., lights could be turned on for four hours from 10:00 P.M. until 2:00 A.M. This means that darkness is provided from 5:00 P.M. until 10:00 P.M. (five hours) and from 2:00 A.M. until 7:20 A.M. (five hours and twenty minutes). In this way, both periods of darkness are shorter than seven hours. Under these conditions, the formation of flower buds is delayed and the plants are allowed to increase in plant size and stem length. If there are other plants growing nearby that would be affected by the extra light, shade curtains should be installed to shield them from the light. In warm, sunny climates where plants naturally grow taller, lighting is not necessary.

For summer flowering, shading is necessary since days are long. Mums set flower buds only when exposed to short days (twelve hours or less) and to night temperatures of 60°F (16°C) or lower. In summertime, black sateen shade cloths are used to shorten the days. Wires strung the length of the growing bed are used to support the cloth. The cloth is pulled over the bed in the evening and left there long enough to shorten the day to twelve hours or less. In very hot weather, the cloth should not be pulled over the plants until 6:00 or 7:00 P.M. If the cloth is pulled over the plants any earlier, heat buildup under the cloth could damage the plants.

OBTAINING STARTER PLANTS

Mums are propagated by softwood cuttings. They are relatively easy to root. There are, however, a variety of diseases that may be passed along from parent plants to the rooted cuttings. For this reason, it is very important to purchase rooted cuttings from specialists who can guarantee that the cuttings are disease free. Contact a local greenhouse operator for a source. The operator may offer to order for you if you need only a few cuttings. Order at least several months in advance to be sure the selected varieties are available. Figure 17-1 shows a typical rooted cutting.

ESTABLISHING A GROWING SCHEDULE

Before ordering rooted cuttings, a growing schedule must be established. The schedule must be planned starting at the end of the schedule, the blooming date.

Example. Assume that Easter Sunday occurs on April 10. If the mums are being grown as a class project and the school closes for the holiday on Friday, April 7, the mums must bloom prior to that date. To be sure that the blooms are available for sale the

Fig. 17-1 Typical rooted chrysanthemum cutting ready for planting in the pot.

Schedule	Short Treatment (2 1/2 weeks of lights)	Medium Treatment 2 weeks of lights)	Tall Treatment (1 week of lights)
Plant	Jan. 4	Jan. 7	Jan. 14
Shade	not needed	not needed	not needed
Lights On*	Jan. 4 – Jan. 21	Jan. 7 – Jan. 21	Jan. 14 – Jan. 21
Pinch	Jan. 22	Jan. 25	Jan. 29
B-Nine 2 weeks after pinching	Feb. 5	Feb. 8	Feb. 12
Bloom date	April 1	April 1	April 1

*From end of Lights on to Bloom Date is 10 weeks (when long nights start)

Fig. 17-2 Growing schedule for Easter flowering (ten-week response variety)

week of April 4 through 7, the blooming date should be set for April 1. (Mums hold their blossoms for a fairly long time.) The schedule would appear as in figure 17-2.

According to variety, mums receive what is called *short treatment, medium treatment* or *tall treatment*. These labels are necessary since some varieties naturally tend to grow short, some medium, and some tall in height. So that the plants grow to the same height and appear uniform, the length of the *vegetative period* (the long day in summer before shade is used, and the short night in fall and winter when artificial light is used) must be regulated differently for each of the three groups.

Example. The mum crop being prepared for the Easter market is grown in winter; therefore, the length of the vegetative period is controlled by using lights to shorten the number of hours of continuous darkness. After planting the rooted cuttings, the lights are left on for two and one-half weeks for the short treatment varieties, two weeks for the medium treatment varieties, and one

week for the tall treatment varieties. Notice that the planting date changes in figure 17-2 according to variety. The schedule shown in 17-2 is for ten-week response varieties; the schedule for eight-week response varieties would be two weeks shorter. These varieties would simply be planted two weeks later.

PLANTING ROOTED CUTTINGS

Container and Media (for potted mums)

A 6-inch plastic pot is a good choice for planting rooted mum cuttings. This size pot requires five cuttings per pot. Varieties which send out more side shoots may be planted with only four per pot. If a 5-inch pot is used, plant four cuttings to each pot. Some markets use a 4-inch pot and place two cuttings in each.

The media should be one that is well drained and yet holds moisture well. If natural soil is used, it must be sterilized first. Peat moss for moisture retention and sand or perlite for drainage should be added to soil in this ratio: one-third soil, one-third sand or perlite, and one-third sphagnum peat moss. There are other growing mixes on the market which contain no soil and are very effective. The media should be such that water added to the surface of the pot drains quickly through the pot and out the bottom. If the soil mix has the proper texture, it will not crack when dry. Be especially careful that the potting media does not dry out after the mums are potted. Mums should never be allowed to wilt from dryness. To prevent this, provide a constantly moist, but not wet, soil. Add water when the top of the potting media first appears to be drying out.

Steps in Planting

As mentioned earlier, mum cuttings should be planted in 6-inch shallow pots (azalea pots), with four or five rooted cuttings

Fig. 17-3 Four mum cuttings planted in a 6-inch pot. Notice that the cuttings are slanted toward the outside of the pot at an angle of about 45 degrees. The roots of the cuttings are barely covered.

per pot. Select cuttings for each pot that are all approximately the same size. This gives more uniformity to the finished product.

1. Fill the pot with planting media to within 1 inch of the desired level after planting.

2. Space the cuttings evenly in the pot. Slant the cuttings at approximately a 45-degree angle so that the top of the cuttings protrude over the edge of the pot, figure 17-3. This angle encourages the growth of more shoots, and thus, better, more compact plants.

3. Place just enough soil in the pot to cover the roots (about 1 inch or less). When the pot is planted and filled, the media level should be about 1 inch below the top of the pot. This makes watering the cuttings an easier job.

4. Water the cuttings until the water runs through the bottom of the pot.

5. Place the pot in the growing area, figure 17-4, page 174. Later, these pots will require more space. Pots should have at least 1 inch of space between them for the first three to four weeks and later, about 8 inches between pot rims. This wider spacing gives the plant room to spread out, resulting in a better shaped plant with more blossoms. Watch for signs of overcrowding within the pots. One of the first signs may be drying of lower leaves. (However, this may also signal under-feeding, disease, or poor drainage.)

CARING FOR POTTED CUTTINGS

Temperature Control

When first placed in the growing area, the greenhouse temperature should be kept at 63°F (17°C) at night. The temperature should be held to 65°F (18°C) on cloudy days and 70°F (21°C)

Fig. 17-4 Pots of mum cuttings placed in the growing area of the greenhouse. It will be necessary to provide more space between the pots in about 3 weeks.

on sunny days. This temperature range is maintained through the first two or three weeks of short days. The higher temperatures promote faster vegetative growth. If the temperature drops below 60°F (16°C) during this period, flower buds may not set evenly. Uneven bud set may be one of the first signs of improperly regulated temperature.

At the end of this period, the temperature should be dropped to 62°F (17°C) at night; to 62° – 65°F (17° – 18°C) on cloudy days; and to 70°F (21°C) on sunny days. For the last three weeks of growth, the temperatures should be dropped to 55°F (13°C) at

night and on cloudy days, and to no higher than 65°F (18°C) on sunny days. This cooler temperature at the end of the growing period tends to harden off the plants.

Fertilizing Cuttings

Mums may be fertilized using one method or a combination of several different methods.

Fertilizing by Constant Application. A very diluted fertilizer may be added to the irrigation water supplying the mums with constant fertilizing. Wait a few days after planting to begin this program. If this form of fertilizing is chosen, a water soluble fertilizer prepared specifically for mums should be used according to directions. The fertilizer should be high in nitrogen and potash. For example, if a 15-10-30 fertilizer is chosen, it should be mixed in a stock solution of 18 ounces per gallon of water and diluted again to 1 part solution to 100 parts water before application or as it is being applied through a siphoning device.

> **Follow directions on the container label.**

Note: During cloudy weather, less fertilizer and water are needed. This fertilizing system works well to reduce both at the same time when necessary.

Fertilizing by Slow Release. A slow release fertilizer may be mixed in the potting media at the time of planting. Use 8 ounces of fertilizer for every bushel of potting media. This fertilizer will furnish all or most of the fertilizer needs for a period of three to four months. If plants appear to need more nitrogen, additional fertilizer may be added in the irrigation water. Do not sterilize a soil mix after a slow release fertilizer has been added.

Time Period	Night	Temperature Cloudy Day	Sunny Day
From planting through the first 2 weeks of short days*	63°F	63°F	70°F
Up to last 3 weeks of growing time	62°F	62°F to 65°F	70°F
Last 3 weeks of growing cycle	55°F	55°F	65°F

*If temperatures are cool in the early stages of short days, an uneven bud set may occur.

Fig. 17-5 Temperature control schedule

Fertilizing by Timed Applications. Fertilizer may be applied once every two weeks in the irrigation water or to each pot as needed. This method is not as effective as either of the other methods listed, and the crop produced may not be as good a one.

Watering

How often mums are watered varies with the temperature and humidity in the room and water-holding capacity of the media. In hot, dry weather, the plants may require watering twice a day. In cool, cloudy weather, once a day or every other day may be sufficient. Check the soil mix frequently for dryness and add water as needed. Do not keep the soil saturated, however; the plant roots may rot. At the same time, allowing the plants to wilt will cause damage to the crop. Add enough water at each watering so that water runs through the pot and out the drainage holes.

Newly planted cuttings require special care and attention the first week. During this time, be especially watchful for signs of dryness.

Pinching and Disbudding

Pinching means simply to break or pinch off tips from plant stems. Hold the tip of the stem and pinch off 1/4 to 1/2 inch of growth, figure 17-6. This causes side shoots to develop, thereby resulting in more branches to set flowers. The final result is a more compact, flowerful plant.

Fig. 17-6 A plant tip is held directly above the mum from which it was pinched. The tip measures about 1/2 inch in height. Because of the pinching, many more side shoots will form, resulting in a bushier, more flowerful plant.

Fig. 17-7 Three buds are present on the shoot being held in this photograph. If larger and fewer flowers are desired, one or two of the smaller buds can be pinched off. Always leave the large terminal, or end, bud. (Richard Kreh, Photographer)

The ideal potted mum should have about ten to fifteen flowers per pot. If more buds are present, remove some or all of the smaller side buds. This will result in larger, more attractive individual flowers, figure 17-7.

Growth Retardants

The chemical B-Nine, a growth retardant, tends to shorten stem growth. This causes the blossoms to grow closer together, resulting in a more compact plant. B-Nine is sprayed on the foliage about two weeks after pinching, or when side shoots are about 2 inches long. Do not wet or water the plant foliage for twenty-four hours after applying B-Nine to prevent the chemical from being washed off. The treatment may be repeated if plants appear to be growing too tall. Follow label directions for second treatment.

Read label directions carefully before using these chemicals.

Controlling Insects and Disease

Preventing attacks on plants by insects and disease organisms is a far better method of plant protection than fighting a serious infestation. Good sanitation and cultural practices are often all the control that is needed. If insects do become a problem, spray with an emulsifiable concentrate spray. Use a recommended fungicide if necessary for fungus diseases.

Note: Wettable powders leave a residue on plants and should not be used within six weeks of sale date. Emulsifiable concentrates do not leave visible residues. (An *emulsifiable concentrate* is a chemical in liquid form that will mix with water.) Emulsifiable concentrates and wettable powders may not mix well if applied in the same solution.

Unless insect or disease problems arise, do not spray. Never use highly toxic substances. When any pesticide is used, read directions carefully and follow all safety precautions.

Space Requirements for Year-round Production

To produce fifty pots of mums every two weeks year-round, about 380 square feet of bench space is needed. Provision must be made for lighting on at least part of the space.

Garden Mums

Many horticulturists prefer to cultivate plants that can be set outdoors after blooming in the pot. This may be done by using a special variety, the garden mum, for the spring crop. After planting outdoors in the spring, the garden mum can be expected to bloom again the following fall.

Plant cuttings of garden mums in 3- to 5-inch pots and grow according to the schedule for that particular variety. Most garden mums are the short response variety and flower in eight to nine weeks of short days. This gives a quick return on a very popular crop.

STUDENT ACTIVITIES

1. Develop a potting soil for chrysanthemum cuttings. Determine by feel and appearance if the potting media has the proper characteristics.

2. Plant at least one pot of four cuttings of mums and care for it throughout the growing period.

3. Develop a complete growing schedule for an Easter mum crop. Compare it with your instructor's schedule for accuracy.

4. Visit a commercial greenhouse operator who produces mums. Observe the operation and discuss a prepared growing schedule with the grower.

SELF-EVALUATION

Select the best answer from the choices offered to complete the statement or answer the question.

1. The mum is a short season crop. This means that it flowers
   a. in a very short period of time.
   b. when days are short and nights are long.
   c. in areas that have a shorter than average growing season.
   d. only on chrysanthemum plants that are very short.

2. Mum cuttings planted in fall and winter require special lighting to lengthen days for the first one to three weeks so that

    a. the plants grow larger.
    b. the plants are more saleable.
    c. the plants do not set flower buds too quickly.
    d. all of the above

3. The amount of or intensity of light necessary to prevent flower bud formation in mums is

    a. about 5 foot candles.
    b. about 15 foot candles.
    c. about 7 foot candles.
    d. about 70 foot candles.

4. Shading of mums in summer is necessary because

    a. days are too long.
    b. temperatures are too high.
    c. heat tends to fade the flowers.
    d. the sunlight is too strong, providing too many foot candles of light.

5. When a mum is labeled an *8-week variety* (response time), this means that

    a. eight weeks is the entire growing time for the crop.
    b. eight weeks is the length of time lighting must be provided for the crop.
    c. eight weeks is the period of time from the start of short days until bloom.
    d. all of the above

6. Mums that are labeled for short treatment are

    a. plants that naturally grow to a shorter than average height.
    b. plants that require more pinching than the average plant.
    c. plants that bloom in a short period of time.
    d. plants that require stronger light, but for a shorter period of time.

7. The suggested potting arrangement for mums is

   a. six cuttings to each 4-inch pot.
   b. one cutting to each 4-inch pot.
   c. three cuttings to each 5-inch pot.
   d. four to five cuttings to each 6-inch pot.

8. When potting mums, the most important characteristic of the potting media is that it

   a. be inexpensive.   c. be locally available.
   b. drain and hold moisture well.   d. contain good, loamy soil.

9. Mum cuttings are equally spaced around the pot and slanted toward the outside of the pot at an angle of

   a. 45 degrees.   c. 60 degrees.
   b. 35 degrees.   d. 25 degrees.

10. After mum cuttings are placed in the pot, how much potting media should be placed in the pot?

    a. 2 inches   c. 3 inches
    b. 1/2 inch   d. 1 inch or less

11. After the first three to four weeks, how far apart should potted mums be spaced (measuring from pot rim to pot rim)?

    a. 1 inch apart   c. 8 inches apart
    b. 2 inches apart   d. 12 inches apart

12. For the first two weeks, the temperature of the greenhouse in which potted mum cuttings are kept should be _____ at night and on cloudy days and _____ on sunny days.

    a. 55°F; 65°F   c. 63°F; 70°F
    b. 45°F; 55°F   d. 70°; 75°F

13. From two weeks after planting until the last three weeks of the growing period, the temperature of the greenhouse in which mums are kept should be _____ at night and on cloudy days and _____ on sunny days.

    a. 62°F; 70°F
    b. 55°F; 65°F
    c. 63°F; 75°F
    d. 70°; 75°F

14. During the last three weeks of the growing cycle, the temperature of the greenhouse in which mums are kept should be _____ at night and on cloudy days and _____ on sunny days.

    a. 62°F; 70°F
    b. 63°F; 75°F
    c. 70°F; 75°F
    d. 55°F; 65°F

15. Mums should be watered when

    a. the top of the media is beginning to appear dry.
    b. media begins to run through the bottom of the pot.
    c. the plants have wilted.
    d. it is convenient.

16. An uneven bud set on mums is usually caused by

    a. improper lighting control.
    b. improper temperature control.
    c. crowding pots together.
    d. insect damage to buds.

Section 7
Pesticides

A *pesticide* is any chemical used to control *pests* (plants or animals that are harmful to human beings or to the crops that they cultivate). There are seven types of pesticides that concern horticulturists, each classified according to the type of pest it destroys.

- *Insecticides* are chemicals used to control insects. The insects are killed by body contact with the chemical or by swallowing the poison.

- *Miticides* are chemicals used to control *mites* (tiny spiderlike animals) and ticks. Mites and ticks are usually killed by coming in contact with the chemical.

- *Fungicides* are chemicals used to control fungus disease. To be effective, the chemical must come in contact with the fungus. Fungicides are usually used to prevent a plant from becoming diseased and are applied before the disease is present.

- *Herbicides* or *weed killers* are chemicals used to kill unwanted plants. *Nonselective* herbicides kill all plants, while *selective* herbicides kill only certain plants.

- *Rodenticides* are chemicals that kill rodents, such as rats and mice. These chemicals are usually applied as bait; the rodents are poisoned by eating the chemical.

- *Nematocides* are chemicals that kill nematodes (tiny hairlike worms that feed on the roots of plants). Nematocides are usually applied in the form of *fumigants*, substances which produce a smoke, vapor, or gas when applied.

- *Molluscicides* are chemicals used to kill slugs and snails (types of *mollusks*). These chemicals are usually applied as bait which attracts the slugs and snails and poisons them.

SELECTING THE PROPER PESTICIDE

The first step in selecting a pesticide is to identify the pest to be controlled and determine if control is necessary. If the pest is doing very little damage, or if the crop is ready for harvest, control may not be needed.

If control is necessary, methods which do not employ chemicals should first be considered. It is possible that the spread of disease can be halted simply by proper *sanitation*, removing and destroying any diseased plant material and disinfecting all tools and work areas. (This procedure should be followed while making cuttings even if there is no sign of disease.)

If the pest to be controlled is an insect, insects which eat other insects (such as the ladybug) may be effective and less expensive than chemicals. Such insects may be purchased and released on the infested plants. It may be unnecessary to introduce the insect to an area; natural parasites (such as the tiny wasp that kills tomato worms) may be present in large enough numbers to minimize attack by the pest. In the same way, mulches, rather than herbicides, may be used for weed control to minimize the use of harsh chemicals.

After it has been decided that chemical control is the only alternative, it is important to select the best pesticide for that particular problem. If correctly chosen, the pesticide will meet the following requirements.

- It kills or controls the pest.

- It does not injure the plant on which it is used.

- Of all chemicals which could be used for the problem at hand, it is the least harmful to the environment.

- It is suitable for use with available equipment.

- The label recommends its use for the plant and pest for which it has been chosen.

OBJECTIVE

To use all pesticides described safely and competently.

COMPETENCY TO BE DEVELOPED

After studying this unit, the student will be able to

- identify the three main routes by which pesticides enter the body.

- examine five pesticide labels and

  - identify the type of each and its degree of toxicity.

  - demonstrate the recommended precautions in the mixing and handling of each.

- list first aid steps to be taken in case of poisoning by one pesticide from each of the three families of pesticides.

unit 18
The Safe Use
of Pesticides

Materials List

pesticide labels or containers with labels

sprayer and duster for hand application of pesticide

all articles of protective clothing specified on pesticide labels including rubber gloves, respirator, rubber boots, waterproof coveralls, goggles, and rubber hat

local pest control schedules and calendars from a local university or extension service

pesticides for application

TOXICITY

Because pesticides can be poisonous to humans and the environment, the United States has established standards for their handling and use through the Environmental Protection Agency (EPA). Other countries have similar regulations. Because of their poisonous nature, pesticides must be used with a degree of caution, depending on how poisonous, or *toxic*, they are.

To poison or injure humans and other animals, pesticides must be present in or on the victim's body. There are three main routes by which poisons enter the body:

- oral contact (by swallowing)

- dermal contact (by contact with the skin)

- inhalation (by breathing)

When dealing with the danger of children coming in contact with poisons, the major concern is preventing the child from taking the poison orally. On the other hand, the individual who applies pesticides is more likely to be poisoned by contact dermally or through inhalation. Chemicals such as organophosphates are absorbed rapidly through the skin. Any chemical that vaporizes, has a strong odor, or is a fine dust or mist can be inhaled and absorbed through the lungs. Pesticides pass very rapidly through the skin on the back of the hands, wrists, armpits, back of the neck, groin, and feet. Cuts or scrapes also allow pesticides to enter the body more easily.

Caution: Never allow a pesticide to come in contact with the skin. If this does occur, wash the affected area immediately with soap and water.

Types of Toxicity

Acute toxicity is a measure of how poisonous a pesticide is after a single exposure. Pesticides are generally rated according to acute toxicity. *Chronic toxicity* is a measure of how poisonous a pesticide is over a period of time and after repeated exposure. Chronic toxicity is a danger of chemicals that accumulate in the body, such as chlorinated hydrocarbons. DDT (dichlorodiphenyl-trichloroethane) is an example of a chemical which builds up in the body. Its use is now banned in the United States.

Measuring Acute Toxicity

Oral and Dermal Toxicity. The method used to measure acute oral and dermal toxicity is LD_{50}. The *LD* stands for *lethal dose* (amount necessary to cause death). The $_{50}$ means that 50 percent of test animals (generally rats) are killed by this dose. The lower the LD_{50} number of a pesticide is, the more poisonous it is. LD_{50} values are given in milligrams of substance per kilogram of test animal body weight. This is a metric measurement which is the same as parts per million.

Figure 18-1 gives LD_{50} values for some important pesticides. The pesticides in the chart are listed with the most poisonous first, as shown by the smaller LD_{50} numbers of these substances.

Inhalation Toxicity. Acute inhalation toxicity is measured by LC_{50} values. *LC* stands for lethal (deadly) concentration. LC_{50} values are measured in milligrams per liter. A *liter* is a volume measurement (in the metric system of measurement) equal to about 1 quart. Again, the lower the LC_{50} number, the more poisonous the pesticide is.

There is no standard measure for chronic toxicity.

Pesticide Name*	Acute Toxicity		Type of Pesticide or Use	Pesticide Name*	Acute Toxicity		Type of Pesticide or Use
	Oral (by mouth)	Dermal (on skin)			Oral (by mouth)	Dermal (on skin)	
TEPP (tetraethyl pyrophosphate)	1.2-2.0	2.4	insecticide	MCPA (Rhomene)	700	––	herbicide
phorate (Thimet)	1.7	4.4	insecticide	2, 4-D	700	––	herbicide
demeton (Systox)	2.5-12.0	8.2-14.0	insecticide	aspirin	750	––	pain reliever
endrin	5-45	60-120	insecticide	malathion (Cython)	1,375	4,100	insecticide
parathion	·8.3	14.0	insecticide	pyrethrins	1,500	1,880	insecticide
azinphos-methyl (Guthion)	16	220	insecticide	linuron (Lorox)	1,500	––	herbicide
methyl parathion	19	67	insecticide	EPTC (Eptam)	1,630	––	herbicide
carbophenothion (Trithion)	20	40	insecticide	ronnel (Korlan)	1,940	––	insecticide
Paris green	22	––	insecticide	atrazine (Aatrex)	3,000	––	herbicide
strychnine	30	––	rat poison	table salt	3,320	––	mineral food item
DNBP (Premerge)	40	––	herbicide	diuron (Karmex)	3,400	––	herbicide
nicotine	50-60	50	insecticide	chloramben (Amiben)	3,500	––	herbicide
warfarin	58	––	rat poison	trifluralin (Treflan)	3,700	––	herbicide
sodium fluoride	75	––	insecticide	resmethrin (SBP 1382)	4,240	over 3,040	insecticide
lindane	88-125	1,000	insecticide	ethyl alcohol	4,500	––	beverage
diazinon	92	678	insecticide	borax	4,980	––	cleaning compound; insecticide
lead arsenate	100	2,400	insecticide	simazine (Princep)	5,000	––	herbicide
PCP (penta)	100	100	wood preservative	bromacil (Hyvar X)	5,200	––	herbicide
crotoxyphos (Ciodrin)	125	385	insecticide	methoxychlor	6,000	––	insecticide
rotenone	133	943	insecticide and fish poison	fenuron (Dybar)	6,400	––	herbicide
caffein (coffee, tea)	200	––	beverage	maneb (Manzate)	6,750	––	fungicide
dimethoate (Cygon)	215	515	insecticide	piperonyl butoxide	7,500	1,880	insecticide
Nemagon or Fumazone	250	1,420	nematocide	captan	9,000	––	fungicide
nabam (Dithane)	394	––	fungicide	benomyl (Benlate)	10,000	––	systemic fungicide
chlordane (Ortho-Klor)	457	765	insecticide	ferbam (Fermate)	17,000	––	fungicide
carbaryl (Sevin)	675	4,000	insecticide	kerosene	28,000	––	fuel

*Pesticides listed in order of oral acute toxicity.

––Where no dermal figure is given, the chemical cannot be absorbed through the skin in dangerous amounts.

Fig. 18-1 LD_{50} **(lethal dose) of common farm chemicals measured in milligrams of substance per kilogram of body weight. (Measurement is same as parts per million.)**

> Before purchasing or using a pesticide, read and understand the label. Watch for signal words indicating toxicity of the chemical. Be prepared to take any precautions listed. If the pesticide is highly toxic, try to purchase a material that is as effective yet less toxic.

Standard Label Information

Labels on pesticide containers always include mixing instructions. These must be followed carefully. Other standard points of information which appear on pesticide containers include:

1. name and address of manufacturer

2. trade name (may or may not be the same as the chemical name)

3. active ingredients, including the official common name of each ingredient. When an accepted common name is not available, the chemical name appears.

4. type of pesticide such as insecticide or fungicide

5. form of substance such as dust, wettable powder, or emulsion

6. EPA registration number

7. storage and disposal precautions

8. hazard statement (Read carefully.)

9. directions for use (Read carefully and follow.)

10. net contents

Signal Words

To alert the user to the toxicity of a pesticide, certain *signal words* appear on the container label. There are four categories of acute toxicity and signal words, figure 18-2.

Categories	Signal Word Required on Label	LETHAL DOSE			Probable Oral Lethal Dose for a Person Weighing 150 lbs.
		LD_{50} mg/kg Oral	LD_{50} mg/kg Dermal	LC_{50} mg/1 Inhalation 24 hr. exposure	
Highly Toxic	DANGER skull and crossbones Poison	0-50	0-200	0-2	a few drops to a teaspoonful
Moderately Toxic	WARNING	over 50 to 500	over 200 to 2,000	over 2 to 20	over 1 teaspoonful to 1 ounce
Slightly Toxic	CAUTION	over 500	over 2,000 to 20,000	————	over 1 ounce to 1 pint or 1 pound
Relatively Nontoxic	CAUTION (or none)	over 5,000	over 20,000	————	over 1 pint or 1 pound

Fig. 18-2 Categories of acute toxicity. (from *Pesticide Applicator Training Manual*, courtesy Northeastern Regional Pesticide Coordinators)

It is essential to read labels to determine the toxicity of pesticides and the precautions necessary in their handling. The label should be examined before the product is purchased. In this way, it can be determined if the chemical is effective against the particular pest which is causing the problem, and if perhaps it is too toxic a substance with which to work.

Examine the partial sample label in figure 18-3. The markings on the label of this pesticide indicate it is in the highly toxic

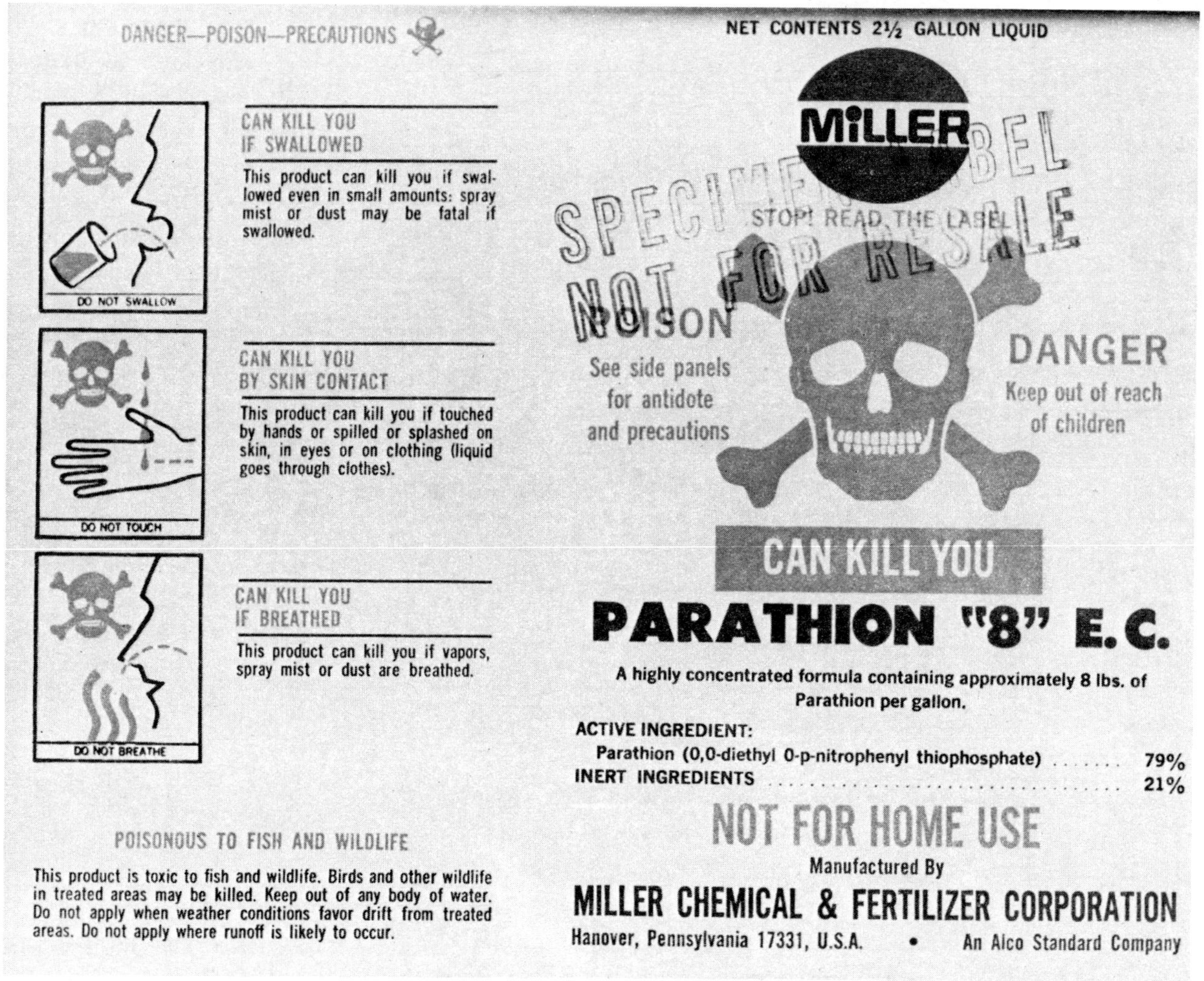

Fig. 18-3 Partial label from container of highly toxic pesticide. Notice the signal word *(DANGER)* and the skull and cross-bones, indicating the extremely high toxicity of this chemical. (Courtesy Miller Chemical & Fertilizer Corporation. Richard Kreh, Photographer)

Fig. 18-4 A relatively nontoxic pesticide (Courtesy Miller Chemical & Fertilizer Corporation. Richard Kreh, Photographer)

category; therefore, it should not be used around the school or home. Compare the information from this label with that found on a relatively nontoxic pesticide, figure 18-4.

> When considering the use of any pesticide, be sure that the intended use appears on the label. It is violation of federal law to use any pesticide in a manner which is inconsistent with the labeling. A use that is *inconsistent* with a label includes use on any pest or on any site that is not specifically given on that label.

SAFETY PRECAUTIONS

Safety precautions must be considered before beginning to mix or apply a pesticide. Adequate safety precautions include:

Reading the label carefully. Preparing to follow all directions and precautions exactly.

Checking the recommended uses to be sure it is the correct substance for the intended use.

Having clean water and detergents available to wash spills.

Wearing protective clothing such as rubber gloves, a respirator, and any other protective gear called for on the label. Some pesticides require little protective clothing, figure 18-5. Other highly toxic materials require complete coverage of the applicator's entire body with waterproof material, figure 18-6.

Fig. 18-5 The chemical being applied here is relatively nontoxic. Note the lack of protective clothing. (USDA photo)

The proper protective clothing can be determined only by reading the label on the pesticide container.

Use extra caution with concentrated chemicals. (Chemicals are considered concentrated prior to mixing or diluting with water or dust.) Protective covering is especially important when handling the concentrated chemical in mixing and filling the spray tank. Always wear rubber gloves and goggles. If a concentrate is spilled or splashed on clothing, wash and change clothes immediately. Always protect the eyes with goggles when handling concentrates; the eyes absorb pesticides rapidly. Any pesticide spilled on the ground or floor should be removed immediately.

Apply the chemical with care. When applying pesticides, be sure that the chemical is directed only at the target area and that as little as possible is allowed to spread anywhere else. Apply only the amount needed to do the job.

Always mix just enough for the job at hand. The label lists the proper amount needed for various crops and pests. The sprayer may need *calibration* (adjustment of the nozzle to a certain size) to ensure that the proper volume of chemical is applied. Applying at the proper rate is especially important when working with herbicides (weed killers).

Guard against inhalation or ingestion (swallowing) of the chemical; do not inhale its fumes, dust, or mist. Never eat or smoke while handling any pesticide.

Consider weather conditions. Pesticides should be applied when weather conditions are favorable. Be aware of forecasted weather conditions before application of the chemical. If used out of doors, application during heavy rain will wash away the chemical. Strong winds prevent applications of sprays and dust from hitting and adhering to the target crop. Some chemicals must be applied only within certain temperature ranges. Read the label and use good judgment. When applying pesticides, avoid contact with the mist or dust that drifts through the air. Do not spray where the pesticide can drift into streams, lakes, or apiaries (bee yards), or near sensitive plants or animals that might be injured.

Store and dispose of chemicals properly. Store all concentrated pesticides in the original container. Keep in a locked area, away from children and animals. Do not store near foods, animal feed, fertilizer, or in a building which houses people or animals. Sprayers and dusters should be treated in the same manner for storage.

Fig. 18-6 The chemical being applied here is highly toxic and requires complete protection. The applicator is wearing a gas mask, rubber head cover, rubber gloves, waterproof clothing, and boots. This prevents the solution from coming into contact with any part of the applicator's body. (USDA photo)

Dispose of all diluted chemical that is in the spray tank. If any is left after spraying or dusting, try to use it on another area or crop where it will not contaminate or damage the environment, livestock, or people. However, the best way to eliminate problems associated with leftover chemicals is to mix only the amount necessary for the job at hand. Empty containers must be disposed of out of the reach of all persons and animals, and in such a way that they will not contaminate the environment. Never reuse a pesticide container. Bury containers at least 18 inches deep in the soil in a location which will not lead to contamination of a stream or water supply, or take them to a sanitary landfill. Never set empty or partly empty containers out to be collected as trash unless they are clearly labeled.

> **Before disposing of any pesticides or empty containers, check local regulations, since the procedures for disposal may vary greatly.**

Clean up thoroughly. The sprayer, all measuring equipment, gloves, and any other protective clothing should be washed thoroughly. The sprayer tank and pump should be rinsed three times with large amounts of water. Some of the water should be run through the sprayer each time to flush the hose and spray pump. The rinse water is then dumped into the mixing tank. Rubber gloves should be washed thoroughly with soap and water while still on the hands and hung up to dry. When the pesticide application is finished and the sprayer or other application equipment is cleaned and safely stored, the applicator should shower and change clothes immediately. Contaminated clothing should be handled carefully and washed separately.

PESTICIDE POISONING

If the warnings mentioned earlier are not heeded, or if accidental spillage results in poisoning, it is important that any symptoms of poisoning be recognized quickly. In cases of spillage or other types of exposure to a toxic chemical, the most important consideration is to remove the contaminated clothing and wash thoroughly with a good soap or detergent and plenty of water. A shower should be used rather than a tub bath because fresh, clean water is continually supplied.

> **Dilution of the poison is the most important first aid practice.**

First Aid for Pesticide Poisoning

First aid is just what the term indicates — the first help given to a victim until medical professionals arrive. If poisoning has occurred, telephone a doctor immediately and read the pesticide label to him or her. Before the doctor arrives, you can do the following.

- For any pesticide spilled on the skin, wash with plenty of soap and water to dilute the chemical.

- remove contaminated clothing. Shower, dry, and wrap or dress the victim in warm blankets or clothing. Cover any chemical burns with loose, clean, soft cloth.

> **Caution: Be careful that you do not come in contact with the pesticide while coming to someone's aid.**

- For eye poisoning, hold the eye open and flush with clean water for at least five minutes.

• For inhaled poisons, carry the victim immediately to fresh air. Do not allow the victim to walk. Loosen clothing and apply artificial respiration if breathing has stopped. Keep the patient quiet and warm. Do not give the victim alcohol.

The label on the pesticide container may give additional first aid information. Use any information available. After notifying a doctor, follow his or her advice. Provide expert medical attention as soon as possible. Do not panic and frighten the victim; instead, remain calm and reassure the victim that help is on its way.

> **Always choose the least toxic chemical that will do the job.**

STUDENT ACTIVITIES

1. Obtain labels from containers of commonly used pesticides. Draw up a list of pesticides, including the following.

   a. name of pesticide
   b. crops on which pesticide is commonly used
   c. pests controlled by the pesticide
   d. toxicity level of the pesticide and necessary precautionary measures including protective clothing

2. Demonstrate the proper procedure for mixing and applying a pesticide.

3. Try on and properly adjust a respirator, rubber gloves, waterproof hat, and goggles.

SELF-EVALUATION

A. Briefly answer each of the following.

   1. List the seven categories of pesticides and define each briefly.

   2. What is the first step in selecting a pesticide?

   3. List the four signal words found on pesticide labels. Also list the category of pesticide that each word signals.

   4. What is a toxic chemical?

   5. List the three routes by which pesticides can enter the body.

B.   Complete each of the following statements with the proper word or words.

1.   The toxicity of a pesticide is determined by reading the _____

2.   Acute toxicity is a measure of how poisonous a pesticide is after _____ exposure.

3.   Chronic toxicity is a measure of how poisonous a pesticide is after _____ exposure.

4.   The more poisonous a pesticide is, the _____(higher, lower) the LD_{50} number is.

5.   If a pesticide is spilled on the skin, it should be washed off immediately with _____and _____.

6.   If a pesticide is splashed in the eyes, they should be flushed with clean water for _____minutes.

7.   When handling concentrated pesticides, always protect the eyes with _____ and wear _____on the hands.

8.   Store all pesticides in a _____storage area.

9.   Leftover spray that has been mixed for use should be disposed of by _____ it on a crop that the pesticide will not damage.

10.   Empty containers should be _____where they will not contaminate the environment.

11.   If a person is poisoned by pesticides, the most important first aid practice is to _____the poison.

12.   In cases of inhaled poisons, the first step in first aid is to carry the victim to

_____.

OBJECTIVE

To identify insect pests and select and apply the appropriate insecticides.

COMPETENCY TO BE DEVELOPED

After studying this unit, the student will be able to

- identify common insect pests and select an effective control method for each.
- list four natural insect control methods.
- describe the six ways in which insects are killed by insecticides and the type of insect against which each is most effective.
- list the names and characteristics of the three major groups of insecticides (according to their chemical makeup).
- compare the six ways in which insecticides are applied.
- explain the relationship between the life cycle of insects and timing of insecticide application.

Materials List

illustrations of insect pests

live insects

recommendations for spraying from local extension service

insecticides with labels (none should be highly toxic)

spraying and dusting equipment

safety gear, including elbow length rubber gloves and respirator

---

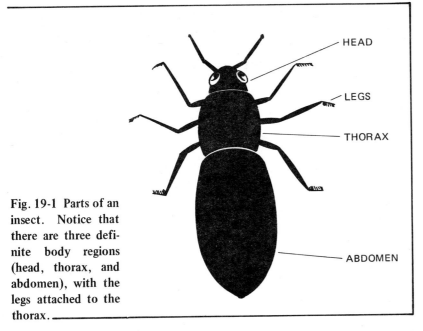

Fig. 19-1 Parts of an insect. Notice that there are three definite body regions (head, thorax, and abdomen), with the legs attached to the thorax.

Insecticides, one of the seven types of pesticides mentioned in the previous unit, are used by most people at one time or another in the control of flies, mosquitoes, and other insect pests. They are used widely by horticulturists to protect plants from insect damage. Although insecticides can effectively and safely control insects, they can be dangerous and even deadly to human beings when used improperly. For this reason, it is extremely important that they be handled with care and only according to directions on the label.

WHAT IS AN INSECT?

An *insect* is a small animal with three clearly defined body regions and three pairs of legs. The three body regions are the head, thorax, and abdomen, figure 19-1. The proper identification

Fig. 19-2 A parasite of the Japanese beetle (*Tiphia*) in its larval form feeding on the larva of a beetle. The United States Department of Agriculture introduced the parasite from Japan to the U.S., where it prevents the Japanese beetle from becoming a serious economic threat. (USDA) photo

of insects is an important step in their control. Only after the insect is identified can the best method of control be selected. Guides to insect identification are available from the United States Department of Agriculture.

CHEMICAL-FREE CONTROL

Insects are controlled to a great extent by natural forces. Natural enemies of insects, such as birds and insects like the ladybug and praying mantis, help to keep the insect population under control. For example, tiny wasps lay eggs on the tomato hornworm;

the eggs hatch into worms which eat the tomato worm from the inside out. Other parasitic wasps kill insects such as the alfalfa weevil and the Mexican bean beetle. If natural enemies are not present in a certain area, they may be introduced to provide the necessary control. Figure 19-2 shows a small worm which was introduced to the United States to aid in the control of the Japanese beetle.

Commercial preparations containing disease organisms which attack insects can also be used for chemical-free control. Naturally occurring disease organisms which attack insects, such as fungi, viruses, and bacteria, are grown on host insects in the laboratory and then prepared for use as pesticides. In this way, the natural diseases of the insects are multiplied to such a great extent that the insect population is greatly decreased or eliminated.

Crop rotation is a method of controlling some types of insects that do not move rapidly from place to place. Crop rotation consists of planting a different crop in a field from year to year, rather than growing the same crop on the same land for several years. Many soil-borne diseases, such as tomato wilts, can only live in the soil for two or three years if the plants on which they feed (host plants) are not present. Thus, if crops are rotated so that the host plant or related plants are not grown in a field more than once in three years, the disease will die out. In the case of tomatoes, this means that tomatoes, eggplant, or potatoes (plants which also support tomato wilt) should be planted in a field only once every three years. A typical rotation might be: tomatoes the first year, beans the second year, corn the third year, and then tomatoes again the fourth year.

Plant breeders are developing more and more resistant varieties of plants. For example, the tomato variety Better Boy is listed as *VFN resistant*. The *N* stands for *nematode*, a root-damaging pest. Nematodes do not like the roots of Better Boy; because of this, nematodes are controlled without chemicals.

These methods of control — the introduction of natural enemies, crop rotation, and the development of resistant varieties of plants — are effective chemical-free ways of keeping certain insects under control. Because they do not employ chemicals, they are not toxic to human beings and animals and do not contaminate the environment. At present, there are not enough of these natural control agents available. Because the food demands of today's world require efficient production of plants free from insect damage, we must sometimes resort to chemical control in the form of insecticides.

HOW INSECTICIDES KILL

To be considered good insecticides, chemicals must kill a specific type of insect while doing little or no damage to the plant. They must also be as safe for the handler and environment as possible.

Stomach Poisons work against insects that actually eat a part of the plant. The chemical is sprayed or dusted on the plant; as the insect eats the plant, it is poisoned through the stomach. Insects that chew their food, such as caterpillars, grasshoppers, and beetles, are controlled by stomach poisons. Paris green, arsenate of lead, and rotenone are considered to be effective stomach poisons.

Contact Poisons kill insects when they are hit by or come in contact with the poison; the insect does not have to consume it. Any type of insect can be controlled by contact poisons, including insects that suck plants, such as aphids and leaf hoppers. Contact poisons kill by upsetting the insect's nervous system or breathing system. An example of a contact insecticide is malathion.

Systemic poisons are chemicals that enter the plant sap and move throughout the entire plant. When insects eat parts of plants or suck juice from plants, the chemical is swallowed. Systemic poisons are effective in the control of insects with either chewing or sucking mouth parts. They are especially effective in controlling insects that hide under leaves or underground that therefore are not affected by contact sprays. If used on plants that are used as food, the chemical must be given time to break down within the plant, and become nonpoisonous before the harvest date for the crop. Two systemic insecticides are dimethoate (Cygon) and demeton (Systox). They are sprayed on the plant foliage or mixed with water and applied to the roots. Both methods result in the chemical spreading throughout the entire plant.

Fumigants are actually contact poisons applied in gaseous form. The gases or fumes kill the insect after entering its system through breathing pores. The insect must actually ingest the poison for this insecticide to be effective. Fumigants are used to control soil-borne insects that damage roots. One example of a fumigant is methyl bromide.

Fumigants are also used in greenhouses. When applied in the greenhouse, the fumigant, consisting of tiny particles of insecticide, settles on the insect and is breathed into its system.

Repellants generally do not kill insects, but, instead, drive them away before they attack the plant. One example of a repellant is aluminum foil which when placed around plants repels flying insects, figure 19-3.

Attractants work in the opposite way from repellants; they lure insects to their death. An example is the Japanese beetle bait used in traps to catch the beetles. Another example is the sex lure used to trap the gypsy moth. The sex lure used in these traps is made from naturally occurring or synthetically produced hormones which attract the adult insect. The male gypsy moth is irresistibly attracted to the sex lure; this attraction is stronger than the normal attraction of the female moth. Since most of the male moths are caught in the trap, the females remain unfertilized and lay sterile, unfertilized eggs which do not hatch. The sex lure acts in the same manner on the Japanese beetle.

THE CHEMICAL MAKEUP OF INSECTICIDES

Insecticides are classified in three major groups according to their chemical makeup.

Fig. 19-3 Aluminum foil strips around squash plants protect the plants from flying aphids by reflecting ultraviolet rays from the sky, making the area undesirable to the aphids. (USDA photo)

Inorganic Compounds

Inorganic compounds are of mineral origin. That is, a mineral is used as the basis for the poison. They usually work in the form of stomach poisons. Sulfur is one type of inorganic insecticide and is also widely used as a fungicide. Lead arsenate is an inorganic insecticide but is not legal in many areas.

Organic Compounds

Organic compounds are those derived from plants. They usually work in the form of stomach poisons or contact poisons. Two examples of organic compound poisons are rotenone and pyrethrum, which are very safe to use.

Synthetic Organic Compounds

These chemicals are of relatively recent origin, many of them having come into use in the last thirty to thirty-five years. They are very effective against insects, but many of them are also toxic to human beings. This large group of insecticides is subdivided into three smaller groups: chlorinated hydrocarbons, organophosphates, and carbamates. They are not found naturally as are the organic compounds listed above, but are produced in the laboratory.

Chlorinated hydrocarbons contain chemicals that have long *residual* control; that is, they continue to kill long after the initial application. DDT is one of the best known chemicals in this group. It is no longer sold in the U.S., however, because of concern about the buildup of the poison in the environment and the bodies of warm-blooded animals. One chemical from this group which is still being used is methoxychlor.

Organophosphates. This group of insecticides is very effective in controlling insects. It also contains some of the chemicals most toxic to warm-blooded animals. TEPP and parathion are so toxic that one drop of the concentrated chemical in the eye can kill a human being. They are also absorbed rapidly through the skin. Malathion, another chemical in this group, is a relatively safe chemical and one that can be used to control many insects. **Read the insecticide label for caution in use of these chemicals.** Organophosphates break down quickly in the environment (within fifteen to thirty days) and do not build up in the bodies of warm-blooded animals.

Carbamates. This group of synthetic organics contains some of the safest insecticides on the market. Carbamates such as carbaryl (Sevin) are very effective in killing some insects although not effective in killing others. They are slightly toxic to warm-blooded animals. Carbamates break down rapidly (in two to seven days) and leave no residue to contaminate the environment. These chemicals do not build up in the bodies of warm-blooded animals.

Chemical Sterilants

The most successful use of sterilants involves the use of gamma radiation treatments which are used to sterilize insects so that they cannot produce offspring. One method is to sterilize male insects and release them to mate with females. Since the eggs are not fertilized, they will not hatch when deposited by the female. This is a very effective method in the control of certain insects, such as the screw worm, which infests cattle. This process is also being tested for use on plant insect pests such as the gypsy moth.

Fig. 19-4 Hand duster being used to apply dust to beans. Note the wide drift of dust away from the target plants. (USDA photo)

APPLICATION OF INSECTICIDES

Dusts

Insecticides are applied with dusters if purchased as dusts. The chemical is already diluted when bought; it requires no mixing before application.

Dusts are easy to apply with inexpensive equipment, but tend to blow or drift from the target, figure 19-4. This drifting can damage other crops or contaminate the environment.

Wettable Powders

Wettable powders (WP) resemble dusts in appearance, but are concentrated and must be diluted with water before application. Read the directions for the amount of insecticide to mix per gallon of water. Wettable powders tend to settle while in solution and must be stirred while being sprayed.

Better coverage of plants is usually possible with a water-mixed spray and less drift occurs, figure 19-5.

Fig. 19-5 Sprayer which is strapped to the shoulders is being used to spray strawberries. Very little drift occurs in this type of application. (USDA photo)

Emulsifiable Concentrates

Emulsifiable concentrates (EC) are liquids which are mixed with water in the same manner as wettable powders. The concentrate is safer to handle since there is no powder or dust to blow during mixing. Emulsifiable concentrates do not settle and separate from the solution, as wettable powders do, and give the same good coverage of plants. Shelf life of these chemicals is generally longer than dusts or wettable powders if the container is kept closed and stored in a cool, dark place.

Granules

Granules are insecticides in the form of small pellets. They are spread on the soil surface where they either penetrate the soil after the application of water, or turn into gases which fumigate the insects. Some of the systemic insecticides mentioned in this unit are sold in the form of granules. For example, the chemical Systox may be applied to the soil in granulated form and watered in, after which it is absorbed by the roots of the plants.

Baits

Baits are poisonous materials which attract insects. The pest eats the bait and is killed by the poison. Slugs, snails, cutworms, grasshoppers, and weevils are examples of pests that are controlled in this manner.

Aerosols

Aerosol insecticides are contained in pressurized cans. They are generally used for small insect control jobs around the house or in the greenhouse. The chemical is already diluted and ready for use when purchased, figure 19-6. Many aerosols in the form of bombs are available for use in the greenhouse. The bombs are punctured and the chemical is slowly released as a fine mist.

EQUIPMENT

Hand-operated Dusters and Sprayers

Application of insecticides on the school grounds or in greenhouses can be done safely using inexpensive hand-operated sprayers and dusters. An explanation of the use of each of these pieces of equipment will be given by the instructor.

Fig. 19-6 Commonly used aerosol sprays. The spray on the left is used outdoors on all types of plants. The aerosol in the center is meant for houseplants. The aerosol on the right is a combination insecticide and fungicide used in the control of both insects and plant disease. (Richard Kreh, Photographer)

Exercise. As part of the study of this unit, use each piece of equipment to properly apply an insecticide. The instructor will demonstrate the proper filling and spraying technique.

Larger Dusters, Sprayers, and Granular Application

Large scale commercial application of sprays, dusts, or granules is accomplished by the use of airplanes, helicopters, tractor sprayers, or trucks with special granular spreaders or sprayers. Figure 19-7 illustrates the use of a large fan jet orchard sprayer.

Airplanes and helicopters are used to spray and dust large acreage of crops, and crops which cannot be treated by ground-operated equipment, figure 19-8. Commercial applicators apply the specified chemical and charge the grower at a cost per acre. This is an economical way to apply insecticides when large areas must be covered quickly.

Fig. 19-7 A large orchard sprayer applying pesticide to a fruit tree. The pesticide is mixed with water and applied in liquid form. A large fan on the back of the sprayer propels a fine mist up and through the tree. (USDA photo)

Applying the Proper Amount

Whatever method of application is used, it is important that the proper amount of the chemical is applied to a given area. This requires calibration of sprayers and dusters. Calibration involves measuring the amount of chemical applied during a specified period of time or a measured area and then adjusting the applicator or speed of application to obtain the proper amount of chemical required.

Dusts and sprays must be applied so that there is total coverage of all leaf surfaces. When dusts are applied to plants, they can be seen settling on the surface of the plants, thereby making it easy to see the extent of the coverage. When sprays are used, leaves of the treated plants should be wet just to the point of dripping.

> Safe, efficient control of insects requires that the proper concentration of the chemical be mixed in sprays. Read the label for proper mixing instructions.

Fig. 19-8 Airplanes are used to apply pesticides on a large scale. Many acres can be covered rapidly by this method. (USDA photo)

IDENTIFYING INSECT DAMAGE TO PLANTS

To identify the insect causing damage to plants, the damage itself must be examined. Chewing insects actually eat away part of the plant. This results in holes in the leaves or missing pieces of bark. Chewing insects are controlled by a stomach poison or a contact spray, figure 19-9.

Sucking insects suck the juices from the leaves of plants. The injury caused by sucking insects is not as easily detected as that which is caused by chewing insects. Some signals of damage by sucking insects include twisted plant tips and rolled leaves. Plants that are infested by sucking insects may show no signs of infestation at all, but simply appear to be less healthy. Figure 19-10, page 202, shows damage to a pea plant by a sucking insect, the aphid. Sucking insects must be controlled with a contact chemical, a fumigant, or a systemic.

Fig. 19-9 Damage to grape leaves by Japanese beetle. These chewing insects are actually eating parts of the leaves. Notice that the tough leaf veins are left uneaten. (USDA photo)

Fig. 19-10 Aphids attacking a pea plant. These insects suck juices from the plant, thereby greatly reducing plant vigor and yield. (USDA photo)

TIMING IN INSECT CONTROL

Insects must be killed when they are actively feeding or moving about on the plant. When in the resting (*pupae stage*) or egg stage, very few chemicals can affect insects. It is important to determine what stage of growth the insect is in before spraying. This is done by examining the insect on the plant.

Insects go through one of the following basic life cycles. The stages during which the insects are controllable are boxed.

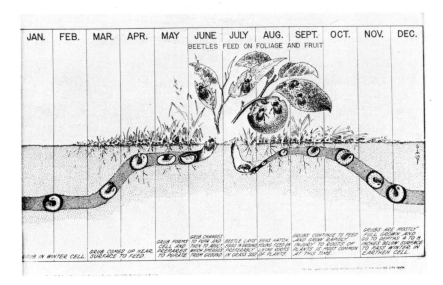

Fig. 19-11 The Japanese beetle is killed with an insecticide applied as a spray only during June, July, August, and September. The rest of its life cycle is spent underground where it does a great deal of damage to grass roots. Insecticides in the form of chemicals or the introduction of natural enemies are used during the larval stage. The insect cannot be killed with sprays during its egg and pupa stages. (USDA photo)

- Complete metamorphosis — egg — larvae — pupae — adult

- Incomplete metamorphosis — egg — nymph — adult

If the entire insect population of a plant is in an inactive stage, insecticides must not be applied until the eggs hatch or the pupae emerges as an adult. If insects in both active and inactive stages are present, an immediate application of insecticide should be made, with a second application made in seven to ten days when the eggs have hatched and before the insect has a chance to mature and lay more eggs. Figure 19-11 illustrates the life cycle of the Japanese beetle.

PLANTS AS INSECT REPELLANTS

According to organic gardening experts, certain plants may be used to prevent attacks by insects on other plants. The following are some of those plants, listed with the insects they repel.

Plant	Repels
mint	flea beetles, cabbage butterflies
onions, garlic*, chives, leeks	aphids
marigolds	root nematodes
nasturtiums	aphids, cucumber beetles
sage	carrot flies, cabbage pests
horseradish	potato bugs, other flying insects

*Garlic also inhibits the growth of asparagus.

THE ORGANIC GARDENER

According to *Webster's Dictionary, organic* is defined as relating to, produced with, or based on the use of fertilizer of plant or animal origin without employment of chemically formulated fertilizers or pesticides. The *organic gardener* incorporates this idea in the production of food. These gardeners rely on insect repellants that do not include chemicals in any form — natural enemies, plants, extracts of ground insects used as sprays, natural insecticides, crop rotation, sanitation, and resistant varieties.

Although there are differing opinions concerning the usefulness of organic gardening, there are some techniques that are practiced widely, even when the gardening being done is not entirely organic. For example, it is a generally accepted fact that the incorporation of organic matter into the soil and the use of mulches results in better plant growth. This is due to better moisture retention in the soil and, possibly, an increased availability of minor plant food elements. This leads to the production of healthier, faster growing plants which are better able to resist attack by insects and disease.

STUDENT ACTIVITIES

1. Observe insect damage to plants. Identify the insect involved and select the proper pesticide for effective control. Record your results in the Insecticide Selection and Application Table, available from your instructor.

2. With guidance from your instructor, demonstrate the proper mixing and application procedure of an insecticide to a live insect pest discussed in the unit and evaluate the results.

3. Obtain the worksheet, Insect Pests Common to Your Area, from your instructor and fill in the blanks with the required information. After completing the project, check with your instructor to be sure the information you have given is accurate.

4. Visit a local greenhouse operation to observe insect control techniques.

SELF-EVALUATION

A. Provide a brief explanation for each of the following.

1. List six ways in which insecticides work to kill insects. Include one insecticide in each category and one insect it controls.

2. List the three major groups or families of insecticides and one chemical belonging to each.

3. List three natural methods of insect control and at least one insect controlled by each.

B. Select the best answer from the choices offered to complete the statement or answer the question.

1. Stomach poison insecticides work best

a. on sucking insects.
b. as a contact insecticide.

c. on chewing insects.
d. on nematodes.

2. An example of an inorganic insecticide is

a. sulfur.
b. Sevin.

c. malathion.
d. rotenone.

3. An example of a natural organic insecticide is

a. lead arsenate.
b. Sevin.

c. malathion.
d. rotenone.

4. An example of a synthetic organic insecticide is

a. lead arsenate.
b. Sevin.

c. malathion
d. rotenone.

5. What kind of insects do systemic insecticides kill better than any other insecticide?

a. sucking
b. chewing

c. active
d. none of these

6. An example of an insecticide that is used as a fumigant is

   a. malathion.
   b. Sevin.
   c. DDT.
   d. methyl bromide.

7. Insecticides that remain active in the environment for the longest period of time are

   a. organophosphates.
   b. carbaryls.
   c. chlorinated hydrocarbons.
   d. none of these

8. The insecticides that are most toxic to human beings are the

   a. carbamates.
   b. chlorinated hydrocarbons.
   c. carbaryls.
   d. organophosphates.

9. Application of dust insecticides requires

   a. a duster.
   b. no dilution.
   c. a quiet day to reduce drift of the insecticide.
   d. all of the above

10. Wettable powders (WP) are powdered insecticides that

    a. must be diluted with water.
    b. are ready to apply with a duster when purchased.
    c. are used only on chewing insects.
    d. all of the above

11. Insecticides which come in the form of small pellets and are scattered on the soil surface are called

    a. pellets.
    b. baits.
    c. granules.
    d. none of these

12. Emulsifiable concentrates (EC) are liquid insecticides that must be

    a. diluted with water.
    b. used as contact sprays.
    c. used for chewing insects only.
    d. all of these

13. An insecticide packaged in a pressurized can is called a(an)

    a. home owner's spray.   c. home and garden spray.

    b. convenience package.   d. aerosol.

14. The proper amount of spray insecticide has been applied to a plant when

    a. the bugs begin to die.

    b. the leaves are damp.

    c. the liquid just begins to drip from the leaves.

    d. none of the above

15. When buying an insecticide to eliminate a specific insect, one should purchase a chemical

    a. that is recommended for that insect.   c. that is safe to use.

    b. only after reading the label.   d. all of these

16. How many pairs of legs do insects have?

    a. six   c. two

    b. three   d. four

17. The legs of every insect are fastened to the body part known as the

    a. thorax.   c. abdomen.

    b. head.   d. shank.

18. A deadly fungus disease is a naturally occurring method used in the control of

    a. tomato hornworms.   c. Japanese beetles.

    b. Mexican bean beetles.   d. all of these

unit 20
Fungicides, Rodenticides, Molluscicides, and Nematocides

OBJECTIVE

To identify fungus diseases, rodents, mollusks, and nematodes, and prescribe a chemical control for each.

COMPETENCY TO BE DEVELOPED

After studying this unit, the student will be able to

- identify fungus diseases on plants and set up a spray or dusting schedule for effective control.

- recognize at least two types of rodents and apply a rodenticide in a bait station.

- recognize slugs and snails on sight and develop a slug control program, including chemicals used in the bait station or as contact poisons.

- identify two nematocides and explain in writing their application to specific situations.

Materials List

plants infected with fungus disease

one live snail and slug

one plant with nematode damage to roots and one healthy plant of the same variety

sample fungicides, rodenticides, molluscicides, and nematocides

---

FUNGICIDES

Fungicides are pesticides used to control plant diseases caused by fungi. *Fungi* are tiny nongreen plants such as rusts, mildew, mold, and smut that lack chlorophyll and live as parasites on green plants. Fungus diseases are spread by *spores* that resemble tiny seeds floating through the air. As fungi grow on the green plants, the plant tissue is killed or the plant is weakened. These diseases must be controlled if attractive, highly productive crops are to be grown.

Chemical Makeup

Fungicides are manufactured from a variety of chemicals. Generally, compounds of copper, sulfur, mercury, and dithio-carbamates make effective fungicides. The following are some of the fungicides in use today.

Bordeaux mixture. The bordeaux mixture is a combination of copper sulfate (blue vitriol) and hydrated lime in water. It is the copper in the bordeaux mixture that kills.

Lime sulfur, a mixture of lime and sulfur, has been used as a fungicide for many years.

Ferbam, maneb, and zineb, classified as dithio-carbamates, are relatively new fungicides that have been used successfully. Sulfur, in the form of finely ground wettable powder, is a very effective fungicide.

Mercury compounds have proven to be good fungicides, but most of these are off the market because of heavy metal contamination of the environment.

Fungicides are most effective when used to prevent germination and/or growth of the fungus spore, thus preventing the attack to the plant. Systemics, such as Benlate (benomyl), work by entering the plant and circulating in the sap. However, they work primarily as other fungicides do; that is to prevent the disease rather than cure it after it is established in the plant.

Fungicides are generally not as toxic to humans and other warm-blooded animals as insecticides are. Review figure 18-1 in unit 18, which lists the LD_{50} for some of the most frequently used fungicides. Compare the toxicity level of fungicides listed in the chart with those of other pesticides. Figure 20-1 illustrates a typical fungicide label.

Fig. 20-1 Typical fungicide label. Notice the different types of information that appear on the label. (Courtesy Miller Chemical Company)

Effects of Fungus

Fungus diseases attack all parts of the plant.

Leaves. Plant leaves are attacked by many so-called *leaf spot diseases.* The term *leaf spot* describes the appearance of the diseased leaf. The leaf is marked by round or irregular spots which are beginning to turn a color different from the rest of the plant, 20-2. The spot or spots on the leaf grow larger and larger as the fungus disease spreads. The entire leaf may be killed, but this is unusual.

Leaf spot diseases kill the plant by destroying the cells that manufacture food. As more cells are killed, the plant is less able to make food and becomes weaker. The appearance of the plant is also affected. A plant sold for its beauty may be damaged so much that it is unsaleable. A plant which produces a crop of fruit or seeds does not produce as large a crop when affected by leaf spot disease because the leaves cannot manufacture enough food for best growth of the fruit or seeds.

Leaves are also attacked by another fungus disease, known as rust disease. Again, the name describes the appearance of the disease; the leaf which takes on a rusty color in the beginning, is gradually killed by the disease. Mildews and molds are other fungus diseases. Their names describe their appearance on the plant.

Fungicides are sprayed on leaves to prevent the entry of the fungus disease into the plant. When the fungus spore sprouts on the plant leaf, the fungicide kills it. A spray or dusting calendar complete with dates must be established so that the fungicide can be applied on a regular schedule. Systemic fungicides such as benomyl (Benlate) which enter the plant sap, do a better job of preventing leaf diseases.

Fig. 20-2 The effects of leaf spot disease on blueberry leaves. (USDA photo)

Most fungicides are used to prevent disease and must therefore be sprayed or dusted on the plant before the disease is present to prevent spread of disease. Some fungicides do cure or eliminate disease already present on the plant, but even these chemicals work best as preventors of disease, and not as a cure.

Figure 20-3, page 210, is a list of fungus diseases and the fungicide used to prevent them.

Stems, fruit, and seed pods of plants are also attacked by fungus diseases. The effect and control are the same as those listed for leaves. Anthracnose, for example, is a fungus disease that attacks the stems of many plants.

High humidity and rain help to spread fungus diseases of stems and leaves. Most fungus diseases grow best in high moisture conditions.

Plant Disease	Symptoms	Fungicide to Use	How to Apply	Remarks
Black Spot	circular black spots up to 1/2″ across, usually on the upper leaf surface	benomyl, Phaltan, or captan	spray or dust	A serious disease of roses. Difficult to control.
Bean Rust	Rusty growth on under-side of leaves; may cause leaves to fall off.	sulfur or maneb	dust	No bean variety is resistant.
Anthracnose (Attacks the black raspberry.)	circular reddish brown sunken spots with purple margins and gray centers	lime sulfur in dormant stage; ferbam	spray	Grows on young shoots; becomes large cankers on older canes.
Powdery Mildew (Attacks the lilac.)	Covers leaves with a thick white coating.	benomyl, Phaltan, or Karathane (best)	spray	worse in dry season
Apple Scab	Appears on leaves as small smoky area, moldy 1/4″ across fruit, raised spots. scablike	ferbam, captan, or Cyprex	spray	Worst apple disease. Attacks leaves and fruit.
Damping-off	Seeds either fail to appear or small seedlings rot off at soil level.	captan-Arasan seed treatment - - - - benomyl	seed treatment or soil drench	Spreads rapidly; prevention is better than cure.
Rhododendron Root Rot	Top of plant starts to wilt on hot afternoons; watering does not help.	none totally effective Nurelle or Dithane Z-78 as a soil drench may work.	soil drench	very difficult to control May be necessary to destroy plant and sterilize soil.
Brown rot (of peach)	Starts as brown spot; spreads rapidly until entire fruit turns brown and dries up.	wettable sulfur	spray	Attacks blossom, stem, and fruit. Easiest to identify on fruit.
Botrytis (Attacks the peony.)	Buds blast (wither) and turn black; leaves get gray mold and rot off.	zineb, ferbam	spray	Attacks many bulb crops.

Note: Most fungus diseases are at least partly controlled by sanitation. If leaves, fruits, stems, or any part infected with the disease are collected and burned at the end of the season, further control measures may not be necessary.

Fig. 20-3 Some common fungus diseases and their control

Roots of plants are attacked by many types of root rot or root mold. *Root rot* and *root mold* describe the appearance of the root after attack by the fungus disease. Root diseases are controlled by drenching the soil with fungicides or by spraying.

Fungus root diseases are most often found in soils that do not drain well and are wet too much of the time. Improved soil drainage generally helps control fungus diseases of roots.

Small seedlings are often attacked by fungus diseases such as *damping-off*. The use of sphagnum moss to grow seedlings often prevents the start of the disease. Dusting or drenching the soil mix with benomyl (Benlate) or captan often gives excellent control of seedling root and stem rot.

Application

Fungicides are sprayed or dusted on plants or poured around the roots as drenches. The methods of application are the same as those for insecticides.

One difference between the application of fungicides and insecticides is in the timing. Insecticides are usually applied after it is determined that an insect is on the plant and is causing damage. Fungicides are applied to prevent attack. The chemical acts as a protective film to kill the fungus spore as it sprouts and before it can penetrate the surface of the plant.

RODENTICIDES

Rodenticides are chemicals used to control rats, mice, other rodents, and bats. Most commonly, rodenticides kill as stomach poisons. They are applied as bait which is attractive to the rodent and is eaten. In limited cases, rodenticides are spread or sprayed on the rodent or in an area through which the rodent travels.

This technique is usually avoided however, because other wildlife or domesticated animals could be poisoned accidentally.

A common chemical used for rats and mice is warfarin. This chemical is applied to grain in *bait stations* where the rodent feeds. Field mice in orchards are often poisoned by a coating of deadly poison on pieces of apple. The apple is placed in bait stations for the mice to eat.

> **Always place poisoned baits in bait stations where other animals cannot reach the poison. Place the bait under a wooden or metal box with a small hole cut in it which is just large enough for mice to enter.**

MOLLUSCICIDES

Molluscicides are chemicals used to control snails and slugs. They are applied as poison baits or contact poisons. These chemicals kill as stomach poisons when eaten or as contact poisons when the slug or snail is covered by the poison or crawls through an area to which the poison has been applied. Slugs and snails usually hide during the daytime; therefore, it is difficult to hit them while spraying or dusting.

Snails and slugs look much alike, except that slugs do not have a shell and snails do, figure 20-4, page 212. Both travel on the soil surface leaving a slimy trail behind, feeding on plant leaves and tender stems at night. Because they feed at night, slugs and snails actually doing damage to plants may be difficult to locate, thereby causing other insects to be blamed for the damage. To pinpoint snails and slugs as the pest, look for the slimy trail.

To find slugs and snails lift stones and mulch and look underneath. Any cool, moist place is a good hiding place for them.

Fig. 20-4 (A) Snail   (B) Slug
(Richard Kreh, photographer)

Control

Rough sand, ashes, gravel, or cinders act as barriers to slugs and snails; the rough surface provided by these materials is difficult for them to crawl on. Another chemical-free control method is to set a shallow pan filled with beer in the soil, with the top of the pan at the soil level. The slugs follow the smell of the beer into the pan and drown.

Various slug killers are sold by pesticide dealers. Some effective chemicals include Slugit, which is a stomach and contact killer; Slug-kill (15 percent methaldehyde dust) which is a contact poison; and Zectran.

NEMATOCIDES

Nematocides are chemicals used to control nematodes. *Nematodes* are small hairlike worms that feed on plant roots.

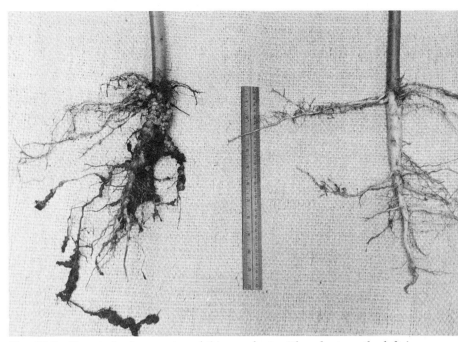

Fig. 20-5 Nematode damage to a hibiscus plant. The plant on the left is badly infested with nematodes and the root system is severely damaged. The plant on the right has a normal root system. (USDA photo)

These tiny worms bore into the root, usually causing knots or bumps to appear on the attacked root, figure 20-5. Nematodes live in the soil and swim in soil water to move from plant to plant. Some nematodes also live inside plant leaves.

Control

Soil fumigants are the most effective treatment for nematodes. The nematocide must contact the nematodes as a gas or liquid passing through the soil. Two chemicals meant specifically for nematodes are Dasanit and Nemagon. Systemic insecticides are effective on leaf nematodes.

STUDENT ACTIVITIES

1. Under the instructor's supervision, apply a fungicide to a plant infected with fungus disease on the school grounds. Use a small aerosol. As the spraying progresses, check for disease control and record the results. If two plants are available, spray one and keep one as a control.

2. Put together a bait station for rodents and slugs. Actually use the station under the supervision of your instructor if circumstances permit.

> **Caution: These activities should be accomplished only under the supervision of the instructor. When applying any pesticide, carefully read and follow all instructions and precautions.**

SELF-EVALUATION

Select the best answer from the choices offered to complete each statement.

1. Fungicides are chemicals used to control

   a. plant rust diseases.   c. plant mildew.
   b. fungus diseases.   d. all of these

2. Fungus diseases are spread by small seedlike structures called

   a. spores.   c. pollen.
   b. roots.   d. nematodes.

3. Fungicides cause a protective film to coat plant leaves and

   a. keep fungus spores from sticking to the leaves.
   b. kill the fungus spore before it can enter the plant cell.
   c. smother fungus diseases on the leaves.
   d. prevent oxygen from reaching the fungus spores so that they cannot grow.

4. Rodenticides generally kill rats and mice

   a. as stomach poisons, so they must be eaten by the rodents.

   b. while they crawl on the poison.

   c. by fumigating them.

   d. all of the above

5. Snails and slugs are not often seen eating plants because

   a. they only eat roots and underground stems.

   b. they feed at night.

   c. they drop to the ground and hide if any vibration occurs.

   d. they are the same color as most plants and therefore difficult to see.

6. Snails and slugs belong to the same family, but have one main difference. This difference is

   a. slugs are larger than snails.

   b. slugs do not have shells; snails do.

   c. slugs leave a slimy trail when they move; snails do not.

   d. slugs have longer antennae than snails.

7. Contact sprays are not completely effective for slugs because

   a. their slimy bodies resist the chemicals.

   b. they form a path with slime on which to walk.

   c. their bodies are slippery so chemicals do not stick to them.

   d. they hide under objects such as stones and mulch and are therefore difficult to hit.

8. Nematodes are tiny hairlike worms that

   a. belong to the insect family.   c. attack animals, including human beings.

   b. feed on plant roots.   d. all of these

9. Nematodes are controlled with fumigants or soil drenches because

   a. they live in the soil where these chemicals are able to reach them.

   b. they do not eat plants, so stomach poisons do not work.

   c. no other chemicals are made for use on them.

   d. all of the above

OBJECTIVE

To select and apply herbicides so that weed control is accomplished without damage to the crop being cultivated or contamination of the environment.

COMPETENCY TO BE DEVELOPED

After studying this unit, the student will be able to

- identify a weed problem and select an herbicide to control the problem.
- define and differentiate between selective and nonselective herbicides.
- describe three ways in which herbicides destroy weeds.
- outline in writing how a sprayer is calibrated.
- list three possible reasons for failure of an herbicide to work properly.
- properly apply a herbicide.

Materials List

weed killers complete with label and recommendations

weed charts for weed identification

sprayer for use in calibration and application of herbicides

an area of lawn or crop land for use in locating and identifying weeds and applying herbicides

---

unit 21
Herbicides

An *herbicide* is a chemical that kills weeds. Herbicides comprise a large percentage of the total pesticides used each year. In

TOXICITY CLASS	COMMON NAME OR DESIGNATION	SOME COMMON TRADE NAMES	ACUTE ORAL LD$_{50}$ mg/kg (rats)	PROBABLE LETHAL DOSE OF UNDILUTED CHEMICAL FOR 150 LB. PERSON
Very Toxic	sodium arsenite	Atlas A	10	2 drops to 1 tsp.
	DNBP (amine)	Premerge, Sinox PE	40	(less than 1/6 oz.)
Moderately Toxic	paraquat	Weedol	150	1 tsp. to 1 ounce
	2, 4, 5-T	various brands	300	
	diquat	Aquacipe	400	
Slightly Toxic	silvex	Kuron, Weedone-TP	500	1 ounce to 1 pint
	2, 4-D	various brands	500	(lb.)
	MSMA	Ansar, Daconate	700	
	cacodylic acid	Phytar 560	830	
	aspirin	(for comparison)	1,240	
	linuron	Lorox	1,500	
	TBA	Trysben 200	1,640	
	DSMA	Ansar, Sodar	1,800	
	norea	Herban	2,500	
	amitrole	Weedazol	2,500	
	borate	Borax, Borascu	2,500	
	dicamba	Banvel	2,900	
	prometon	Pramitol	2,980	
	DCPA	Dacthal	3,000	
	atrazine	Aatrex	3,080	
	table salt	(for comparison)	3,320	
	diuron	Karmex, Krovar 1	3,400	
	monuron	Telvar	3,600	
	chloroxuron	Tenoran	3,700	
	prometryn	Caparol	3,750	
	AMS	Ammate	3,900	
	TCA	various brands	5,000	
Almost Nontoxic	siduron	Tupersan	5,000	
	simazine	Princep	5,000	
	sodium chlorate	Atratol	5,000	
	propazine	Miloguard	5,000	
	bromacil	Hyvar X, X-L, X-P Krovar 1	5,200	1 pint to 1 quart or 1 lb. to 2 lbs.
	picloram	Tordon	8,200	
	dalapon	dowpon	9,300	
	benefin	Balan	10,000	

Fig. 21-1 Relative toxicity of some herbicides to mammals

the past twenty years, many new weed killers have been introduced on the market and each year, new uses are found for traditional weed killers. It is estimated that in the United States, weeds cause damage and loss of production to crops in excess of $4.5 billion each year. Herbicides are used to kill weeds in crops, on lawns, along roadways, and in many other places.

A *weed* is considered to be any plant growing where it is not wanted. Grass which is growing in a lawn is not a weed, but is considered a weed if it is growing in the garden.

Weed control can be accomplished without the use of chemicals. In some cases, mulching and the cultivation of crops may be more effective and just as economical to control weed growth. For example, it would be very difficult to spray weed killers in a small garden without damaging other plants. It is in large acreages of single crops that selective weed killers can be used to greatest advantage.

There are many different types of chemicals used as weed killers. A partial list of herbicides and their toxicity levels is given in figure 21-1. Notice that some of these pesticides may be very poisonous if swallowed (oral). They are generally not as toxic dermally (on the skin).

TYPES OF HERBICIDES

Nonselective Herbicides

Nonselective herbicides kill all plants to which they are applied. These herbicides are used in places where no plant growth is wanted. Lumber yards, railroad tracks, fence lines, driveways, and parking areas are a few examples of areas where these chemicals are used. Examples of some weed killers in this category are Atratol, Aatrex, and sodium arsenite. Sodium arsenite is a deadly poison used to kill all vegetation and will continue to keep the soil sterile for up to five years.

Nonselective weed killers sometimes may also be used as selective weed killers. Examples of this type of substance include Atratol, and atrazine (Aatrex). Used at a rate of two to four pounds of active chemical per acre, it kills only small weeds or selected weeds; if application is increased to from five to ten pounds per acre, it kills all plants. Sodium arsenite, however, is used only when all vegetation is to be eliminated.

Selective Herbicides

Selective herbicides are just what the name implies — chemicals which kill some plants, but not others. This is by far the largest group of weed killers. Weed control in a particular crop can be accomplished safely and effectively with selective weed killers. These chemicals are designed to kill the weed but not the crop.

Generally, a selective weed killer which kills broadleaf plants will not kill grasses which have narrow, bladelike leaves. This makes it possible to spray a lawn with 2,4-D to kill dandelions which have a broad leaf and not kill the grasses which have narrow leaves. Since corn is a grass, broadleaf weeds in a corn field can be killed with a selective herbicide without damaging the corn. The herbicide 2,4-D is one of the best known selective weed killers used to kill broadleaf plants without killing grasses.

Herbicides that have the potential to kill a particular crop at high concentrations can be used at low concentrations to kill young weeds that are growing in the crop and not damage the crop. By applying the weed killer at a low concentration, young tender weeds are killed and the older mature crop plant is not harmed. Strawberries are protected from weeds in this way.

THE EFFECT OF HERBICIDES

Herbicides work by upsetting the *metabolism*, or life functions, of the plant. The plant either starves to death, or wears itself out due to the increased rate of activity caused within its system by the chemical. Herbicides are designed to affect the metabolism of some plants and not others.

The following are some specific examples of the effect of herbicides.

Atrazine

Atrazine is widely used to kill both broadleaf weeds and grassy weeds in corn and turf grass sod as well as those in tree and shrubbery plantings.

Application. Atrazine may be applied *preplanting* (before planting has taken place); *preemergence* (before emergence of seedlings aboveground); or *postemergence* (after seedlings have grown aboveground but before the crop matures). However, it must be applied before the weeds reach a height of more than 1 1/2 inches. The chemical is usually applied with an ordinary sprayer.

Behavior in the Plant. The chemical is absorbed through both roots and foliage. It moves through the plant in the xylem or wood and accumulates in the stem tips and leaves of plants. Atrazine kills the plant by preventing photosynthesis from occurring. Since the plant cannot manufacture food, it starves to death. Plants such as corn are not killed by atrazine when applied in usual dosages because they are able to break down the chemical before any harm is done to them. Cyanazine (Bladex) also works as a photosynthesis inhibitor.

2,4-D

2,4-D is a systemic herbicide and is widely used to control broadleaf weeds. Most dicot (broadleaf) plants are susceptible to the chemical at normal application rates.

Application. Application is usually by spray directed on the plant foliage. Both plant leaves and roots absorb 2,4-D.

Behavior in the Plant. The chemical moves through the plant phloem or bark. Movement is more rapid in fast growing plants and therefore plants in active growth are killed more easily. The chemical 2,4-D causes abnormal growth in the weed and affects respiration, food reserves, and cell division.

Dacthal

Dacthal is used to control annual grasses and certain annual broadleaf weeds.

Application. Application is generally at planting time before the weeds come up. The chemical is usually diluted with water and applied as a spray.

Behavior in the Plant. Dacthal is not absorbed by leaves and it does not move through the plant. Rather, it kills sprouting seeds. The exact way it acts is not yet known.

The weed killer Treflan kills weeds in a similar manner.

The three herbicides just discussed kill weeds in three different ways:

atrazine ⟶ prevents photosynthesis

2,4-D ⟶ upsets cell division, respiration, and food reserves

Dacthal ⟶ destroys sprouting seeds (specific action not yet known)

Weed killers do not all fall into these three groups, but they are representative of a great percentage of herbicides.

APPLICATION OF HERBICIDES

Herbicides are either sprayed with ordinary sprayers or applied already mixed in fertilizers or in granules. The same type of equipment is used as with insecticides. However, a sprayer that is used to apply herbicides should not be used to apply insecticides or fungicides unless it is thoroughly cleaned. This does not mean simply rinsing out the sprayer; it must be washed with a substance that neutralizes the herbicides, such as ammonia. Read and follow the directions carefully; otherwise, enough herbicide may remain in the sprayer to damage desirable plants sprayed for insects or diseases. When using small hand sprayers, a separate sprayer should be used for the application of herbicides.

Time of Application

Application of herbicides must be made at one of three times to avoid injuring desirable plants.

Preplanting herbicides are mixed into or sprayed onto the soil or seed bed. Some herbicides are effective only when incorporated with or mixed into the soil; others need no mixing. Read the label before buying.

Preemergence treatment is made before any plant growth appears or after the crop comes up but before the weeds appear. Read the label to see if the chemical should be applied before the crop emerges or after the crop emerges and before the weeds emerge. When used at the concentration recommended, these chemicals usually prevent seeds from germination or kill only small tender weeds without damage to desirable plants.

Postemergence applications must be made very selectively in terms of the chemical used.

> The horticulturist must be very careful to choose an herbicide that kills the weeds but not the crop since at the time of application, both weeds and crop are aboveground.

Calibration of Sprayers

It is extremely important that the proper amount of actual chemical be applied to a particular area when spraying herbicides. Even if the right mixture or concentration of spray is in the spray tank, it is possible to apply the wrong amount of weed killer. To avoid this problem, the sprayer should be properly *calibrated*, or adjusted, before use. The correct amount of active chemical must be applied on a measured area for the most effective control.

Large Sprayers. The first step in the calibration of large sprayers is to measure and mark off an acre. Spray the acre with water and refill the spray tank, measuring exactly how many gallons of water are required to refill the tank, the amount of water applied to the acre. This is the rate of application per acre. If 10 gallons of water is applied per acre and 2 pints of chemical is to be applied per acre, then 2 pints of chemical is added to each 10 gallons of water. **Example.** For a 100-gallon tank, add 20 pints, or 2.5 gallons, of chemical.

> Caution: The rate of travel over the area and the sprayer pressure must be the same as in the trial run.

Small Hand Sprayers. Follow these steps to calibrate a small hand sprayer.

1. Lay out an area measuring 10 feet x 100 feet (1000 square feet).

2. Fill the sprayer with water.

3. Spray the 1000-square foot area.

4. Measure the number of pints of water required to refill the sprayer to the same level as before. This is the number of pints applied per 1000 square feet.

If 8 pints (1 gallon) is sprayed on the 1000 square feet of land, and 2 1/4 teaspoons of chemical is required per 1000 square feet, add 2 1/4 teaspoons of chemical to each 1 gallon of water. This is the same rate as 1 pint per acre. (One acre is 43,560 square feet.)

> Caution: Spray with the same pressure and speed of application as in the trial run.

Another way to determine the amount of water that was applied to the 1000-square foot area is to measure the length of time required for the application. Spray into a container for the same length of time and measure the amount of water collected.

> Caution: Use the same spray pressure as when spraying the 1000-square foot area.

Drift

Any substance that is sprayed can blow or drift in the wind. If an herbicide drifts onto a nontarget plant, there is a chance that this plant will also be damaged or killed. When spraying herbicides, use spray nozzles that apply large droplets of spray. These larger droplets do not blow or drift as much as a fine mist and greatly reduce the danger of damage to other crops nearby. Tomatoes and tobacco are injured especially easily if 2,4-D drifts over them. Injury of other crops by drifting chemicals is a greater problem with herbicides than with other types of pesticides.

BREAKDOWN OF HERBICIDES

Most herbicides are broken down in the soil by soil microorganisms and eventually become harmless materials. The length of time required for this breakdown varies from several weeks to several years. Read the label for recommendations. If an herbicide persists or stays in the soil for a year or more, it may kill or damage the crop planted in that soil the following year or later in the same season. Crops which are not damaged by the chemical should be planted there until the soil is clean enough for crops which could be damaged by the herbicide.

EXPLANATION OF SOIL FUMIGANTS CHART

The following herbicide applications developed by Dr. Francis R. Gouin, are based on manufacturers' and university recommendations and field trial results. All recommendations are based on active ingredients per acre. For the reader's convenience, these levels have been calculated in pounds, ounces, gallons, or quarts of formulated, commercially available materials per acre (1,000 square feet [ft^2] or per cubic yard [yd^3]). When two levels are recommended, the lower concentration pertains to sandy or gravel soils; the higher concentration pertains to heavy silt or clayey loam soils.

Since preemergent herbicides are only effective in controlling germinating weed seeds, they must be applied on clean, cultivated soils before weed seeds germinate. Of all the preemergent herbicides recommended in this chart, only Chloro-IPC, Casoron, and

Premerge will control existing chickweed as well as germinating chickweed seeds.

These recommendations should be used only as a guide in selecting the proper test herbicide. Under no circumstances should an herbicide be applied over the entire nursery without first spending one growing season testing it in a limited area under the same soil conditions and the same growing program.

Herbicides are noted for not being equally dependable under all growing conditions.

Always read herbicide labels for instructions, including the species that have been cleared for use for that herbicide. Even when materials are used as recommended, the manufacturer cannot be held responsible for injury when it is applied to ornamentals not listed on the label.

SOIL FUMIGANTS RECOMMENDED FOR FIELD NURSERY CROPS*

To control weeds, insects, diseases and nematodes

Material	Formulation Rec./A.	Soil Temp. °F.	Exposure Days	Aeration Period	Method of application
Brozone	500-600 lbs.	50-80	2-6	2 wks.	Chisel-type applicator
Dowfume MC-33	250-350 lbs.	50-80	1-2	2 wks.	Chisel-type applicator
Picfume	35 gal.	60-85	1	2 wks.	Chisel-type applicator
Vapam	40-100 gal.	60-90	2-7	2-3 wks.	Chisel-type applicator, soil incorporation or irrigation.
Vorlex	25-50 gal.	40-80	2-7	3-5 wks.	Same as Vapam
Vorlex 201	25-60 gal.	40-80	2-7	3-5 wks.	Same

SOIL FUMIGANTS RECOMMENDED FOR SMALL AREAS OR POTTING MIXES

Material	Formulation Rec./1,000 ft.^{2}	Formulation Rec./yd.^{3}	Soil Temp. °F.	Exposure Days	Aeration Period	Methods of application
Dowfume MC-2	9.5 lbs.	1/2 lb.	60-90	1	2-3 wks.	Use gas-proof cover over moist loose soil and punch 1″ holes through the soil pile every 6 to 8 inches.
Picfume	3.25 qts.	1 lb.	60-85	1	2 wks.	Same as above
Vapam	2.5 gal.		60-90	2-3	2-3 wks.	Soil incorporate or drench with water and seal with water or plastic.
Vorlex	3.5 qts.		40-80	2-7	3-5 wks.	Same as above
Vorlex 201	3.5 qts.		40-80	2-7	3-5 wks.	Same as above

All of these materials are highly toxic and should be used with extreme caution. Always read the labels thoroughly before putting these materials into use. The labels are your best source of information for methods and rates of application, and the pests they will control.

*Courtesy Dr. Francis R. Gouin, Horticulture Department, University of Maryland.

PRE-EMERGENT HERBICIDES FOR NURSERY BEDS, GROUND COVERS, AND SUMMER ANNUALS

Materials	Active ingre-dients/acre	Formulated herbicide/A.	Formulated herbicide/ 1,000 ft.^{2}	Time of application	Comments
DCPA (Dacthal W-75) (Dacthal 5G)	10 lbs.	13 lbs. 200 lbs.	5 ozs. 4.5 lbs.	Mar.-Oct.	Control only annual weeds, must be applied at 4 to 6 weeks intervals.
Diphenamid (Dymid 80W) (Enide 50W) (Dymid 5G)	3 lbs.	3.75 lbs. 6 lbs. 60 lbs.	1.25 ozs. 2 ozs. 1.5 lbs.	Oct.-Apr.	Avoid foliage contact, irrigation or shallow cultivation improves activity
EPTC (Eptam 7-E)	4 1/4-6 1/2* lbs.	5 1/2-6 3/4 pts.	2-2.5 fl. oz.	Spring	*levels for perennial grass control. Apply after growth starts in the spring.
Nitrofen (Tok E-25) (Tok WP-50)	2-6 lbs.	1-3 gal. 4-12 lbs.	1/3-1 1/10 cup 1.5-4.5 ozs.	Apr.-Sept.	Controls annuals only, but delay application 7 days after transplanting.
Ornamental Weeder	4.3	100 lbs.	2.3 lbs.	Apr.-Nov.	Controls only annual weeds, irrigate within 7 days.
Trifluralin (Treflan 4-E) (Treflan 5G)	4 lbs.	4 qts. 80 lbs.	5 tbs. 2 lbs.	Apr.-Nov.	Controls only annuals, Improved by incorporation or heavy irrigation.

HERBICIDES FOR BULB CROPS

Materials	Active ingredients/A.	Formulated herbicides/A.	Formulated herbicide/ 1,000 sq. ft.	Time of application	Weeds Controlled	Remarks
CIPC (Chloro-IPC 4E) (Chloro-IPC 10G)	4-6 lbs.	 1-1.5 gal. 40-60 lbs.	 1.25-2.5 tsp. .75-1.5 lbs.	Sept.-Apr.	Annuals	Repeat every 6 weeks during growing season.
DCPA (Dacthal W-75) (Dacthal 5G)	10-15 lbs.	 13-19 lbs. 200-300 lbs.	 5-7 ozs. 4.5-7 lbs.	Mar.-July	Annuals	2 applications are necessary for complete summer control.
Diuron (Karmex)	1.25-2.5 lbs.	1-2 lbs.	1/2-1 oz.	Mar.-May	Annuals	*Recommended for gladiolus only.* Do not apply after bulb sprouts have appeared above.
DNBP (Premerge dinitro weed killer) (Premerge selective weed killer)	7.5-10.5 lbs.	 2.5-3.5 gal. 3 qts.	 3/4-1 1/4 cups 4 tbs.	Mar.-May	Annuals	Same as above. Apply when glad leaves are still short and straight.
Ornamental Weeder	4.3 lbs.	100 lbs.	2.3 lbs.	Apr.-Nov.	Annuals	Irrigate within one week of application.
Trifluralin (Treflan 4-E) (Treflan 5G)	4 lbs.	 4 qts. 80 lbs.	 5 tbs. 1 3/4 lbs.	Apr.-Nov.	Annuals	Irrigate within a few hours after surface application.

PRE-EMERGENT HERBICIDES FOR ESTABLISHED DECIDUOUS TREES & SHRUBS

Materials	Active ingredients/A.	Formulated herbicide/A.	Time of application	Weeds	Remarks
CIPA (Chloro-IPC 10G) (Chloro-IPC 4E)	8-10 lbs.	40-60 lbs. 2-2 1/2 gal.	Sept.-Apr.	Annuals	Particularly effective against chickweed both mature & seeds.
DCPA (Dacthal W-75) (Dacthal 5 G)	10-15 lbs.	13 1/4-19 1/2 lbs. 200-300 lbs.	Mar.-Apr.	Annuals	Light cultivation 7 to 10 days after application will improve its effectiveness. Poor control of Ragweed.
Dichlobenil (Casoron 4G)	4-6 lbs.	100-150 lbs.	Nov.-Apr.	Annuals Perennials	Use higher levels for control of perennial weeds and lower level for annual weeds.
Diphenamid (Enide 50W) (Dymid 80W) (Dymid 5G)	4-6 lbs.	8-12 lbs. 5-7 1/2 lbs. 80-120 lbs.	Nov.-Apr.	Annuals	Very effective on grasses. Its activity is increased by shallow cultivation 7 to 10 days after application.
EPTC (Eptam 7-E) (Eptam 10-G)	4.25-6.75 lbs.	5.75-6.75 pts. 40-60 lbs.	May-Sept.	Annuals Perennials	Apply after growth starts in the spring and 1 to 2 wks. after transplanting. Incorporate for perennial weed control.
Simazine (Princep 4G) (Princep 80W)	2-3 lbs.	50-75 lbs. 2 1/2-3 3/4 lbs.	Sept.-Apr.	Annuals	Suggest split applications with 1-1 1/2 lbs. applied in the spring and 1-1 1/2 lbs. applied late summer or early fall.
Ornamental Weeder	4.3 lbs.	100 lbs.	Apr.-Nov.	Annuals	Irrigate within one week of application.
Trifluralin (Treflan 4E) (Treflan 5G)	4 lbs.	4 qts. 80 lbs.	Apr.-Nov.	Annuals	Irrigate within a few hours after surface application.
Amitrol-Simazine (Amizine)	1 & 3 lbs.	7 lbs.	Apr.-Aug.	Annuals Perennials	Use only directed spray. Weeds must be actively growing Pre and Post emergent control.

PRE-EMERGENT HERBICIDES FOR ESTABLISHED NARROW-LEAF AND BROADLEAF EVERGREENS

Materials	Active ingredients/A.	Formulated herbicide/A.	Time of application	Weeds	Remarks
DCPA (Dacthal W-75) (Dacthal 5G)	10-15 lbs.	13-19 lbs. 200-300 lbs.	Mar.-July	Annuals	W.P. can be applied through the irrigation system. 2 to 3 applications are necessary for complete summer control.
Dichlobenil (Casoran 4G)	4-8 lbs.	100-200 lbs.	Nov.-Apr.	Annuals Perennials	8 lb. levels are recommended for control of nutsedge, quackgrass, & mugwart. Applied from Dec.-Jan.
Diphenamid (Enide 50W) (Dymid 80W) (Dymid 5G)	4-6 lbs.	8-12 lbs. 5-7 1/2 lbs. 80-120 lbs.	Nov.-Apr.	Annuals	Activity is increased by shallow cultivation 7 to 10 days after application.
EPTC (Eptam 10G) (Eptam 7E)	4.25-6.75 lbs.	40-60 lbs. 3.75-6.75 pts.	May-Sept.	Annuals Perennials	Apply after new growth starts or 2 weeks after transplanting. Soil incorporation is necessary.
Simazine (Princep 4G) (Princep 80W)	3-4 lbs.	75-100 lbs. 3 3/4-5 lbs.	Sept.-Apr.	Annuals	Suggest split applications with 1 1/2 to 2 lbs. applied in the spring and 1 1/2 to 2 lbs. applied in the late summer or early fall.
Ornamental Weeder	4.3 lbs.	100 lbs.	Apr.-Nov.	Annuals	Irrigate within one week of application.
Trifluralin (Treflan 4E) (Treflan 5G)	4 lbs.	4 qts. 80 lbs.	Apr.-Nov.	Annuals	Irrigate within a few hours after surface application.
Amitrol-Simazine (Amizine)	1 & 3 lbs.	7 lbs.	Apr.-Aug.	Annuals Perennials	Use only directed spray. Weeds must be actively growing. Pre and post emergent control.

POSTEMERGENT HERBICIDES FOR DIRECTED SPRAY APPLICATIONS

Materials	Active ingredients/A.	Formulation/ A./gal.	Formulation/gal. (spot treatment)	Time of application	Remarks
Translocated Weed Killers					
Amino-triazole (Amitrol-T) (Amizol) (Weedazol)	4-6 lbs.	2-4 gal./100 4 lbs./100 8 lbs./100	1/2 cup 6 level tbs.	Spr. & Sum.	Thoroughly wet the foliage. Spray when most weeds are 6 to 10 inches high. On broadleaf weeds spray when leaves have reached full size.
Dalapon (Dowpon) (Dowpon-M)	7 lbs.	8 lbs.	3 level tbs.	Spr. & Sum.	Specific for grasses. Apply in the spring and summer when grasses are growing rapidly. Wait 1 week before plowing.
Contact Weed Killers					
Brominal	1 lb.	2 qts./200	1 tbs.	Spr. & Sum.	Mix with adjuvant and wet foliage thoroughly.
Diquat	1/2 lb.	1 qt./50	4 tsp.	Spr. & Sum.	Mix with 4 ozs. of X77/50 gal. or 1/2 tsp./gal.
Paraquat	1/2-1 lb.	1-2 qts./50	4-8 tsp.	Spr. & Sum.	Same
Phytar 560	2.48-4.96 lbs.	1-2 gal./100	5 tbs.	Sum.	Do not add wetting agents. Apply on a warm sunny day and avoid spraying foliage, green bark, graft unions or scuffed trunks of ornamental plants.

FOAMING AGENTS

Material	Manufacturer	Rates/100		Comments
		Ground	Air	
Fomex	Kalo Lab. Inc.	2-3 qts.	2-4 qts.	Must be applied with specially designed foaming nozzles.
Triton A-F	Rohm and Haas	2-6 pts.	2-4 pts.	

FOAMING NOZZLES

Name	Manufacturer	Remarks
Flat Spray Foamjet	Spraying Systems Co.	Four nozzle sizes available.
Dela-Foam Nozzles	Delavan Manufacturing Co.	Five nozzle sizes available.

BRUSH CONTROL AROUND THE NURSERY

Treatment	Materials	Application Rates Water	Time	Comments
Foliage	Ammonium Sulfamate *(Ammate X) *(Ammate X-NI)	60 lbs./100 gal. 60 lbs./100 gal.	June-Sept.	Best results when applied under high temperature and humidity. Recommended for vegetation control near streams, and ponds.
	2,4-D **(Esteron 44 Improved) **(Weedar 64)	3-4 qts./50 gal. 2-4 qts./30 gal.	May-Sept.	Controls many broadleaf weeds, herbaceous perennials and woody plants susceptible to 2,4-D. Plants must be actively growing.
	2,4-D & 2,4-DP ***(Weedone 170)	1-1 1/2 gal./100 gal.	May-Sept.	
	2,4-D & 2,4,5-T **(Esteron Brush Killer) **(Weedone IBK)	3-4 qts./100 gal. 1 gal./100 gal.	May-Sept.	For difficult to control weeds. Spray when brush is 6-8 ft. tall and foliage is well developed. Spray both leaves and stems.
Stump	*(Amate X) **(Esteron Brush Killer) ***(Weedone IBK)	Crystals 1 1/2 pts./3 gal. oil 1 pt./3 gal. oil		Sprinkle crystals on freshly cut stumps. Spray cut surface, bark and exposed roots. Spray cut surface, bark, and exposed roots.
Tree Injection	**(Formula 40) **(Tordon 101) ***(Weedar Amine BK) ***(Weedar 64)	Undiluted 1:1 with water 1:19 with water 1:19 with water	March-Dec. Jan.-Jan. May-Oct. May-Oct.	1 ml. every 1-3″. 1 ml. every 3″. One injection for every inch of trunk diameter. One injection for every inch of trunk diameter.

*   E.I. DuPont De Nemours & Co.
**   The Dow Chemical Company
***   Amchem Products, Inc.

INDUSTRIAL HERBICIDES FOR NON-CROP

Materials	Formulation per 1,000 sq. ft.			Formulation per acre		
	Annual	Perennial	Total Vegetation	Annual	Perennial	Total Vegetation
Atratol 8P	2.5-5 lbs.	5-10 lbs.	10 lbs.	108-215 lbs.	251-430 lbs.	430 lbs.
1 ** Atratol 80W	2.2-4.6 ozs.	4.6-9.2 ozs.	9.2-18.4 ozs.	6-12 1/2 lbs.	12 1/2-25 lbs.	25-50 lbs.
2 ** Fenac	1 pt.	1 1/2 pt.	- - - - -	10 gal.	12 gal.	- - - - -
2 ** Fenamine	1 pt.	1 pt.	- - - - -	9 gal.	9 gal.	- - - - -
2 ** Fenavar	1 cup	1 1/2 cups	1 1/2 pts.	3 gal.	5 gal.	10 gal.
2 Fenavar Gr.	1.3 lbs.	1.8 lbs.	2.2 lbs.	40 lbs.	75 lbs.	150 lbs.
3 ** Hyvar X-L	3-6 ozs.	6-12 ozs.	12-24 ozs.	1-2 gal.	2-4 gal.	4-8 gal.
3 Hyvar X-P	3/4 lb.	1 1/2 lbs.	2 1/2 lbs.	30-60 lbs.	60-120 lbs.	120-240 lbs.
3 ** Karmex	7-14 ozs.	14-28 ozs.	2 3/4 lbs.	20-40 lbs.	40-60 lbs.	80 lbs.
4 ** Kenapon	12 ozs.	16 ozs. Grasses		4 gal.	5 gal. Grasses	
1 ** Pramitol 25E	1-1 1/2 pts.	4-5 1/2 pts.	- - - - -	5-7 1/2 gal.	20-30 gal.	- - - - -
1 Pramitol 5P	5-10 lbs.	20 lbs.	- - - - -	215-430 lbs.	860 lbs.	- - - - -
5 Tandex 4G	1 1/2-2 3/4 lbs.	3 1/4-5 1/2 lbs.	7-14 lbs.	60-120 lbs.	140-240 lbs.	300-600 lbs.
5 ** Tandex 80WP	1 cup	1 1/2 cups	2 1/2 cups	3-6 lbs.	7-12 lbs.	15-30 lbs.
4 ** TCA Pellets	1.7 lbs.	2.5 lbs. Grasses		75 lbs.	100 lbs. Grasses	

* These materials are to be used only around buildings, storage areas, parking lots, fence-rows, driveways and when specified under asphalt. These materials should not be used near desirable trees and shrubs.

** To insure proper coverage dissolve recommended amounts in approximately 50-100 gallons of water per acre. For areas of 1,000 square feet dissolve recommended amounts into approximately 2 to 5 gallons of water.

1 Geigy Agricultural Chemicals
2 Amchen Products Inc.
3 E.I. DuPont deNemours & Co. (Inc.)
4 The Dow Chemical Company
5 FMC Corporation

PERENNIAL WEED CONTROL IN THE NURSERY

(Pounds of formulated material per acre)

Materials	Artemesia	Johnsongrass	Nutsedge	Quackgrass	Time
Amino-triazole* (Weedazol)	16 lbs./100 gal.		8 lbs./50 gal.	8 lbs./40 gal.	May-July
Dichlobenil (Casoron G-4)	150-200 lbs.			150-200 lbs.	Dec.-Mar.
Dalapon* (Dowpon C and		15 lbs./100 gal.		15 lbs./100 gal.	May-Aug.
(Dowpon M)**		15 lbs./100 gal.		15 lbs./100 gal.	May-Aug.

* Use only directed sprays and avoid wetting the foliage of desired plants. For best results eradicate these weeds one year before planting. If drift is a problem refer to page 220 for information on foaming agents and foaming nozzles.

** If you are not getting proper wetting of the foliage add a surfactant [wetting agent], but in limited amounts, to avoid excessive wetting.

WEED CONTROL FOR CONTAINER ORNAMENTALS

Materials	Active ingredients/A.	Formulated herbicide/A.	Formulated herbicide/ 1,000 sq. ft.	Remarks
DCPA (Dacthal W-75) (Dacthal 5-G)	10 lbs.	13 lbs. 200 lbs.	5 ozs. 4 1/2 lbs.	W-75 can effectively be applied through overhead irrigation systems. Repeated application must be made at 6 week intervals.
Trifluralin (Treflan 5G)	4 lbs.	80 lbs.	1 3/4 lbs.	Allow 2 weeks for the plants to become established.
Black Plastic Disc				Available from nursery supply dealers.
Weed Check Discs				Available from nursery supply dealers. Beware of splitting stems when over-wintering such species as cottoneaster, euonymus, & weigela.
Bark Mulch				

WHY HERBICIDES SOMETIMES FAIL

The question of why herbicides are sometimes ineffective is a very complex one. The following are some of the more common reasons for failure.

Incorrect Selection of the Herbicides

Unsatisfactory weed control may result from use of the wrong herbicide. It has been noted that many weed killers are selective, that is, they only kill certain weeds. Therefore, it is most important that the right chemical be selected to control a specific weed. This requires accurate identification of the weed to be controlled. To use herbicides effectively, the applicator must be able to accurately identify weeds.

Improper Application of the Herbicide

Herbicides are designed for application in a variety of ways and at various stages of the growing process. Poor application or poor timing of application are reasons for herbicide failure. For example, preemergence herbicides such as Dacthal and Lasso must be spread evenly on the surface of the soil. To contact with sprouting weed seeds, the entire soil surface must be covered with the chemical; if not, the herbicide could be rendered ineffective.

Lasso also must be washed into the germination zone by rainfall or irrigation within seven to ten days after application. If no rain occurs or irrigation is not possible, poor weed control will result. Preplant weed killers such as Eptam must be mixed into the soil immediately after application; if not, they evaporate into the air. Much of the active chemical is thus lost and control is poor. Postemergence weed killers such as 2,4-D must be sprayed onto the plant to be killed or over the roots of the plant so that the chemical can be absorbed into the plant. The chemical 2,4-D also works most effectively if the weeds are in active growth.

Sprayer calibration is also very important for proper application of the herbicides. This may involve tests to determine the proper amount of chemical to apply per acre of land. Too low a rate results in poor weed control; too high a rate could kill the crop as well as the weeds.

Environmental Variables

The maturity of the weeds is an important variable; most weeds are killed more easily when they are young and in active growth. In fact, this is the only time many weed killers work effectively. An old, hardened plant requires much more chemical, or may even require a different type of weed killer.

Rain which falls just after application of the herbicide can be an advantage with some preemergence chemicals. However, excessive rain can leach or wash the chemical through the soil and dilute it so much that weeds are not controlled.

Soil type and organic matter content of the soil affect the amount of active chemical that must be applied. Some chemicals are absorbed or held on the surface by clay particles and organic matter. The University of Maryland weed control table gives two dosages for some chemicals. For example, Dacthal is recommended in doses of 10-15 pounds of active ingredient per acre. Ten pounds is recommended for sandy soils, and 15 pounds for silt and clay soils. A soil high in humus absorbs and makes some of the weed killer inactive, requiring a higher rate of chemical application.

Not all chemicals are affected by organic matter in the soil to the same degree. For example, dalapon is not absorbed at all, while Dacthal is easily absorbed and would be greatly affected by organic matter level of the soil.

Weed control with chemicals is not a simple job; many variables must be considered to achieve effective control. Herbicides have limitations and are not a magic cure-all for weed problems, but if used carefully and accurately, good results can be expected.

STUDENT ACTIVITIES

1. Demonstrate the proper calculation and mixing of at least one herbicide, working in groups of three to five students.

2. If land is available, mark off plots of 1 square rod (16 1/2 square feet). Apply the herbicide mixed in activity 1 to an assigned area. Work in the same groups.

   a. Identify weeds present and estimate number of each.
   b. Select the proper weed killer.
   c. Mix and apply the chemical.
   d. Record the results.

> Complete both projects only under supervision of your instructor.

SELF-EVALUATION

Select the best answer from the choices offered to complete each statement.

1. A weed is a plant
   a. growing in the lawn.
   b. growing where it is not wanted.
   c. growing in the wild.
   d. that is not a food item.

2. Weeds can be controlled without the use of herbicides by
   a. hand weeding and cultivation.
   b. mulches and cultivation of crops.
   c. hoeing and weeding.
   d. all of these

3. Selective weed killers can be used to greatest advantage
   a. on large acreages of a single crop.
   b. in the home vegetable garden.
   c. where there is a mixture of weeds.
   d. where the land is level.

4. Nonselective weed killers

   a. should not be used to kill weeds.
   b. can be used in any weed problem situation.
   c. kill all of the plants to which they are applied.
   d. are the best weed killers for the home garden.

5. The herbicide 2,4-D controls the serious weed problem caused by

   a. crabgrass.   c. orchard grass.
   b. bluegrass.   d. dandelions.

6. The weed killer atrazine kills plants by

   a. starving them to death.
   b. preventing photosynthesis.
   c. accumulating in the leaves and buds and causing damage therein.
   d. all of the above

7. The herbicide 2,4-D, a systemic herbicide, is widely used to control

   a. grasses.
   b. broadleaf weeds.
   c. weeds in the garden.
   d. weeds in commercial tomato plantings.

8. Dacthal is a weed killer that is generally applied

   a. at planting time.
   b. at the time weeds are 1 1/2 to 2 inches tall.
   c. at postemergence.
   d. to control perennials.

9. Postemergence application of an herbicide means that the chemical is applied

   a. before the crop is up.   c. after the crop is up.
   b. at planting time.   d. after the crop is harvested.

10. To be sure an herbicide is recommended for use on a specific weed and crop

   a. test the chemical on a small sample area first.
   b. read the label for recommendations.
   c. ask your neighbors if they have used it.
   d. ask the salesperson for advice.

11. Preemergence application of herbicides means that the chemical is applied

   a. before the crop is up.
   b. at planting time.
   c. after the crop is up.
   d. after the crop is harvested.

12. Herbicides kill plants by all the methods listed below *except* for

   a. preventing photosynthesis.
   b. upsetting cell division and respiration.
   c. preventing seed germination.
   d. drying them up.

13. A sprayer used to apply herbicides should not be used to spray insecticides on crop plants because

   a. it is difficult to clean all of the weed killer out of the sprayer.
   b. insecticides and weed killers do not mix.
   c. weed killers may damage the sprayer.
   d. you may forget to clean the sprayer.

14. The three different times for applying weed killers are

   a. spring, summer, and fall.
   b. preplant, preemergence, and postemergence.
   c. preplant, postplant, and after harvest.
   d. preplant, postplant, and postemergence.

15. If an herbicide is to be applied at 2 pints to the acre and the sprayer applies 10 gallons of water to the acre, how much chemical should be mixed in each gallon of water?

   a. 2 pints
   b. 20 pints
   c. 0.2 pints
   d. 5 pints

16. Even if the proper amount of chemical is mixed per gallon of water, the wrong amount of chemical can be applied per acre

   a. if the wrong amount of spray solution is applied per acre.
   b. if the wrong amount of chemical is added to each gallon of water.
   c. if the wrong amount of chemical is put in the spray tank.
   d. if the weeds are too old and tough.

17. Drift of weed killers to nontarget plants can be reduced by

   a. using a nozzle that sprays larger droplets.
   b. spraying downwind.
   c. using the proper chemical.
   d. mixing the proper amount of chemical per gallon of water.

18. Most weed killers are broken down into harmless materials in the soil

   a. by freezing and thawing.   c. by soil microorganisms.
   b. by rainwater.   d. by plant uptake and detoxification.

19. A soil high in organic matter tends to

   a. hold a great deal of water, thereby diluting herbicides.
   b. absorb herbicides and cause them to be ineffective.
   c. grow weeds so quickly that a higher dosage of herbicide is needed.
   d. hold herbicides in the soil, thereby making them more effective.

20. An example of a nonselective weed killer is

   a. sodium arsenite.   c. Dacthal.
   b. 2,4-D.   d. atrazine.

Section 8
Container
Grown Plants

OBJECTIVE

To construct and care for terrariums.

COMPETENCY TO BE DEVELOPED

After studying this unit, the student will be able to

- prepare a terrarium for planting by mixing a suitable media and placing it with the proper drainage material in the container.
- identify ten plants suitable for use in terrariums.
- plant a terrarium.
- identify five cultural conditions that affect the growth of plants in terrariums.
- list four problem symptoms of terrarium plants.
- prune two terrarium plants to control growth.
- list and identify five insects that affect terrarium plants.

unit 22
Terrariums

Materials List

2 1/2-inch tall potted foliage plants

plant mister

single-edge razor blades

bamboo stakes

6-inch wooden pot labels

containers (brandy snifter, bottle, Plexiglas box, etc.)

sphagnum moss or peat moss

small pebbles or crushed rock

---

activated charcoal

teaspoon

number 4 wooden pencil (very sharp)

16-gauge florist wire

small funnel

small paint brushes in assorted sizes

lazy Susan

sand in assorted colors

construction grade sand

The original *terrarium* (container, usually covered, in which plants are grown) was developed by Dr. Nathaniel Ward, a nineteenth-century English surgeon. It was quite by accident, and in an experiment unrelated to horticulture, that Dr. Ward made his terrarium discovery. He had been attempting to grow bog ferns, but found that the polluted air from the London factories near his home inhibited their growth. During this same period, Dr. Ward investigated many aspects of nature, including the life cycle of butterflies. As part of this work, he placed the pupa (cocoon) of a butterfly in a glass jar to keep it alive, along with some soil from the same area. After several weeks, he noticed a seedling growing inside. He found that when the seedling lost moisture from its leaves (transpiration), it collected on the walls of the terrarium. As the water formed and moved down the sides of the container, it was reabsorbed by the plant and soil. This established a rain cycle for the plant inside the container, making it an ideal place to grow healthy plants with a minimum of attention. Despite the rain cycle that developed, Dr. Ward found it necessary to occasionally add a few drops of water to prevent the plant from drying out. After several weeks, the plant continued to grow without any particular attention.

Dr. Ward went on to experiment with growing plants in containers of various sizes. As a result of this work, the *Wardian case* was developed. This enclosed case containing media and moisture in which plants could be started and grown evolved into the modern day terrarium. It was originally used to transport sensitive tropical plants which needed protection from sea air and the rough weather of sea voyages. Over the past 150 years, the case has continued to be used for both growing and transporting plants. Today, many nurserymen, horticulturists, and plant enthusiasts use the case for starting and displaying plants.

THE TERRARIUM CONTAINER

Selecting the Container

There are various containers suitable for the construction of terrariums, figure 22-1. When choosing a container, several things should be considered.

Location of Terrarium. The terrarium should harmonize with, as well as accent, the furnishings and materials already present in the surroundings. For example, a modernistic terrarium would not be a good choice for a room furnished with antiques.

Size of Terrarium. Terrarium size should be chosen in relation to the size of the area in which it is to be located. A terrarium which is too large or too small is less effective than one which is in proportion with the rest of the room.

Color of Glass. The chosen container must allow sufficient light to reach the plants so that photosynthesis can occur. Dark tinted

ANTIQUE BOTTLE

BRANDY SNIFTER

WINE BOTTLE

1-GALLON JUG

GLOBE-SHAPED BOWL

5-GALLON WATER JUG

Fig. 22-1 Various terrarium containers

glass should be avoided since it filters out too much sunlight; plants are unable to grow in this environment.

Size of Container Opening.   The size of the opening in the container limits the type of plant which may be used. For example, if a one-gallon bottle is selected, small plants which will fit through the narrow neck of the container must be chosen. A wider variety of plants is possible when a container with a large opening, such as a brandy snifter, is used.

Condition of Container.   Check to be sure the container is free from cracks or scratches. It should also be clean and dry. Water spots on the inside or outside should be removed so that the finished terrarium appears clean and attractive. If a closed container is chosen, it must be watertight. This helps to maintain the necessary humidity in the terrarium to allow best growth of plants.

Constructing a Plexiglas Terrarium

Many people enjoy constructing their own terrarium containers. One way in which this can be done is by using Plexiglas and special Plexiglas glue, both of which can be purchased at most hardware stores.   To make a 2 1/4-gallon terrarium, follow the directions given below.   The diagram for assembly is shown in figure 22-2, page 240.

1. Cut pieces the following sizes from a sheet of Plexiglas, leaving its protective paper covering attached:

    two pieces 4 inches by 5 inches for end walls
    two pieces 8 inches by 5 inches for sides
    two pieces 8 inches by 4 1/8 inches for top and base

2. Glue the end pieces to the base of the terrarium. Set one end in place and carefully brush the glue onto the seam. Allow a few minutes for drying; the glue seals immediately.

3. Repeat this procedure for the other end and sides.

4. After allowing the sides to dry for a few minutes, place the top on the terrarium. Do not glue the top to the rest of the structure.

5. After planting, seal the top with masking tape.

PREPARING THE CONTAINER FOR PLANTING

Several layers of material are placed in the terrarium in preparation for planting, figure 22-3. These include a layer of drainage material, sphagnum moss, charcoal, and soil mix.

Drainage Material

Proper drainage is important to allow the movement of excess water through the root zone area. It also helps to maintain the proper balance between air and water in the container. Excess water in the root zone of terrariums makes it impossible for plants to absorb oxygen from the air in the soil mixture. The average soil mixture contains about 25 percent air, 25 percent water, 45 percent mineral matter, and 5 percent organic matter. If the plants in a terrarium are not provided with the best possible drainage in the root zone, they will turn yellow and eventually die from excess water and lack of oxygen.

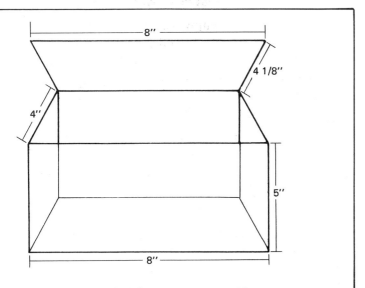

Fig. 22-2 Plexiglas terrarium assembly

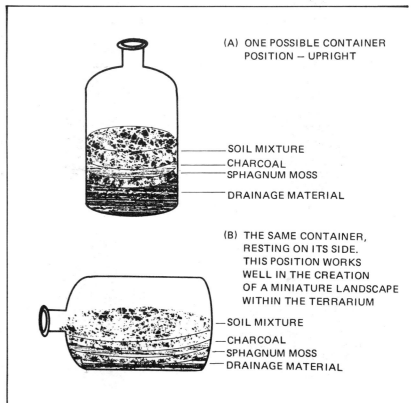

Fig. 22-3 Two terrarium containers prepared for planting. Container (B) illustrates an interesting variation in container position.

Fig. 22-4 A 1/4-inch layer of sphagmum moss is added to prevent the potting soil from seeping into the drainage material (in this case, a sand design). Richard Kreh, photographer)

Fig. 22-5 A layer of charcoal helps in the absorption of bacterial odors in the terrarium. (Richard Kreh, photographer)

Drainage in terrariums can be provided by crushed stone, pea gravel, washed river gravel, or pieces of broken clay pots. These materials are available at florists, nurseries, and garden centers. The material selected is the first to be placed in the container. It is spread directly on the bottom of the terrarium.

Sphagnum Moss and Charcoal

Sphagnum moss, composed of sterile leaf and stem tissue from acid bog plants, is spread in a layer over the drainage material in the container. The main purpose of the moss layer is to prevent the soil mixture from seeping into the drainage layer, figure 22-4.

A thin layer of activated charcoal is spread over the sphagmum moss to absorb any odors in the terrarium, figure 22-5. These odors usually develop from bacteria which results from breakdown of soil particles in the terrarium.

Soil Mixture

A good soil mixture placed over the charcoal layer is important since a wide variety of plant materials are usually used in terrariums. A good general potting mixture consists of two parts loamy soil, one part organic matter (peat moss), and one part horticultural grade perlite. If this mixture is used, it must be sterilized to kill all weed seeds and harmful disease organisms which might be present. To sterilize the soil, spread a layer of the mixture in a shallow pan, place in an oven, and bake at 180°F for 30 minutes.

Commercial medias also work very well in terrariums. If only a small amount of media is required for a terrarium, it is most practical to purchase commercial terrarium soil from a florist, nursery, or garden center. Fertilizer should not be added to closed terrariums. When the fertilizer breaks down, ammonia is released and causes the plants to burn and eventually die.

SELECTING PLANTS FOR THE TERRARIUM

The process of selecting plants for a terrarium can be a very interesting and enjoyable experience. Usually, the first consideration in plant selection is the appearance of the plant and how it will fit into the overall design of the terrarium. One very important plant characteristic is leaf color. Some plants have solid foliage colors, while others are *variegated* (have more than one color). For example, a leaf may have a dark green center and a white outer edge. A combination of solid colored and variegated plants creates an attractive contrast in a terrarium.

The texture, shape, and finish of the leaves and the shape and size of the plant are also important factors. Plants selected for terrariums are usually small, relatively inexpensive, and easy to work with when planting. Small, newly rooted cuttings or small potted plants produce the best results. However, larger plants work well if a large container with a good-sized opening is selected.

Terrariums vary in their environmental requirements, depending on the type of plants that are used. Since the entire terrarium provides the same environment, it is wise to choose plants that have similar basic needs in terms of light, soil, and moisture. Terrariums are often constructed by selecting one of the groups described below and using a variety of plants from that group.

Types of Plants

Cacti and succulents require natural desert conditions and full sun. They do not require a great deal of humidity, and therefore work well in open containers. During the spring and summer, cacti and succulents are watered to promote active growth. They enter a period of rest in winter, during which watering is reduced. The media previously recommended (one part loamy soil, one part organic matter, and one part horticultural perlite) with the addition of one part coarse sand is recommended for these plants.

Woodland plants are native to forests or woods. When grown away from their natural environment, they require similar conditions to those present in a typical woodland setting. The woodland terrarium requires dense shade, soil with a high organic content, and a good supply of moisture. The soil mix previously mentioned (one part loamy soil, one part organic matter, and one

part horticultural grade perlite) usually works well in woodland terrariums.

Familiar woodland plants include ferns, mosses, yellow root, ground ivy, jack-in-the-pulpit, lichens, oxalis, partridgeberry, trailing arbutus, and violets (Johnny-jump-ups). These plants grow well in closed terrariums because of the humidity created by the accumulation of moisture in the closed container. Generally, woodland terrariums are easy to manage and keep clean.

Tropical plants respond favorably to a more humid type of terrarium than either cacti or woodland plants since their natural environment is the moist tropical regions. Light requirements for this group vary with different plants. It is advisable to use those plants which require low light. Although there is a great variety of large tropical plants, the smaller varieties are best for terrariums since they are easier to work with and less expensive.

PLANTING THE TERRARIUM

Planting Equipment

There are a few simple tools which simplify planting terrariums. A funnel is used to work the soil mixture into the container. In this way, soil can be evenly distributed in the terrarium. When performing this operation, it is best to avoid getting soil on the sides of the container, since it is difficult to remove after planting. A wooden dowel is used to form planting holes in the soil mixture. One end is sharpened for digging while the other end is used to tamp the soil around the plant roots. When planting in a narrow-necked bottle, a bamboo stick slit about 2 inches on one end can be used to hold plants. The plant is slipped into the slit in the stick to hold it securely while planting. A bent spoon is helpful in adding soil around the plants. After the terrarium is planted, a mister is used to spray the foliage.

Planting Technique

The following steps comprise the planting process in the construction of terrariums.

(1) Push the wooden dowel into the soil mixture to form holes in which to place the plants.

(2) With the plant held by the split in the bamboo stake, transplant the plant into the bottle. (Gently bending the leaves does not harm the plant.)

(3) After the plant is in position, carefully remove the plant from the bamboo stake with the dowel. Lightly tamp the soil around the plant roots with the dowel to help hold the plant straight and firm.

(4) Add plants until the terrarium design is complete.

(5) Water the completed terrarium with a mister. Spray the foliage until water droplets form on the leaves. Finally, seal the terrarium and enjoy watching it grow.

Plants give off moisture. During this process (transpiration), moisture is given off from small openings on the underside of the leaves known as *stomata* (plural of *stoma*). When plants are in a closed terrarium, the moisture collects on the sides of the container and drops back into the soil so that the plants may use it again, figure 22-6. If the moisture is concentrated in the terrarium, the container may fog up. In this case, remove the top and allow the moisture to escape. This establishes the balance of air and moisture in the terrarium. After the fogging has cleared, replace and seal the container top.

If a terrarium lacks moisture, the leaves of the plants may droop, and the soil may appear lighter than normal. When this occurs, spray the terrarium with a mist of water until beads of

Fig. 22-6 Terrarium with established rain cycle. Notice the condensation forming on the glass.

Fig. 22-7 Terrarium after fogging has cleared. This terrarium is made in a brandy snifter with a plastic lid. It contains common house plants growing in a mixture of equal parts of garden soil, sand, and peat moss. A 1-inch layer of charcoal has been placed in the bottom of the container for drainage. The sides of the snifter below the soil level have been lined with moss. (USDA photo)

water form on the leaves of the plants. Seal the terrarium and allow the rain cycle to become reestablished. The plants can then continue their natural growth process, figure 22-7.

SAND DESIGN

Unique effects can be developed in terrariums by creating designs in the sand which is used as the drainage material. This process, known as *sand art*, is not new but has recently gained popularity in terrarium construction.

Materials and Equipment

Most of the materials and equipment required for sand design are easy to obtain. The most important item is colored sand. It is best to buy it already colored rather than attempt to color it. The amount of colored sand used will vary with the size of container selected. A construction grade sand is also necessary. Its use as filler material reduces the amount of the more expensive colored sands which are needed.

Basic equipment for sand design consists of an old metal spoon, a piece of 16-gauge florist wire, a 4H wooden pencil, a funnel with a small piece of plastic hose, and several small paint brushes. The metal spoon is bent at a right angle and is used to place small quantities of sand in the container. Larger amounts of sand are added to narrow-necked containers by using the funnel and piece of plastic hose. Various designs are created with the aid of the pencil and florist wire. The paint brushes are used to replace sand which falls outside the design. If a lazy Susan is available, place the terrarium on it while the design is being created. This permits turning of the terrarium without disturbing the sand.

Design Technique

The first step in sand design is to decide upon the type of design which is to be created. This may range from simple layering of colored sand to the creation of a landscape or picture in the terrarium. A layered design is the easiest type to construct and the best choice for the beginner in sand design.

Almost any clear container can be used for sand design, but the wider the container opening, the easier it is to create the design. A brandy snifter is an ideal container.

When using colored sand, small quantities can be placed in plastic drinking glasses or similar containers. Taking the sand from these smaller containers when the designer is ready to use it in the terrarium helps to avoid wasting sand.

Creating a Layered Design

If a layered design is desired, it is best to select three to five colors of sand. The first layer is placed as a base in the bottom of the container. It should completely cover the bottom. To place this sand, use the bent metal spoon described in the section on equipment.

After the base color is in place, use the spoon to add construction grade sand in the center of the colored sand. The construction sand should come within a 1/4 inch of touching the glass sides of the container. In other words, the colored sand need only be a 1/4-inch layer at the outside of the design. It is held in place against the glass by the construction sand and appears to be the only material used.

The layers may consist of a variety of colors and may be placed in any sequence which gives a pleasing appearance. Interest may be added by varying the thickness of the different colored layers, figure 22-8. *Caution:* Do not shake or tip the terrarium

Fig. 22-8 Creating a pattern with layers of colored sand is probably the best method of sand design for the beginner. (Richard Kreh, photographer)

while working on the design. If the terrarium is shaken but the colors are not mixed, more colored sand can be added to fill in spots where the sand has fallen to the center. Additional filler sand is then added to hold the colored sand in place.

When the colored sand is built up to the desired level, add the prepared soil mixture in a 2-to-3-inch layer over the sand design, figure 22-9. Be sure to keep the glass clean as the soil mixture is added.

Fig. 22-9 A 2- to 3-inch layer of moist soil is placed over the sand design. (Richard Kreh, photographer)

Creating Landscape Scenes

To create a landscape in sand design, first imagine how the scene will appear in the terrarium. A good practice is to draw a picture to follow when placing the sand. Land forms such as snowcapped mountains, valleys, rivers, seashores, and deserts can be created.

Snowcapped mountains are made by forming the base of the mountains in black or dark brown sand and topping them with white sand to make the snowcaps. Once the base of the mountain has been formed, carefully push a sharpened wooden pencil or florist wire down the side of the glass to make an indentation in the black sand. The white sand falls down into the indentation as the pencil or wire is slowly withdrawn. By repeating this procedure several times, a series of peaks and valleys is created. This pattern may be used on all sides of the container.

In one valley, create a sunset or sunrise by filling in the valley with red or red softened with red-yellow sand. The appearance of the sunset is up to the individual designer. To form the sunset, carefully place the empty spoon in the area where the sunset is to be located. Press firmly to prevent the sand from sliding out of place. Take a small sample of the sand to be used for the sunset and add it to the terrarium. If grains of red sand drop out of place, brush them back to the center with a small paint brush. Continue adding sand until the desired sunset is created.

A river or ocean can be created in another valley by using a deep blue sand and following the same procedure as was used to form the sunset.

The sky is made by mixing equal amounts of blue and white sand. Using the spoon, spread this mixture over the entire landscape, making it thicker over the water and sunset. This gives depth and contrast to the mountains.

Birds in flight are designed by taking the spoon and firming the sand in the place where the bird is to be located. Place about one-half teaspoon of black sand in the sky. Then push a sharp wooden pencil down through the center of the black sand about one-quarter inch into the sky. The black sand flows down into the indentation made by the pencil to form the shape of the bird. Any color sand may be used for birds. To show birds which appear to be in the distance, make them smaller. This technique adds depth to the design, figure 22-10.

SEALING THE SAND DESIGN

If it is important that the colored sand be prevented from fading, the sand design can be covered with a sealer. Sealer can be applied in several ways. A commercially prepared sealer, available at most florists, garden centers, and large department stores, should be applied according to manufacturer's direction. A home-made sealer consists of melted paraffin (wax) which is poured in a thin layer over the sand. However, this method is difficult and hazardous and should be avoided by the novice sand designer. If it is used, extreme caution should be exercised.

If colored sand is sealed, it does not serve as drainage material, but strictly as a decorative feature. To prepare a terrarium for planting in which a sand design has been created, follow the same procedure as that for preparing a terrarium without a sand design. Place the drainage material, sphagnum moss, and soil mixture on top of the sealed design. Refer to figure 22-3 for proper placement of these materials.

CARING FOR THE TERRARIUM

After the terrarium has been planted, it should be checked periodically. It is especially important to see that it continues to have the proper light, moisture, and temperature.

Problem Symptoms

Musty odor. This can be corrected by the addition of activated charcoal to the terrarium. The charcoal absorbs the bacterial odor and aids in drainage.

Fig. 22-10 Sand design showing mountain, birds, and sky. (Richard Kreh, photographer)

Chloric (yellowish) and Stunted Growth of Plants. This condition could result from improper drainage, improper ventilation, or improper soil mixture.

Decaying Foliage. This may be the result of several things. Improper ventilation, one possible cause, may be corrected by removing the sealed lid to ventilate the terrarium. High nitrogen fertilizers may be present in the soil mixture, giving off ammonia and thus burning the plant leaves. In this case, it is best to replace the soil mixture.

Brown Foliage. This indicates that the terrarium has not received enough water.

Pruning

In some cases, the growth of plants in the terrarium is so great that they become too big for the container. This may be due to the plant type or very favorable cultural conditions. Whatever the reason, the plants must be pruned to remove excess foliage. Pruning plants causes them to branch out and become fuller.

Use a single-edge razor blade for pruning. If the terrarium container is a long-necked bottle, insert a single-edge razor blade into a bamboo stake.

Remember to remove any material that is pruned from the terrarium to prevent it from decaying inside the container.

Insect Pests

Insect pests are controlled most efficiently by planting the terrarium with clean, insect-free plants. At times, however, pests may be present in the soil and not noticed when the terrarium is planted. Some insecticides may be used for insect control in terrariums; consult the local extension service for recommendations.

Specific insects that plague terrarium plants include the whitefly, mite, aphid, mealybug, and scale.

Whiteflies have very small, white, wedge-shaped wings. When the plant is disturbed, these insects fly around the plant's leaves where they can be seen. Whiteflies feed on the underside of the leaf. Control: Spray with malathion.

Mites are very small insects and difficult to see. If it is suspected that a plant has mites, take a sheet of white paper, insert it under the leaf, and tap the leaf lightly. If mites are on the plant, they will fall onto the paper. Check the paper for small moving insects. Draw a finger across the insect and the paper; if mites are present, there will be a red mark on the paper. Control: Apply a systemic insecticide to the soil. (*Systemic* insecticides are absorbed into the plant's system through its leaves and roots. As the insect sucks the plant juices containing the insecticide, the body system of the insect is affected, thus causing death. Systemics are used to control sucking insects.)

Mealybugs are small, flat, white insects that resemble a small spot of cotton on the plant. Mealybugs suck the plant juices until the plant dies. Control: Swab the areas of the leaves where the bugs are located with alcohol.

Scale insects harm plants by sucking the juices from them. Scale insects are small, gray to brown in color, and flat, with a soft or hard shell. They can be found sticking to the stems of plants. After scale attacks, the plant becomes yellow, then turns brown, and finally dies. Control: Spray with malathion.

Ants. Tiny black ants are drawn to some plants by the presence of honeydew on their leaves. *Honeydew* is sap which is secreted from the leaves of plants after an attack by aphids, scales, or, sometimes, a fungus. These ants usually attack the root system of the plant. Control: Spray with malathion.

Watering

If moisture beads are present on the sides of the terrarium container, there is a good chance that the rain cycle has been established. Be cautious of overwatering. If the terrarium is overwatered, the plants may die from poor drainage and a lack of air circulation in the soil. In the case of overwatering, it is best to open the top of the terrarium and allow it to dry out somewhat. When enough water has evaporated, reseal the terrarium to reestablish the rain cycle.

Temperature

The ideal temperature of terrariums depends to some extent upon the type of plants they contain, but a temperature of 65° to 75°F is suitable in most cases. Remember that the temperature is greatly influenced by the location of the terrarium. For example, if it is located in or very close to a window, it will experience greater changes in temperature than a terrarium which receives indirect light.

Light

Different types of plants require different amounts of light. When selecting plants for terrariums, refer to Section 10 of this text for correct light exposures. Remember that plants in terrariums located near windows grow toward the light source (phototropism).

Artificial Light. Terrariums are excellent decorative additions to the home, office, or school. The sunlight in these areas may be very limited, however, making artificial light necessary. There are special lights on the market today that are especially good for this purpose. A fluorescent light is a better choice than an incandescent light (standard light bulb).

Various manufacturers of fluorescent lights publish information concerning the placement of the lights for best results in the terrarium. On the average, it is recommended that they be located about 15 to 18 inches away from the terrarium. The light should be provided from 12 to 14 hours per day. Flowering plants require more light than others.

Fertilizers

If plants are grown in a good media, additional fertilizer is not necessary. Fertilizer tends to stimulate excessive growth of terrarium plants, causing the terrarium to appear unkempt and cramped.

If the plants must be fertilized, use an organic fertilizer such as bone meal. Simply sprinkle a small amount of the fertilizer in the terrarium. Organic fertilizers release slowly and do not burn plants, as inorganic fertilizers tend to do.

Preventive Disease Control

When the terrarium has been completed, preventive measures should be taken to control disease organisms that may be present in the terrarium. One chemical that is commonly used is Benlate. Follow manufacturer's directions for its use. Use a spray bottle to ensure even distribution of the chemical on the foliage and soil.

STUDENT ACTIVITIES

1. Ask a local florist to visit the class and demonstrate the various techniques used in terrarium construction and planting.

2. Organize a field trip to a woods area and collect native plant materials for terrarium plantings.

3. Propagate plants for the terrarium. For woodland plants, students might visit a nearby forest or woods and collect seeds, sow them, and select the terrarium plants from the results.

4. As a class project, construct and plant three terrariums, using the following medias: (1) soil; (2) two parts loamy soil, one part organic matter, and one part horticultural perlite; and (3) Pro-Mix. Compare the growth in each terrarium. (Students might consider constructing and selling other terrariums as a fund-raising project.)

5. As the plantings in the three terrariums grow, observe and record the results. Care for the plants, employing the techniques found in the text. Prune when necessary.

6. Construct stands with artificial lighting for the terrariums.

SELF-EVALUATION

Select the best answer from the choices offered to complete the statement or answer the question.

1. When selecting a container for terrariums, one must consider
   a. size.
   b. color of glass.
   c. desired environmental conditions.
   d. all of these

2. Charcoal is used in a terrarium for its
   a. color.
   b. drainage ability.
   c. ability to absorb bacterial odors.
   d. none of these

3. Crushed rock or pebbles are used in terrariums for
   a. soil mixture.
   b. added weight.
   c. drainage material.
   d. none of these

4. A musty odor in a terrarium can be eliminated by adding

   a. stone.
   b. peat moss.

   c. more soil.
   d. activated charcoal.

5. Terrarium plants which are chloric (yellow) and are stunted in growth may have this condition because of

   a. improper drainage.
   b. improper ventilation.

   c. improper soil mixture.
   d. all of these

6. The use of high nitrogen fertilizers in terrarium soil results in the release of

   a. moisture.
   b. new plants.

   c. ammonia gas.
   d. none of these

7. The pruning of terrarium plants is necessary to

   a. remove excess growth.
   b. remove dead leaves.

   c. remove broken plants.
   d. all of these

8. Terrarium plants that have small, white, wedge-shaped winged insects on the underside of the leaves are probably infested with

   a. whiteflies.
   b. horseflies.

   c. ants.
   d. none of these

9. The ideal temperature for terrariums depends upon the

   a. amount of soil used.
   b. size of the terrarium.

   c. number of plants in the terrarium.
   d. type of plants in the terrarium.

10. Plants tend to grow toward light sources. This is known as

    a. photosynthesis.
    b. transpiration.

    c. the rain cycle.
    d. phototropism.

11. Flowering plants in terrariums usually require

    a. more light than foliage plants.
    b. less light than foliage plants.
    c. the same amount of light as foliage plants.
    d. none of the above

12. Plants requiring a natural desert condition and full sun are

    a. woodland.   c. cacti and succulents.
    b. tropical.   d. all of these

13. Plants which are native to the forest or woods are

    a. woodland.   c. cacti and succulents.
    b. tropical.   d. none of these

14. Ferns, mosses, ground ivy, and lichens are classified as

    a. tropical.   c. cacti and succulents.
    b. woodland.   d. none of these

15. The loss of moisture from plant leaves is known as

    a. condensation.   c. tropism.
    b. evaporation.   d. transpiration.

16. Moisture in plants is lost through small openings on the underside of the leaves known as

    a. air holes.   c. stems.
    b. stomata.   d. none of these

17. Once a terrarium has the correct balance of air and moisture, it has established its

    a. air cycle.   c. transpiration.
    b. rain cycle.   d. life cycle.

18. If there is too much moisture in a terrarium, it will

    a. dry out naturally.   c. pop its top.
    b. fog up.   d. none of these

19. A sealer over a sand design in a terrarium will prevent _____ of the colored sand.

   a. fading
   b. erosion
   c. transpiration
   d. none of these

20. Ideally, a woodland terrarium should have

   a. dense shade.
   b. soil which is high in organic matter.
   c. a good supply of moisture.
   d. all of the above

OBJECTIVE

To construct and maintain an attractive, healthy bonsai planting.

COMPETENCY TO BE DEVELOPED

After studying this unit, the student will be able to

- explain the purpose and general technique of bonsai.
- list the steps in the construction of a bonsai planting. This may be done orally or in writing, followed by a demonstration of the technique.
- describe the aftercare of bonsai plantings.
- explain how to repot bonsai plantings.

Materials List

bonsai containers

9- to 20-gauge copper wire (for shaping)

screen wire (to cover drain holes)

potting media and gravel

chopsticks or similar sharp sticks

pruners (some should be for fine work).

potted plants (to be trained to bonsai)

---

unit 23
The Art of Bonsai

Bonsai, a word of Japanese origin meaning "tray planting," is the art of dwarfing plants by growing them in shallow pots and trays. The end result is a plant which resembles a large tree in miniature. By applying the techniques of bonsai, the horticulturist

is able to produce an 18-inch tree which resembles an 80-foot tree in all aspects except size. Part of bonsai tradition is to make these plants appear old and rugged, just as they appear in nature after the passing of many generations, figure 23-1. Bonsai designs may be simple or complex, ranging from a simple one-tree planting to a miniature forest, figure 23-2.

Fig. 23-1 A 100-year-old Japanese hemlock. Plants used in bonsai design are frequently hundreds of years old. The height of this plant is 31 inches. (from the collection of the National Arboretum, Washington, D.C.)

Fig. 23-2 A bonsai tray planting designed to resemble a group of rugged trees growing on a rocky hill. This planting is about 21 inches tall. (from the collection of Cliff Pottberg, Richard Kreh, photographer)

Fig. 23-3 A 250-year-old Sargent juniper trained to resemble an old, twisted tree. The juniper is 29 inches tall. (from the collection of the National Arboretum, Washington, D.C.)

PLANTING THE BONSAI

Choosing the Plants

Plants used in bonsai may be evergreen or deciduous. The plants that are selected should have small leaves. This quality helps them appear more as reduced size models of larger trees or plants. Plants selected should also be able to grow in crowded conditions. Flowers, berries, and fruit can add to the attractive quality of bonsai specimens. The following species are some of the best choices for bonsai plantings.

Evergreens
Japanese white pine (*Pinus parviflora*)
Japanese larch (*Larix kaempferi*)
Hinoki cypress (*Chamaecyparis obtusa*)
Azaleas (Satsuki and Kurume)

Deciduous plants
Japanese maple (*Acer palmatum*)
Flowering peach (*Prunus persica*)
Japanese flowering apricot (*Prunus mume*)
Japanese flowering quince (*Chaenomeles lagenaria*)

Source of Plants

There are two sources of bonsai plants.

Naturally Dwarfed Plants. It is sometimes possible to find old plants growing in the wild that have been twisted and dwarfed by cold or dry conditions. These plants may be many years old, yet only a few inches tall. The artistic beauty of bonsai may already be present with very little pruning or training needed to result in a good specimen. A great deal of care must be taken in the digging and transplanting of these specimens so that the roots are not torn.

These wild plants are becoming very scarce and are expensive. The bonsai planting in figure 23-3, found in the wild, is 250 years old. It has been in training in a pot for seventy years.

Some examples of plants that are often collected from the wild are:

Sargent juniper (which lives on high cliffs)
Rhododendron obtusum and *keusianum*
Japanese black pine
Yeddo spruce

Artificially Dwarfed Plants. The second source of plants is a nursery, where potted plants that have the potential to be pruned to the desired shape may be purchased. Plants with small leaves or needles and that have short, twisted growth should be chosen. They should also be selected according to their ability to grow well in the area.

Plants for bonsai may also be grown from seed or cuttings. These plants are trained from the very beginning by restricting their growth and by pruning to obtain the desired shape. The first step is selection of the plant. Figure 23-4 shows a container grown plant selected for a bonsai.

Containers

The bonsai container should

- have one or more drain holes.
- be shallow.
- be one-quarter to one-sixth the height of the plant placed in it.
- not detract from the beauty of the plant placed in it. Subdued earthy colors are best; browns, greens, and grays are most natural.

Fig. 23-4 Mr. Cliff Pottberg, an expert in the art of bonsai, holds a plant chosen for bonsai. When selecting container grown plants for bonsai, the desired shape must be visualized and one branch selected for the main trunk. All other branches are pruned. Mr. Pottberg points to the branch he will save for the main trunk of the new bonsai plant. (Richard Kreh, photographer)

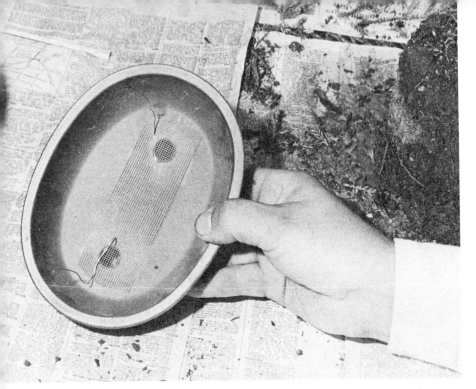

Fig. 23-5 One type of bonsai container. It is shallow, unglazed, and has two large drainage holes. The screen wire placed over the bottom prevents loss of potting mix. The wires protruding through the bottom will be used to hold the plant firmly in the tray. (Richard Kreh, photographer)

- have a simple form (round, square, or rectangular).
- be made of pottery and unglazed on the inside.

Figure 23-5 illustrates one type of bonsai container.

Tools

Various tools used to cut and prune plants in preparation for bonsai are shown in figure 23-6. Not all of these tools are essential, however. The large pinchers (the first tool on the left) are used to remove the tap root (the long, central root of the plant). The remaining tools in the front row are used for pruning and cutting. The ordinary slip joint pliers in the back row are used to remove bark from limbs. The copper wire of various thicknesses is twisted around branches to hold them in place as the branches are shaped.

DWARFING PLANTS

Determining Plant Shape. The first step is to visualize the bonsai design in its final form. Will it have one stem or several? Will there be more than one plant in the container? Is the plant to be set straight or slanted in the container? Will roots show above the soil level or be covered? These are some of the questions that must be answered before pruning is begun. When pruning, remember that the trunk and branches of the plant are the focal points of the design.

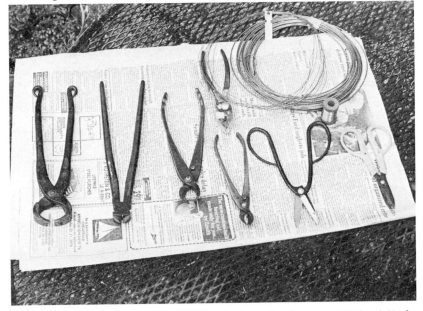

Fig. 23-6 Tools used in preparation of plants for bonsai. (Richard Kreh, photographer)

Fig. 23-7 Plant after major pruning to the top. (Richard Kreh, photographer)

Fig. 23-8 Pliers are used to loosen the bark on branches that are to represent dead limbs. The cambium layer must be scraped off to ensure that the branch dies. (Richard Kreh, photographer)

pruning is done. Either or both of the two short stubs may be stripped of bark (as has been done to the twig on the right) to represent old, dead limbs. The bark on the twig was loosened by gripping the branch with pliers and twisting it, figure 23-8. The branch is then scraped to remove all of the cambium layer, ensuring that the branch dies.

Pruning the top of the tree is essential in developing the rugged but miniature shape which is characteristic of bonsai plants. Any active shoots must be shortened and undesirable buds or shoots removed. There should be little space between limbs or side branches to give the effect of a tall tree.

Initial Pruning. Remove all of the unwanted branches. This step must be done very carefully to ensure that enough branches remain to form the desired shape for the entire plant. Figure 23-7 shows the same plant as the one in figure 23-4 after the first

Wiring. The remaining branches are wired with copper wire. Branches are wired to force growth in a direction other than the direction in which they naturally grow. Copper wire is desirable because it does not rust and is not highly visible after it weathers. Heavy wire is used for the larger branches and trunk and smaller copper wire for small branches. The size of the wire used varies from 9 to 20 gauge. The wire must be pliable enough to bend and yet rigid enough to hold the plants in shape. Cut the wire a few inches longer than the combined length of the trunk and branch to be wired. Twist the wire around the trunk with both ends protruding, figure 23-9. Wrap the wire over itself, locking it in place. Wrap the branches with spirals about 1 inch apart, being careful not to tear the bark on the twigs. Avoid wrapping plant foliage under the wire since this foliage may die and disfigure the plant. When the wrapping is finished, cut the end of the wire off close to the branch so that the ends do not show.

When two branches are to be pushed apart, one end of the wire is used on each branch. The first branch is wired and pushed down, and then the second branch is wired, figure 23-10. Notice that the two pieces of wire are twisted in opposite directions. After the wire is twisted on both branches, the branches are pushed in the desired direction. The wire holds them in place. The ends of the wire are cut off close to the branches so that pieces of wire do not protrude from the plant. The wire should not be left on long enough to girdle the branches as they grow.

> Be careful not to twist the wire too far and break the branches or to twist the branches from the trunk.

Additional Pruning. The top of the plant is now pruned to remove some of the small branches and shorten others. Figure 23-11 shows special pruners being used to remove small branches not needed or desired. The branches on top are pruned to form a flat, wide shape and are arranged at desired levels. (Figure 23-17 shows the extent of top pruning.)

Root Pruning. After the top is pruned, attention is focused on the roots. Dig the soil ball apart by pulling down through roots which make up the root ball with a chopstick or other pointed stick, figure 23-12. Continue until most of the soil is removed, and the roots are exposed. Expect many roots to be torn off in this process. The large pinchers are used to cut off the center tap root, figure 23-13. Many of the side roots must also be cut back so that the

new plant fits in the bonsai container.

Placing the Plant in the Container. The plant is now ready to be placed in the bonsai container. The container is prepared by placing a screen wire over the drainage holes in the bottom. A thin layer of gravel is then spread over the screen wire to ensure proper water drainage, figure 23-14. The plant is set on top of the gravel and pushed down so that the root level is higher than the edge of the container, figure 23-15. Notice that the plant is placed one-third of the container diameter from the left side and two-thirds from the right side. The plant leans towards the center of the container balancing its appearance. Bonsai plants are generally not placed in the exact center of their containers.

Fig. 23-9 The copper wire is twisted around the plant where the two branches are to be pushed apart. (Richard Kreh, photographer)

Fig. 23-10 The wire is twisted around both branches when the objective is to push the branches apart. The lower branch on the right will form a side limb. Wire is being twisted around the second branch to push it upward, forming part of the top of the plant. Notice that the wire is held at the end, allowing it to be wrapped loosely around the plant. (Richard Kreh, photographer)

Fig. 23-11 Detailed pruning of small branches after wiring (Richard Kreh, photographer)

Fig. 23-12 Soil is carefully removed from the root ball with a pointed stick. (Richard Kreh, photographer)

Fig. 23-13 The center root of the plant is removed with the pinchers. (Richard Kreh, photographer)

Fig. 23-14 Gravel is placed in the bottom of the container for proper drainage. (Richard Kreh, photographer)

Fig. 23-15 The plant is placed in the bonsai container. Notice that the root system has been pruned to fit the container. The anchor wire that was inserted earlier in the process is twisted over the top of the main roots to hold the plant firmly in place. (Richard Kreh, photographer)

Fig. 23-16 The media is worked through the roots and firmed with a chopstick. The media used here is clay kitty litter. (Richard Kreh, photographer)

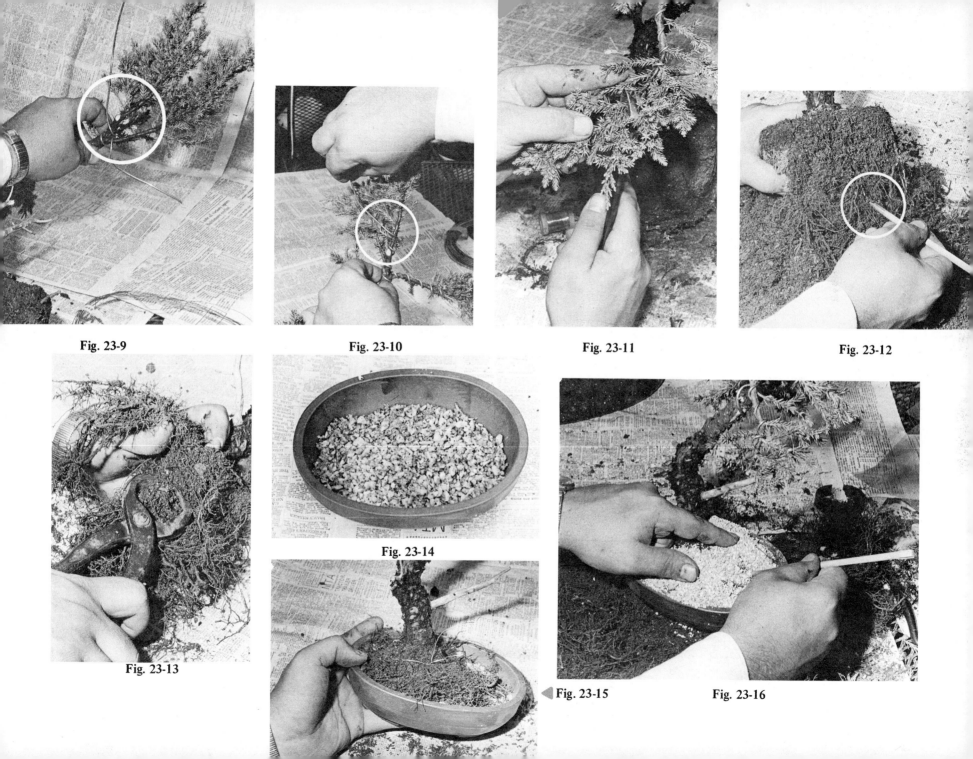

Fig. 23-9

Fig. 23-10

Fig. 23-11

Fig. 23-12

Fig. 23-13

Fig. 23-14

Fig. 23-15

Fig. 23-16

Fig. 23-17 The finished bonsai planting. In a few years, the plant will develop more foliage and look much more like a miniature tree. (Richard Kreh, photographer)

Adding Potting Media. A potting media with good drainage is now placed on top of the root system and gently worked through the roots with a pointed stick to fill any air pockets, figure 23-16. The media should cover all of the roots except for large roots near the trunk, which are sometimes left exposed for effect. Exposed roots cause the plant to look like an older tree whose large roots have gradually been exposed.

Fill the container to a point within 1/4 inch of the top of the edge and form a gradual mound toward the center of the container to cover the plant roots. Firm the media so that there are no large air pockets.

Watering and Fertilizing. Water the new plant with a diluted solution of a complete fertilizer mix made specially for bonsai, or use one of the water-soluble plant fertilizers for potted plants. Add water until it runs from the drainage holes in the container.

Figure 23-17 shows the finished product. Notice that the larger dead limb was removed because it was out of scale with the rest of the plant. The new plant should be protected from wind and sun for at least two weeks until it recovers from the severe pruning.

CARE AND MANAGEMENT OF BONSAI

Aftercare of the plant itself consists of continually pruning the roots so that they fit the container and shaping the top to obtain the desired effect.

Watering

Keep the soil moist; the bonsai plant has a small root system and must not be allowed to dry out. Add water until it runs through the bottom of the container.

Location

During summer months, keep the bonsai outdoors in light shade. Protect it in a cold frame or greenhouse in winter months to prevent the roots from freezing and winds from drying the plant out. Storage in a cool basement in winter is adequate. The plant must be watered and cared for while it is inside. A bonsai should not be kept inside except when protection from heat or cold is necessary.

The Root Bound Plant

When trees become root bound, they must be replanted. (Lift the plant out of the container to check for the pot bound state. If the plant is not pot bound, replace it as it was.) Prune roots in the spring.

Fertilizing

Organic materials such as cottonseed meal, well-rotted animal or vegetable manure, or the newer liquid fertilizers work well. Sprinkle on the surface (2 teaspoons per plant of organic fertilizer every six weeks from midspring to August 1). Liquid fertilizers may be applied with watering. Apply them about three times a year, beginning when active growth starts, and repeat every four weeks until August 1.

REPOTTING AND TRAINING PROCEDURE

As the new plant grows, additional pruning is required. Repotting should be done whenever the plant becomes root bound or roots have curled around the root ball. As a general rule, repotting should be done every two or three years for deciduous trees and every three to five years for conifers. The best time to repot is in early spring just before new growth begins. When repotting bonsai plants, root pruning is the most important consideration. Roots must be pruned to bring about the growth of new roots. Many small feeding roots will emerge from the cut ends of large roots.

Repotting Procedure

1. Remove the plant from the container.

2. Straighten out roots if curled in a ball.

3. Remove one-third of the soil from the outside of the root ball, being careful not to break up the rest of the root ball. This process is much easier if the soil is dry.

4. Trim the roots back to the soil left in the root ball with sharp scissors or shears. (One-third of the root length is cut off.)

5. The tree is now repotted, using the same procedure as was used when the bonsai was originally potted. The new root ball should have 2 inches of space between it and the outside edge of the pot. New soil is worked firmly into this area with a small stick. The 2 inches of new soil provide room for new active feeding roots to develop and keep the root system active. Soil is packed tightly against all roots so that no air space is left. Take the time to do this job thoroughly. The soil added should be dry so that it sifts down around the roots.

6. If support is necessary, tie the plant to the container by guiding a piece of copper wire up through a drain hole and over part of the root system. The wire will not show, since it is covered with soil. Do not wrap the wire completely around the roots, or they will be girdled.

Bonsai can be both a rewarding hobby and business, and can be accomplished by anyone with the patience to work with the plant and wait for its beauty and character to develop. It is not necessary to wait fifty to one hundred years for an attractive bonsai. The small plant used for demonstration purposes in this unit will develop shape, character, and beauty in just a few years.

STUDENT ACTIVITY

Select a container and suitable plant and prepare a bonsai planting following the steps in the text. This may be done in groups of two to four students or individually.

SELF-EVALUATION

A. Select the best answer from the choices offered to complete each statement.

1. Plants selected for bonsai should have small leaves because
   a. large leaves hide too much of the trunk.
   b. large leaves get in the way when wiring and shaping.
   c. they make the plants more closely resemble dwarfed trees, and the scale of all parts appears more nearly correct.
   d. large leaves lose too much moisture.

2. The bonsai container should
   a. be shallow and well drained.
   b. have a simple form and not detract from the plant placed in it.
   c. be made of pottery and unglazed on the inside.
   d. all of the above

3. Copper wire is used for shaping plants because
   a. it is inexpensive.
   b. it is pliable but strong and does not rust.
   c. it is the only wire available in the sizes needed.
   d. it does not girdle plants.

4. The large center root of plants used in bonsai is cut off

   a. to dwarf the plant.
   b. to cause root branching.
   c. to make the root system shallow enough to fit in the container.
   d. to simplify the root pruning job.

5. The new bonsai plant is wired in the container because

   a. the shallow root system may not be able to anchor the plant and hold it up.
   b. it helps to restrict root growth.
   c. the potting media is too light for good anchorage.
   d. all of the above

6. The new bonsai plant must be watered regularly because

   a. it has a small, shallow root system.
   b. the small container cannot hold much moisture.
   c. any drying out would damage a valuable plant.
   d. all of the above

7. The bonsai should be kept

   a. outside most of the time in light shade.
   b. inside year-round in a protected area.
   c. in a cool basement.
   d. either inside or outside in a very dark place.

8. Bonsai plants generally must be repotted after three to five years

   a. so that new soil can be added, encouraging new root growth.
   b. to invigorate the root system.
   c. because the roots become pot bound in this length of time.
   d. all of the above

B. Briefly list the eight steps in the construction of a bonsai planting.

OBJECTIVE

To grow and maintain healthy, happy indoor plants.

COMPETENCY TO BE DEVELOPED

After studying this unit, the student will be able to

- list the four major concerns in caring for houseplants.
- mix a media to use for growing various houseplants.
- describe the best cultural conditions (water, light, and soil requirements) for ten selected houseplants of his or her choice.
- list the four methods of watering houseplants.

Materials List

fifteen houseplants from the plant list in the text

one bag of soluble fertilizer

one bag of slow-release fertilizer

peat moss, soil, construction grade sand, clay pots, and pieces of clay pots

---

unit 24
Houseplants

Plants have been a special part of people's lives throughout time. Of course, plants that bear food and materials for making articles have always been essential to life, but the plant that simply adds beauty to its surroundings serves a special purpose. Since the initiation of the environmental protection and "green survival" movements, the popularity of houseplants has grown rapidly. In 1971, $38 million worth of foliage houseplants was sold. In 1976, the figure jumped to $260 million, an increase of nearly 600 percent.

This increased interest in live indoor plants is reflected in both public places and private homes. Shopping malls, office buildings, and motel complexes are now using live plants to landscape indoor areas. The ancient practice of building homes around an *atrium* (indoor garden accessible from all main rooms of a house) is receiving renewed attention. People who live in crowded urban areas often find the idea of a private garden of houseplants particularly appealing. Houseplants can be used solely for the interest they create or to serve a specific function. Plants act functionally when used as room dividers, screens, and to accent pieces of furniture.

CARING FOR HOUSEPLANTS

To care for houseplants properly, the following questions must be answered for each type of plant being grown.

- What is the proper way to water the plant?
- How much light does it need?
- What is the best temperature for optimum growth?
- What type of soil mixture is necessary?

WATERING

To be sure that houseplants are watered properly, there must be careful observation of the plant's behavior. There are several signs of improper watering. Drooping leaves or leaves that do not seem as full as they should may indicate a lack of water. Too much water can cause leaves to wilt or turn yellow and finally drop off the plant. Other factors besides the amount of water given a plant are important to consider in moisture control, such as using the proper media and providing good drainage.

There are several methods of watering houseplants.

Drench/let dry. Enough water is added to moisten the media which is then permitted to dry before watering again. In the drenching method, the desired amount of water is added all at one time.

Drench/let dry slightly. Enough water is added to moisten the media. It is then permitted to dry slightly so that it is damp to the touch.

Keeping media constantly moist. Enough water is added so that the media is damp to the touch at all times.

Temperature

Many times, houseplants are exposed to a great range of temperatures within one 24-hour day. For this reason, plants must sometimes be given special protection. For example, plants located in windows during winter months may receive full sunlight during the day. During the night, however, the temperature next to the window will be much cooler. If the temperature is below $32°F$ $(0°C)$ and the plants are very close to the windowpane, the plants may freeze. In these cases, it may be enough to place a shade or curtain between the pane and the plant. The temperature recommended for houseplants ($50°$ to $85°F$) is the daytime temperature; the night temperature can generally drop 10 to 15 degrees below this temperature without harming the plants.

The common use of air conditioners in homes and offices can be a source of trouble for plants. The plants can be protected, however, by locating them away from drafts and direct air flows.

Lighting

As watering requirements vary with individual plants, so does the amount of light which they require. All plants need at least some light, natural or artificial, so that their leaves can manufacture food (the process of photosynthesis).

Light meters (small machines that measure illumination, or light) can be used to determine if the location selected for the plants is suitable. Light meters are sold at florists, garden centers, and department stores with plant shops. Light intensity requirements, which vary from plant to plant, are classified into the following groups.

Direct Sun. Plant should receive full intensity of natural sunlight.

Partial Sun or Weak Sun. Plant should receive less than 50 percent of natural sunlight.

Indirect or Filtered Light. Plant should receive no direct sunlight at all.

In general, the more light that is available, the greater is the variety of plants which may be grown. In most cases, foliage plants require less light than flowering plants.

Artificial Lighting

The advent of the fluorescent light greatly influenced the art of growing houseplants. The fluorescent light enables the horticulturist to:

- root cuttings of houseplants more easily.

- start seeds of new and different houseplants.

- grow tropical plants, cacti, and succulents with greater ease.

Humidity

Humidity is the percentage of moisture in the air — an important factor to consider when growing houseplants. Humidity varies with seasons; it is lower than usual in the winter; therefore, additional moisture is needed in the atmosphere. During the summer the humidity is usually higher than in the winter.

Humidity around houseplants can be increased by

- spraying a fine mist of water on the foliage of the plant with an inexpensive plastic spray bottle or any type of mister.

- using a watertight tray filled with small pea gravel to hold the potted plants. Water is added to the tray to come to the top of the gravel and plants are set on the gravel making certain that the pots are not actually in the water.

- using a humidifier to provide moisture in the air.

- placing the plants in that part of the house that has the highest humidity. For example, areas of highest humidity are probably the bathroom or the kitchen just above the sink.

FERTILIZERS

There are two types of houseplant fertilizers on the market today — slow release and soluble. *Slow-release fertilizer* is in the form of small beads and is applied to the soil mixture. As it comes in contact with normal soil moisture, plant nutrients are steadily released. The plant absorbs the released nutrients as needed, thus eliminating the danger of overfeeding or fertilizer burn.

Soluble fertilizer is available in liquid or solid form. Solid soluble fertilizer must be dissolved in water before being applied to houseplants. Applying this fertilizer dry may burn the roots of the plant. Soluble fertilizers in liquid form are usually concentrated, requiring that water be added to the mixture before application.

> **Always follow the manufacturer's recommendations when using any fertilizer.**

SOIL MIXTURE

The composition of the soil mixture varies with the requirements of the particular houseplant. One type of soil mixture is not necessarily good for all plants. A good general potting mixture includes two parts loamy soil, one part organic matter such as peat moss, leaf mold, or well-rotted compost, and one part coarse sand or a horticultural grade perlite. When using sterilized soil, the addition of trace elements is usually not necessary.

A good loamy soil consists of equal parts of sand, silt, and clay. Organic matter, rich in carbohydrates, helps to improve the water-holding capacity of the mixture. It makes the mixture easy to work with and helps to keep the soil crumbly. Peat moss, leaf mold, and well-rotted compost act as fertilizers and soil conditioners. The use of coarse sand or horticultural grade perlite helps to keep the mixture loose and provides good drainage. This allows better movement of air and nutrients carried by the water.

CONTAINERS

There are many beautiful ceramic and pottery plant containers on the market today. Plants grow in almost every kind of container, provided that drainage is available. However, if there are no drainage holes in the container, the soil mixture will become waterlogged. Plants cannot survive under these conditions. In addition to drainage, there must be some air spaces in the mixture so that the plant roots breathe properly. Proper air circulation depends upon the ingredients of the soil.

Clay has long been the standard type of container for growing plants, but many greenhouse growers have now changed to other types of containers such as plastic, glass, fiberglass, and metal. All plant containers should be set in some type of saucer to catch excess drainage water. Since clay is porous, water can escape through the clay saucers to table surfaces. This problem can be eliminated by coating the saucer with plastic or paraffin wax.

POTTING PLANTS

After the proper container and soil mixture have been selected, potting or repotting houseplants is a simple process. First, some type of drainage must be provided in the bottom of the pot. For clay pots, a crock may be used. A *crock* is a broken piece of clay pot. Place the crock so that it is curved upward over the drainage hole in the bottom of the pot. If it is placed with the curve down, it may seal the drainage hole and cause problems by not allowing excess water to drain out of the pot. Stones may also be used in containers. If they are used, be sure that the stones don't completely close the drainage holes in the bottom of the pot.

Add the selected potting mixture to the container so that it covers the drainage stone or crock. Place the plant in the container, making sure to check the depth of the plant. If the plant is below the rim of the pot, raise it so it is 1/2 inch below the top of the pot. This allows for an area at the top of the pot to hold additional water while the plant is being watered. If no space is allowed, the water will run off over the edge of the pot. Not having received much of the water added, the plant will dry out much more quickly, thereby inhibiting growth.

Repotting

Houseplants should be repotted when they become pot bound. To determine if the plant is pot bound, remove the plant from the original pot. If the roots of the plant are growing around the root ball of the plant, the plant should be repotted. If the plant is pot bound, the roots must be separated; this allows the roots to develop and make new growth. Repotting should be done

A. Select a pole 3 to 5 times taller than the container in which the totem is to be used.

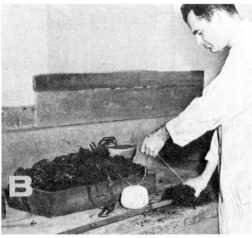

B. Wrap the pole with a layer of sphagnum moss 2 to 3 inches deep and bind the moss with string.

C. Push the pole into the soil.

D. Wind the vine around the totem.

E. Use hairpins to fasten the vine to the totem.

F. Keep the moss damp; roots will grow into the moss and the leaves will form a solid mass.

Fig. 24-A Making a totem (USDA photo)

before new growth starts. If the plant is flowering, wait until after flowering is completed.

GENERAL CARE

The leaves of houseplants should be cleaned occasionally with a fine spray of water at room temperature. Wipe the leaves on both the upper and underside. This is a good way to remove dust and insects, help maintain humidity, and keep the houseplant fresh looking and attractive, figure 24-B.

THE HOUSEPLANT CHART

Following is a houseplant chart which gives the essential information for growing forty-two houseplants. Each plant is illustrated with a picture to aid in identification. The common name of the plant is given first and the scientific name given second.

- *Watering* tells how to water each plant.

- *Sunshine* gives the amount of light exposure each plant needs, ranging from full sun to indirect light.

- *Temperature* indicates the optimum temperature at which to grow each individual plant.

- *Soil mixture* gives a general recommendation for the plant using a mixture of sand, peat moss, and soil.

- *Ease of growth* is a rating given for each plant on a scale of 1 to 4. A number 1 means that the plant is very easy to grow and requires little care. A number 2 means that the plant is easy to grow but needs some special attention. A number 3 means that the plant is very difficult to grow. A number 4 means that the plant is very difficult to grow and must have very special care. This rating may be helpful when selecting plants to grow.

 INDICATES PLANT WILL GROW IN A CLOSED TERRARIUM

 INDICATES PLANT WILL GROW IN A HANGING BASKET

 INDICATES PLANT WILL GROW IN A DISH GARDEN OR POT

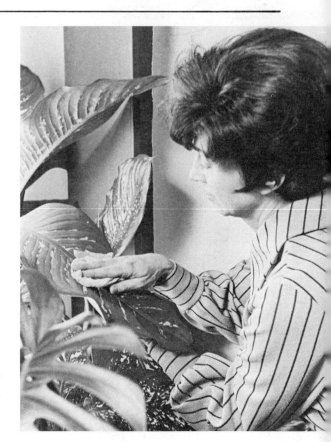

Fig. 24-B Cleaning the leaves of dieffenbachia to remove the dust. (USDA photo)

_____ HOUSEPLANT CHART _____

Asparagus Fern (*Asparagus sprengeri*)

Watering Drench/let dry slightly
Sunshine Direct sun
Temperature 50° to 70°F
Soil Mixture 1 peat, 1 sand, 2 soil
Ease of Growth 1 (Easy)

Cyclamen (*Cyclamen persicum*)

Watering Keep evenly moist
Sunshine Partial shade
Temperature 50° to 55°F (night)
              65° to 75°F (day)
Soil Mixture 2 peat, 1 sand, 1 soil
Ease of Growth 3 (Difficult)

Fuchsia (*Fuchsia x hybrida*)

Watering Drench/let dry slightly
Sunshine Direct sun
Temperature 50° to 65°F
Soil Mixture 1 peat, 1 sand, 2 soil
Ease of Growth 2 (Easy)

Geranium (*Pelargonium x hortorum*)

Watering Drench/let dry slightly
            Mist often
Sunshine Direct sun
Temperature 50° to 60°F
Soil Mixture 1 peat, 1 sand, 2 soil
Ease of Growth 1 (Easy)

Gardenia (*Gardenia jasminoides 'Veitchii'*)

Watering Drench/let dry slightly
            Mist often
Sunshine Direct sun
Temperature 60° to 75°F
Soil Mixture 1 peat, 1 sand, 1 soil
Ease of Growth 4 (Difficult)

Gloxinia (*Sinningia speciosa*)

Watering Keep soil moist
Sunshine Weak (partial) sun
Temperature 55° to 70°F
Soil Mixture 1 peat, 2 sand, 1 soil
Ease of Growth 2 (Easy)

HOUSEPLANT CHART

  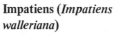

Impatiens (*Impatiens walleriana*)

Watering Keep soil moist
Sunshine Weak (partial) sun
Temperature 50° to 60°F
Soil Mixture 1 peat, 2 sand, 1 soil
Ease of Growth 1 (Easy)

Blue Hens-and-Chickens (*Echeveria secunda*)

Watering Drench/let dry
Sunshine Direct sun
Temperature 50° to 65°F
Soil Mixture 1 peat, 2 sand, 1 soil
Ease of Growth 1 (Easy)

Variegated Wax Plant (*Hoya carnosa variegata*)

Watering Drench/let dry slightly
Sunshine Weak (partial) sun
Temperature 50° to 65°F
Soil Mixture 1 peat, 1 sand, 2 soil
Ease of Growth 2 (Easy)

  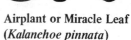

Airplant or Miracle Leaf (*Kalanchoe pinnata*)

Watering Drench/let dry slightly
Sunshine Direct sun to partial sun
Temperature 55° to 65°F
Soil Mixture 1 peat, 2 sand, 1 soil
Ease of Growth 2 (Easy)

Aluminum Plant (*Pilea   cadierei*)

Watering Drench/let dry slightly
Sunshine Partial to Direct sun
Temperature 50° to 65°F
Soil Mixture 3 peat, 2 sand, 1 soil
Ease of Growth 2 (Easy)

Azalea (*Rhododendron*)

Watering Drench/let dry slightly
Sunshine Weak (partial) sun
Temperature 50° to 65°F
Soil Mixture 1 peat, 1 sand, 1 soil
Ease of Growth 3 (Difficult)

HOUSEPLANT CHART

Boston Fern (*Nephrolepis exaltata bostoniensis*)

Watering Keep mixture moist
Mist often
Sunshine Partial sun
Temperature 50° to 70°F
Soil Mixture 1 peat, 2 sand, 1 soil
Ease of Growth 1 (Easy)

Dracaena - Ribbon Plant (*Dracaena sanderiana*)

Watering Keep evenly moist
Sunshine Partial shade
Temperature 62° to 80°F
Soil Mixture 1 soil, 1 peat, 1 sand
Ease of Growth 2 (Easy)

Norfolk Island Pine (*Araucaria heterophylla*)

Watering Drench/let dry slightly
Sunshine Direct sun
Temperature 55° to 75°F
Soil Mixture 3 peat, 2 sand, 1 soil
Ease of Growth 1 (Easy)

Coleus (*Coleus blumei*)

Watering Keep mixture moist
Sunshine Full sun
Temperature 50° to 70°F
Soil Mixture 1 peat, 1 sand, 2 soil
Ease of Growth 1 (Easy)

Dumb Cane (*Dieffenbachia picta 'Superba'*)

Watering Drench/let dry slightly
Sunshine Avoid direct sun
Temperature 60° to 75°F
Soil Mixture 3 peat, 2 sand, 1 soil
Ease of Growth 2 (Easy)

Emerald Ripple Peperomia (*Peperomia caperata*)

Watering Drench/let dry slightly
Sunshine Weak (partial) sun
Temperature 50° to 65°F
Soil Mixture 3 peat, 2 sand, 1 soil
Ease of Growth 1 (Easy)

HOUSEPLANT CHART

Peperomia (*Peperomia obtusifolia*)

Watering Drench/let dry slightly
Sunshine Weak (partial) sun
Temperature 50° to 65°F
Soil Mixture 3 peat, 2 sand, 1 soil
Ease of Growth 1 (Easy)

Prayer Plant (*Maranta*)

Watering Keep moist
Sunshine Weak (partial) sun
Temperature 60° to 70°F
Soil Mixture 1 peat, 2 sand, 1 soil
Ease of Growth 3 (Difficult)

Rubber Plant (*Ficus elastica decora*)

Watering Drench/let dry slightly
Sunshine Weak (partial) sun
Temperature 50° to 75°F
Soil Mixture 3 peat, 2 sand, 1 soil
Ease of Growth 1 (Easy)

Umbrella Tree (*Schefflera*)

Watering Drench/let dry slightly
Sunshine Weak (partial) sun
Temperature 50° to 70°F
Soil Mixture 1 peat, 2 sand, 1 soil
Ease of Growth 1 (Easy)

Spider Plant (*Chloro-phytum comosum*)

Watering Drench/let dry slightly
Sunshine Weak (partial) sun
Temperature 50° to 70°F
Soil Mixture 1 peat, 1 sand, 2 soil
Ease of Growth 1 (Easy)

Snake Plant (*Sansevieria*)

Watering Drench/let dry
Sunshine direct to indirect
Temperature 50° to 75°F
Soil Mixture 1 peat, 2 sand, 1 soil
Ease of Growth 1 (Easy)

HOUSEPLANT CHART

Zebra Plant (*Aphelandra squarrosa*)

Watering Drench/let dry slightly
Sunshine Weak (partial) sun
Temperature 60° to 75°F
Soil Mixture 1 peat, 1 sand, 2 soil
Ease of Growth 2 (Easy)

Lipstick Vine *(Aeschynanthus lobbianus)*

Watering Keep soil uniformly moist but not wet
Sunshine Direct light (except in summer or outdoors)
Temperature 70° to 80°F
Soil Mixture 1 peat, 1 sand, 1 soil
Ease of growth 2 (Easy)

Bird's-Nest Fern (*Asplenium nidus*)

Watering Keep soil moist, mist foliage
Sunshine Direct sun to weak sun
Temperature 50° to 70°F
Soil Mixture 3 peat, 2 sand, 1 soil
Ease of Growth 2 (Easy)

English Ivy (*Hedera helix*)

Watering Drench/let dry slightly
Sunshine Partial sun
Temperature 50° to 65°F
Soil Mixture 1 peat, 2 sand, 1 soil
Ease of Growth 1 (Easy)

Wandering Jew (*Zebrina pendula*)

Watering Keep soil moist
Sunshine Indirect sun
Temperature 50° to 65°F
Soil Mixture 1 peat, 1 sand, 1 soil
Ease of Growth 1 (Easy)

Swedish Ivy (*Plectranthus australis*)

Watering Keep soil moist
Sunshine Indirect sun
Temperature 50° to 65°F
Soil Mixture 3 peat, 2 sand, 1 soil
Ease of Growth 2 (Easy)

HOUSEPLANT CHART

Pineapple (*Ananas comosus*)

Watering Keep soil moist
Sunshine Partial sun
Temperature 50° to 70°F
Soil Mixture 3 peat, 2 sand, 1 soil
Ease of Growth 2 (Easy)

False Aralia (*Dizygotheca elegantissima*)

Watering Keep soil moist
Sunshine Direct sun to partial sun
Temperature 55° to 75°F
Soil Mixture 3 peat, 2 sand, 1 soil
Ease of Growth 1 (Easy)

Ivy-Leaved Geranium (*Pelargonium peltatum*)

Watering Drench/let dry
Sunshine Direct sun
Temperature 50° to 60°F
Soil Mixture 1 peat, 1 sand, 1 soil
Ease of Growth 2 (Easy)

Velvet Plant (*Gynura 'Sarmentosa'*)

Watering Keep soil moist
Sunshine Partial sun
Temperature 65° to 75°F
Soil Mixture 3 peat, 2 sand, 1 soil
Ease of Growth 2 (Easy)

Banana (*Musa maurelii*)

Watering Keep soil moist
Sunshine Partial sun
Temperature 60° to 85°F
Soil Mixture 1 peat, 1 sand, 1 soil
Ease of Growth 3 (Difficult)

Bird of Paradise (*Strelitzia reginae*)

Watering Keep soil moist — Summer
  Drench/let dry — Winter
Sunshine Direct sun
Temperature 50° to 75°F
Soil Mixture 1 peat, 1 sand, 1 soil
Ease of Growth 3 (Difficult)

HOUSEPLANT CHART

Philodendron (*Philodendron cardin*)

Watering Drench/let dry slightly, mist leaves often
Sunshine Indirect
Temperature 50° to 75°F
Soil Mixture 3 peat, 2 sand, 1 soil
Ease of Growth 1 (Easy)

Jade Plant (*Crassula argentea*)

Watering Drench/let dry
Sunshine Partial sun
Temperature 60° to 70°F
Soil Mixture 1 peat, 1 sand, 1 soil
Ease of Growth 1 (Easy)

Medicine Plant (*Aloe vera*)

Watering Drench/let dry
Sunshine Direct sun
Temperature 55° to 70°F
Soil Mixture 1 peat, 2 sand, 1 soil
Ease of Growth 1 (Easy)

African Violet (*Saintpaulia sp.*)

Watering Keep soil moist
Sunshine Weak (partial) sun
Temperature 65° to 75°F
Soil Mixture 3 peat, 2 sand, 1 soil
Ease of Growth 2 (Easy)

Amaryllis (*Hippeastrum sp.*)

Watering Keep soil moist
Sunshine Direct sun
Temperature 55° to 65°F
Soil Mixture 1 peat, 1 sand, 1 soil
Ease of Growth 3 (Difficult)

Pothos (*Scindapsus aureus*)

Watering Drench/let dry
Sunshine Indirect
Temperature 50° to 75°F
Soil Mixture 3 peat, 2 sand, 1 soil
Ease of Growth 1 (Easy)

(Plant photos by Joseph Tardi Associates)

SELF-EVALUATION

1. What are the four questions that must be answered when considering the purchase of a specific houseplant?

2. What is the makeup of a complete houseplant soil mixture?

3. What are the characteristics of a houseplant that has been overwatered?

4. Make a chart listing the cultural conditions for ten houseplants that interest you.

5. List two types of fertilizers used on houseplants.

6. Identify the three methods of watering houseplants.

Section 9
Using Plants in
the Landscape

▲ Color Plate 1

One of the newer rhododendron hybrids (Mary Belle) showing some yellow color. Rhododendrons are good specimen plants and, in large areas, are also very effective in mass plantings.

Color Plate 2 (Upper right)

The red rhododendron Jean Marie De Montague creates a brilliant flash of color when placed against the background of a white house. This plant is hardy to 0°F, and, like all rhododendrons, thrives in acid soil and partial shade.

▶ Color Plate 3

Trees provide the background and vertical depth for a beautiful scene. This landscape demonstrates how a rough or rolling terrain can be developed into a formal rock garden through the use of evergreens, bulbs, perennials, and herbaceous annuals. A natural walk meanders through this rock garden located in Longwood Gardens, Pennsylvania.

◀ Color Plate 4

An 80-year-old bonsai tree from the collection at the National Arboretum in Washington, DC (#35). This Japanese Witch Hazel (Mansaku) tree is growing in a very shallow container. It was a gift to the United States from the government of Japan and has been in training as a bonsai for 50 years.

Color Plate 5 (Lower left)

An excellent example of the use of plants for fall color contrast. The rhododendrons in front and underneath provide deep green color, the dogwood adds red, while the yellow maple in back and the larger oak trees form a background for the scene. Notice the three layers of plants: an understory of rhododendron, a subcanopy of dogwood and maple, and a canopy of oak.

▼ Color Plate 6

Pachysandra used as a ground cover along a brick walk in a heavily traveled area. This plant is a very reliable ground cover for partially or densely shaded areas and will thrive where grass could not survive. Pachysandra, like most ground covers, requires very little care.

▲ Color Plate 7
   The new, red azalea Stewartonian used in an informal planting. Notice
ow the azalea blends naturally with the other plants.

Color Plate 8 (Upper right)
   Annuals (lantana) used as perimeter plants around a large circular lawn
rea. These plantings add color and break up the solid green lawn area.
Notice the different shapes and textures of the trees.

▶ Color Plate 9
   Euonymus as the center of a corner planting. Three red rhododendron
nd the crocuses provide spring color. This planting hides a high foundation
nd ties together the masonry wall and steps, the sidewalk, and the lawn
rea.

▼ Color Plate 10
Fresh garden peas ready to be hulled and cooked for eating. This variety, Blue Bantam, derives its name from the color of its foliage (shown behind the basket), and its shorter height, which eliminates the need for staking or tying.

▲ Color Plate 11
A new tomato hybrid, Big Girl VF. A major factor in this plant's ability to produce high-quality fruit is its resistance to verticillium and fusarium wilt. Disease-resistant varieties should be used whenever these diseases are a problem.

Color Plate 12

A new watermelon variety, Sugar Bush, which has smaller fruit (6 to 8 pounds each) on a more compact plant, is ideal for use in small home gardens. The vines are from 3 to 3 1/2 feet long.

Color Plate 13

Early Sunglow, a new, early maturing variety of sweet corn which is ready for eating in 63 days. Notice the well-filled ear, picked at just the right maturity for best flavor and tenderness.

◀ Color Plate 14

Stark Red Giant Strawberry ready for picking. Notice the straw mulch used for weed control, to conserve soil moisture, and to help keep the berries clean.

Color Plate 15 (Lower left)

A nursery worker asexually propagating apple trees using the whip or tongue graft method. A very sharp knife is required to obtain the smooth, straight cuts which allow the root and scion to fit tightly together. The propagator's fingers are wrapped with tape as a safety precaution. Note that the root section is about the same diameter as the scion. The cambium must match on at least one side in order for the graft to grow.

▼ Color Plate 16

Grape variety Buffalo ready for harvest. This fine crop is the result of proper pruning, fertilizing, and spraying.

▲ Color Plate 17

A blueberry plant showing a large crop of high-quality berries. The vigorous wood necessary for such an excellent crop is produced by proper pruning and fertilizing. The sawdust mulch around the plant controls weeds, holds soil moisture, and keeps the roots cool.

◀ Color Plate 18

A rooted chrysanthemum cutting. Note the compact, well-developed root system.

Color Plate 19 (Lower left)

Four rooted chrysanthemum cuttings planted in a 6-inch pot. Note that the cuttings are planted near the perimeter of the pot and slanted toward the outside. This gives the plant more spread.

▼ Color Plate 20

Garden mums ready for sale. Mums are perennial plants in most areas of the United States. If planted in an outdoor garden, they will bloom each fall for many years.

OBJECTIVE

To plant and maintain annual and perennial flowers in the landscape.

COMPETENCY TO BE DEVELOPED

After studying this unit, the student will be able to

- identify four uses of annual and perennial flowers.
- design two bed layouts using both annual and perennial flowers.
- explain the steps in preparing the soil for annual and perennial flowers.
- list the six steps in the aftercare of annual flowers.
- demonstrate proper transplanting techniques for annual and perennial flowers.

Materials List

sources of color pictures of annuals and perennials, such as nursery catalogs

seed flats and growing media in which to start plants

seed and nursery catalogs (one for each two students)

planting area in which to design a flower bed, using annuals and perennials

all necessary ingredients for preparation of flower beds (peat moss, fertilizer, etc.)

Fig. 25-1 A planting of petunias adds a solid mass of color to the landscape.

Annuals are plants that complete their life cycle in one year. The plant starts from seed, grows, blooms, sets seed, and dies in one season. (The meaning of the word *annual* is "one year.") *Perennials* persist from season to season without replanting. (The root system remains hardy, but the aboveground growth is usually killed by frost.) *Biennials*, another plant classification, are plants that require two years to complete their life cycle from seed to seed. This unit concentrates on the two more commonly occurring types of flowers, annuals and perennials.

Both vegetable plants and flower plants can be classified as annuals and perennials. While both types bear flowers, the flower plant is usually used for landscaping purposes.

FLOWERING ANNUALS

Flowering annuals are easy to grow and lend color to the landscape. They are usually started from seed indoors early in the spring and transplanted to the garden. The most popular annual flowering plants include marigolds, petunias, zinnias, ageratums, celosias, coleus, portulaca, pansies, and snapdragons. Figure 25-1 shows mass planting of petunias.

Flowering annuals are generally used to

- provide a mass of color around a house foundation, in flower beds, or in front of evergreens, figure 25-2.

- fill spaces between shrub plantings or other perennials and give color when these plants are not blooming.

Fig. 25-2 Dwarf marigolds used as a border planting in front of small evergreens. (Richard Kreh, photographer)

- provide color in bulb beds after the bulbs have bloomed.
- supply cut flowers.
- plant along fences or walks.
- cover bare spots between or in front of larger shrubs.
- act as a decoration in hanging baskets, figure 25-3.

Selecting Flowering Annuals

When deciding upon which annuals to grow, first consider the purpose of the plant and where it is to be planted. Be sure to consider height, keeping shorter plants in front of beds. Select plants with colors that will blend in well with one another. Figure 25-4 details other information concerning various annuals.

Great improvements have been made in the quality and vigor of annual flowering plants. One example is the Triploid hybrid marigold which blooms within six weeks from seed. It sets no seed and blooms longer than the traditional marigold, figure 25-5. If seed is allowed to set on annual plants, the seed production causes strength to be sapped from the plants and reduces blooming. For this reason, faded blooms should be cut from all annual plants.

Growing Flowering Annuals

For a summer of color with annuals:

1. Decide which annuals to grow. Outline the plan for the flower bed in writing. Probably the easiest plants for the beginner to grow are marigolds and zinnias.

2. Seed six to eight weeks prior to outside planting date (frost-free date).

3. Prepare the soil well ahead of the planting date. Add peat moss, fertilizer, lime, or sand as needed.

4. Plant according to the landscape plan after the frost-free date.

5. Water, weed, and mulch as needed.

6. Pinch off all faded blossoms.

7. Spray if necessary for insect control.

8. Enjoy cut flowers.

Planning and Designing the Flower Bed. Flower beds for annuals range in size and design according to individual needs and preferences.

Fig. 25-3 Transplanting marigolds into a hanging basket. After a hole is dug in the soil, the plant is set in the basket at its original soil level. (Richard Kreh, photographer)

Name	Height	Spacing	Color	Uses	How to Start	Remarks
Ageratum (ageratum)	6″-12″	6″-9″	blue, pink, white	edging	**seed and transplant	compact, excellent bloom Needs full sun.
Antirrhinum (snapdragon)	12″-36″	6″-8″	yellow, pink, white, blue, red, salmon	cut flowers, flower bed	seed and transplant	Must be staked for straight flower spikes. Needs full sun.
Begonia (begonia)	6″-12″	6″-10″	pink, red, white	edging, potted plant, window box	cuttings or seed	May be used as a house plant. Needs direct light.
Browallia (browallia)	12″-18″	8″-10″	blue, white	hanging basket, window box	cuttings or seed	Makes an attractive house-plant. Needs full sun.
Calendula (potted marigold)	12″-24″	10″-12″	yellow, orange	flower bed	seed and transplant	Flower petal used in cooking stews to add color; needs full sun.
Callistephus (China aster)	12″-24″	10″-12″	blue, purple, white, yellow	flower bed, cut flowers	seed and transplant	Gives excellent summer color; needs full sun.
Celosia (cockscomb)	12″-30″	10″-12″	red, orange, yellow, pink	flower bed, cut flowers	seed and transplant	Is excellent dried flower; needs full sun.
Cleome (spider plant or spiderflower)	24″-36″	18″-36″	pink, white	flower bed	*seed directly	Makes an attractive house-plant.
Coleus (coleus)	12″-24″	10″-12″	red, bronze, yellow, pink foliage	flower bed	seed and transplant	Needs full to partial sun. beautiful foliage plant
Coreopsis (coreopsis)	12″-24″	6″-10″	yellow, red	flower bed	seed directly	good for cut flowers
Cosmos (cosmos)	48″-72″	12″-18″	red, pink, yellow	flower bed	seed and transplant	Keep near back of beds.
Dahlia (dahlia)	12″-24″	12″-15″	red, pink, yellow, rose	flower bed	seed and transplant	profuse bloomer For maximum bloom, sow several weeks before other annuals.

Fig. 25-4 Bedding Plant Chart for Annual Flowers

Name	Height	Spacing	Color	Uses	How to Start	Remarks
Delphinium (larkspur)	18″-24″	8″-10″	blue, pink, white	flower bed, cut flowers	seed and transplant	Grows best in peat pots; difficult to transplant.
Dianthus (pink)	6″-18″	6″-8″	white, pink	edging, cut flowers	cuttings or seed	very fragrant flowers
Dimorphotheca (cape marigold)	12″-18″	8″-18″	yellow, orange, white	flower bed	seed and transplant	Grows well in dry areas.
Gaillardia (gaillardia)	12″-24″	8″-10″	red, orange	cut flowers, flower bed	cuttings	Loves seashore conditions; Does well in dry areas.
Gomphrena (globe amaranth)	9″-24″	8″-10″	blue, pink, white	cut flowers, flower bed	seed and transplant	Makes good dried flower. Collect and hang in dry, dark place.
Gypsophila (baby's breath)	12″-18″	10″-12″	pink, white	flower bed, cut flowers	seed directly	This is the annual form; there is also a perennial plant.
Helianthus (sunflower)	18″-60″	12″-36″	yellow, brown	flower bed	seed directly	Seeds are edible.
Helichrysum (strawflower)	24″-36″	8″-10″	yellow, red, white	flower bed	seed and transplant	Makes good dried flower. Cut; hang in dry dark place.
Iberis (candytuft)	10″-18″	6″-8″	white, pink, red	edging	seed and transplant	Makes attractive ground cover.
Impatiens (impatience)	6″-18″	12″-18″	pink, red, white, multicolor	edging	seed and transplant	Does best in shaded area.
Ipomoea (morning glory)	60″-120″	10″-12″	blue	vine — trellis	seed	vine
Lantana (lantana)	12″-36″	2″-15″	yellow, blue, red	hanging basket, flower bed potted plant	cuttings	Makes excellent topiary plant for patio.
Lathyrus (sweet pea)	36″-60″	6″-10″	pink, blue, white	cut flowers	seed	Grows best in cool conditions.

Fig. 25-4 **Bedding Plant Chart for Annual Flowers (continued)**

Name	Height	Spacing	Color	Uses	How to Start	Remarks
Limonium (statice or sea lavender)	18″-30″	10″-12″	blue, pink, white	flower bed, cut flowers	seed and transplant	Makes good dried flower.
Lobelia (lobelia)	3″-6″	4″-6″	blue, pink, white	edging, hanging basket, potted plant	seed	very attractive in hanging basket
Lobularia (sweet alyssum)	4″-6″	4″-6″	white, pink, blue	window box, edging	seed and transplant	Blooms all summer.
Matthiola (stock)	12″-36″	8″-10″	white, pink, yellow, red	cut flowers, flower bed	seed and transplant	fragrant flowers
Mirabilis (four-o'clock)	24″-36″	10″-12″	yellow, red, white	flower bed	seed and transplant	Withstands city conditions.
Myosotis (forget-me-not)	6″-12″	6″-8″	blue, pink	edging, cut flowers	seed directly	Makes attractive cut flowers.
Nicotiana (flowering tobacco)	18″-24″	10″-12″	rose, white	cut flowers, flower bed	seed and transplant	fragrant flowers after dark
Papaver (poppy)	18″-36″	10″-12″	red, yellow, white	cut flowers, flower bed	seed directly	Grows in masses.
Pelargonium (geranium)	12″-18″	10″-12″	red, white, pink, lavender	flower bed, window box, hanging basket	cuttings and seeds	excellent flowering plant
Petunia (petunia)	6″-18″	10″-12″	pink, red, white, blue	edging, flower beds, hanging basket	seed and transplant	most widely used annual
Phlox (phlox)	6″-18″	6″-8″	pink, blue, white, red	cut flowers, flower bed, window box	seed	very intense colors

Fig. 25-4 **Bedding Plant Chart for Annual Flowers (continued)**

Name	Height	Spacing	Color	Uses	How to Start	Remarks
Portulaca (rose moss)	4″-6″	3″-6″	red, orange, yellow	edging	seed	Reseeds by itself.
Salvia (scarlet sage)	10″-36″	6″-12″	blue, red, white	cut flowers, flower bed	seed and transplant	very showy color
Salpiglossis (giant velvet flower)	18″-36″	10″-12″	gold, scarlet, rose, blue	flower bed, cut flowers	seed and transplant	very small seed
Scabiosa (pincushion flower)	30″-36″	12″-14″	multicolor	cut flowers	seed and transplant	variety of plants from which to select
Tagetes (marigold)	6″-48″	6″-12″	orange, yellow	flower bed, edging, potted plant	seed and transplant or seed directly	wide range of varieties no insect problems
Thunbergia (black-eyed Susan vine)	24″-60″	10″-12″	orange	climbing vine	seed and transplant	
Tropaeolum (nasturtium)	12″-18″	10″-12″	orange, yellow	flower bed, window box	seed and transplant or seed directly	Flowers and leaves are edible.
Verbena (verbena)	6″-12″	10″-12″	purple, red, white	flower bed, edging	seed and transplant	good rock garden plant
viola (pansy)	6″-8″	6″-8″	multicolor	edging, flower bed, cut flowers	seed and transplant	Gives excellent color in summer.
Zinnia (zinnia)	8″-36″	6″-12″	multicolor	edging, flower bed, cut flowers	seed and transplant or seed directly	Gives excellent color in summer.

*seed directly — sow the seed directly into the soil where the plants are desired.

**seed and transplant — start the seed indoors (in greenhouse, hotbed, or portable seed germination case). As the seed develops, they are transplanted from the seed flat to other containers where they grow until they are ready to be set out.

Fig. 25-4 Bedding Plant Chart for Annual Flowers (continued)

Example. One way to beautify a landscape is to plant a flower bed between a walkway and a home. This planting could be designed according to the following plan.

←————— 20 feet —————→

10	9	8
5	6	7	5
3	4	3
2
1

(Numbers refer to the plants listed below.)

1. white sweet alyssum
2. blue ageratum
3. white begonias
4. red petunia
5. deep red cockscomb
6. white geraniums
7. red geraniums
8. blue snapdragons
9. red snapdragons
10. white snapdragons

Preparing the soil. It is best to prepare the soil in the fall before planting the next spring. The soil should be spaded or tilled to loosen it. The addition of peat moss or sand may be necessary. Peat moss adds organic matter to the soil and aids in moisture retention. Sand gives better drainage and aeration to the soil. If peat moss or sand is needed, it should be added in the fall prior to planting.

When planting in beds that contain bulbs, do not dig deeply enough to damage the bulbs. Be sure that the bed does not dry out completely when dug. This drying could also damage the bulbs.

Fig. 25-5 The Triploid hybrid marigold. The marigold is one of the easiest annuals to grow. (Richard Kreh, photographer)

Before preparing new beds, test the soil for drainage. To do this, dig a 10-inch hole and fill it with water. Fill it with water again the following day. If all the water has not seeped into the soil within ten hours on the second day, drainage is a problem. To improve drainage, raise the bed level above the normal soil level and dig a shallow trench around it.

Fertilizing. To fertilize the bed, add about two pounds of a 5-10-10 fertilizer per 100 square feet in the spring as the bed is being prepared. If the soil pH is below 6.0, add ground limestone at the rate of five pounds per 100 square feet and rake it into the soil surface. If the plants do not have a healthy, dark green color, sprinkle a 10-10-10 fertilizer lightly around each plant on the soil surface. Do not place fertilizer on or against the plant.

Many people are now growing beautiful annuals with soluble fertilizers. To mix a soluble fertilizer, mix a complete fertilizer

Fig. 25-6 To start seed indoors, fill the flat with properly prepared seeding media. Small, square peat pots have been set in this flat, with one or two seeds placed in each pot. The peat pots are separated and transplanted individually to the flower bed after the danger of frost has passed. Very small seeds are planted in flats without peat pots and are transplanted to the pots at a later date. (USDA photo)

(20-20-20) with water at the rate of one tablespoon per gallon of water or according to directions. Fertilizer applied to plants in a solution is more readily available to the plants.

Seeding. As mentioned before, most annuals are seeded indoors and set outside as plants. Figures 25-6, 25-7, and 25-8 detail this process. Some annuals may be seeded directly outdoors. Some examples of these are baby's breath, cornflower, globe amaranth, poppy, cleome, strawflower, sweet alyssum, and sweet pea. Seeds of these plants should be sown as early in spring as possible.

Fig. 25-7 Annual seed should be sown carefully to obtain only one to two plants per peat pot. (USDA photo)

The plants are set outside in the flower bed as soon as the danger of frost has passed. Wait until the soil warms to about 60°F before setting the plants outdoors, since germination of seed is very slow until this time. Seedlings in peat pots are set directly into the soil. Very small seeds which are started in flats will probably require transplanting to a larger container before setting outside. These may be replanted to peat pots so that they can establish better growth prior to being planted outside, figure 25-9.

Plant seedlings at about the same depth as the depth at which they grew in the starting container. Firm the soil around the roots and water well. It is helpful to add a starter solution of one

Fig. 25-8 There are two methods for watering: (1) Sprinkle the flat as shown or (2) Set the entire flat in water and allow it to soak up from the bottom. Both methods allow even distribution of water on the seeds. Very small seeds such as petunia must not be watered from the top with any force, or they will be washed down into the planting media too deeply. (USDA photo)

tablespoon of a water soluble, high phosphate fertilizer in one gallon of water after transplanting. Space the plants according to the Bedding Plant Chart (figure 25-4).

Caring for the Plants

Care for the rest of the season is relatively simple.

1. Water if the soil becomes too dry.

2. Pull weeds or mulch for weed control.

3. Remove all faded blossoms.

4. If necessary, spray with malathion or Sevin, depending upon the insect to be controlled. If a chemical is used, read the label carefully before use.

Fig. 25-9 To transplant the small seeded annual to a larger pot, first grasp the seedling by its true leaves. Gently dislodge it from its original pot and place it in the new pot. Fill the pot with soil mixture to the soil level of the original pot. Firm the soil by pressing down lightly with the thumb and index finger. (Richard Kreh, photographer)

5. Fertilize as necessary to keep plants growing actively.

6. Cut flowers as desired for use in the home.

7. Prune or pinch when necessary for better growth. When the plants are herbaceous (soft stemmed), the stem may be shortened by pinching with the fingers. The pinching will result in a thicker, more attractive plant.

TRANSPLANTING SMALL PLANTS

CARE OF PLANTS BEFORE PLANTING

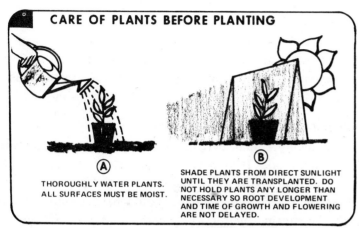

Ⓐ
THOROUGHLY WATER PLANTS. ALL SURFACES MUST BE MOIST.

Ⓑ
SHADE PLANTS FROM DIRECT SUNLIGHT UNTIL THEY ARE TRANSPLANTED. DO NOT HOLD PLANTS ANY LONGER THAN NECESSARY SO ROOT DEVELOPMENT AND TIME OF GROWTH AND FLOWERING ARE NOT DELAYED.

INSERTING PLANTS IN SOIL

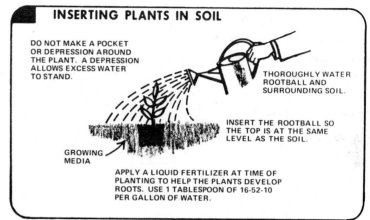

DO NOT MAKE A POCKET OR DEPRESSION AROUND THE PLANT. A DEPRESSION ALLOWS EXCESS WATER TO STAND.

THOROUGHLY WATER ROOTBALL AND SURROUNDING SOIL.

INSERT THE ROOTBALL SO THE TOP IS AT THE SAME LEVEL AS THE SOIL.

GROWING MEDIA

APPLY A LIQUID FERTILIZER AT TIME OF PLANTING TO HELP THE PLANTS DEVELOP ROOTS. USE 1 TABLESPOON OF 16-52-10 PER GALLON OF WATER.

PREPARATION OF PLANTS FOR PLANTING

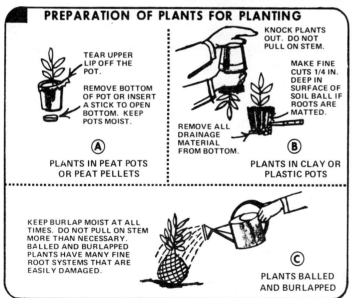

TEAR UPPER LIP OFF THE POT.

REMOVE BOTTOM OF POT OR INSERT A STICK TO OPEN BOTTOM. KEEP POTS MOIST.

KNOCK PLANTS OUT. DO NOT PULL ON STEM.

MAKE FINE CUTS 1/4 IN. DEEP IN SURFACE OF SOIL BALL IF ROOTS ARE MATTED.

REMOVE ALL DRAINAGE MATERIAL FROM BOTTOM.

Ⓐ
PLANTS IN PEAT POTS OR PEAT PELLETS

Ⓑ
PLANTS IN CLAY OR PLASTIC POTS

KEEP BURLAP MOIST AT ALL TIMES. DO NOT PULL ON STEM MORE THAN NECESSARY. BALLED AND BURLAPPED PLANTS HAVE MANY FINE ROOT SYSTEMS THAT ARE EASILY DAMAGED.

Ⓒ
PLANTS BALLED AND BURLAPPED

CARE OF PLANTS AFTER PLANTING

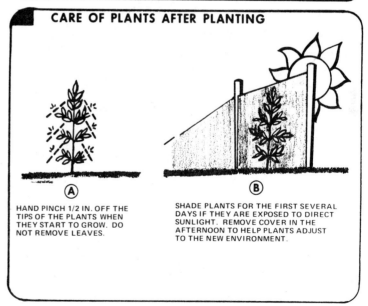

Ⓐ
HAND PINCH 1/2 IN. OFF THE TIPS OF THE PLANTS WHEN THEY START TO GROW. DO NOT REMOVE LEAVES.

Ⓑ
SHADE PLANTS FOR THE FIRST SEVERAL DAYS IF THEY ARE EXPOSED TO DIRECT SUNLIGHT. REMOVE COVER IN THE AFTERNOON TO HELP PLANTS ADJUST TO THE NEW ENVIRONMENT.

Fig. 25-10 Care of plants before and after planting. (USDA photo)

PERENNIALS

Perennials are plants that live from year to year and that therefore do not require replanting. The tops may or may not die back in the winter or dry season. Some perennials bloom their first year, but most produce larger, more attractive flowers and develop stronger root systems in successive seasons. Perennials may be small flowering plants, shrubs, or flowering trees.

Uses of Flowering Perennials

Some ways in which flowering perennials are used to accent landscapes include:
- mass plantings in flower beds.
- as edges around larger shrubbery or evergreens.
- as accents in evergreen plantings.
- to form screens that also add color, thereby creating a functional item that also beautifies.

Selection

Flowering perennials should be selected according to personal preference and local growing conditions. Examine plants grown in the area or ask for advice from a nursery, extension service, or horticulture instructor. The color pictures in seed and nursery catalogs are another source that can be helpful in plant selection. Check figure 25-11 to determine if the plants will grow under the light and soil conditions in the area in which they are to be planted, and to be sure they will flower during the season when flowers are needed to add color to the landscape (mid to late summer). Plant flowers that bloom at the same time together for best display of color.

Soil Preparation and Transplanting

Remember that perennials live in the same soil for more than one year. It is necessary, therefore, to prepare the soil well. Organic matter should be added and drainage improved if necessary. Adequate fertilizer should be dug into the soil along with the organic matter to a depth of at least 1 foot. Careful preparation prior to planting results in a much more attractive floral display.

It is usually best to start perennial seeds indoors six to eight weeks prior to the transplanting date. Plants that do not transplant well must be started in peat pots or direct seeded in the permanent location. The procedures for starting seeds and transplanting perennials are the same as those used for annuals.

STUDENT ACTIVITIES

1. With the assistance of the instructor or a local landscaper, outline on paper a layout for a flower bed using annuals.
2. Start flowering annuals or vegetable annuals from seed according to directions in the text.
3. Plant the flowering annuals started on the school grounds or at home. Plant the annuals according to a plan approved by the instructor.
4. Draw a layout for a perennial flower bed to be planted on the school grounds or at home. If possible, plant the flower bed. Observe and record changes in the perennials as they enter the winter season.

Plant	When to Plant Seed	Exposure	Germination time (days)	Spacing	Height	Best Use	Color	Remarks
Achillea mille folium (yarrow)	early spring or late fall	sun	7-14	36"	24"	borders, cut flowers	yellow, white	Seed is small. Water with a mist. easy to grow.
Alyssum saxatile (golddust)	early spring	sun	21-28	24"	9"-12"	rock garden edging cut flowers	yellow	Blooms early spring. good in dry and sandy soils
Anchusa italica (Alkanet)	spring to September	partial shade	21-28	24"	48"-60"	borders, background, cut flowers	blue	Blooms June or July. Refrigerate seed 72 hrs. before sowing.
Anemone pulsatilla (windflower)	early spring or late fall for tuberous	sun	4	35"-42"	12"	borders, rock garden, potted plant, cut flowers	blue, rose, scarlet,	Blooms May and June. Is not hardy north of Wash., D.C.
Anthemis tinctoria (golden daisy)	late spring outdoors	sun	21-28	24"	24"	borders, cut flowers	yellow	Blooms misdummer to frost. Prefers dry or sandy soil.
Arabis alpina (rock cress)	spring to September	light shade	5	12"	8"-12"	edging, rock garden	white	Blooms early spring.
Armeria alpina (sea pink)	spring to September	sun	10	12"	18"-24"	rock garden, edging, borders, cut flowers	pink	Blooms May and June. Plant in dry sandy soil. Shade until plants are well established.
Aster alpinus (hardy aster)	early spring	sun	14-21	36"	12"-60"	rock garden, borders, cut flowers	white,	Blooms June.
Astilbe japonica (astilbe)	early spring	sun	14-21	24"	12"-36"	borders	pink, red, white	Blooms July and August. Gives masses of color.

Fig. 25-11 Various Flowering Perennials (from USDA bulletin 114)

Plant	When to Plant Seed	Exposure	Germination time (days)	Spacing	Height	Best Use	Color	Remarks
Begonia evansiana (hardy begonia)	summer in shady, moist spot	shade	12	9″-12″	12″	flower bed	yellow, pink, white	Blooms late in summer. Can be propagated from bulblets in leaf axils.
Candytuft (*Iberis*)	early spring or late fall	sun	20	12″	10″	rock garden, edging, ground cover	white	Blooms in late spring. Prefers dry places. Cut faded flowers to promote branching.
Canterbury bells (*Campanula medium*)	spring to September (Do not cover.)	partial shade	20	15″	24″-30″	borders, cut flowers	white, pink, blue	Divide mature plants every other year; best as a biennial.
Carnation (*Dianthus caryophyllus*)	late spring	sun	20	12″	18″-24″	flower bed, borders, edging, rock gardens	pink, red, white, yellow	Blooms in late summer. Cut plants back in late fall and hold in cold frame.
Cerastium tomentosum (snow-in-summer)	early spring	sun	14-28	18″	6″	rock garden, ground cover	white	Blooms in May and June. Forms a creeping mat and is a fast grower. Prefers a dry spot.
Chinese lantern (*Physalis alkekengi*)	late fall	sùn	15	36″	24″	borders, specimen plant	orange	Lanterns are borne the second year in the fall.
Columbine (*Aquilegia*)	spring to September	sun or partial shade	30	12″-18″	30″-36″	borders, cut flowers	wide color range	Bloom in late spring. Best grown as a biennial.
Coreopsis grandiflora	early spring	sun	5	30″	24″-30″	borders	yellow	Blooms from May to fall if old flowers are removed. Grown as a biennial.

Fig. 25-11 Various Flowering Perennials (from USDA bulletin 114) (continued)

Plant	When to Plant Seed	Exposure	Germination time (days)	Spacing	Height	Best Use	Color	Remarks
Daisy, Shasta (*Chrysanthemum maximum*)	early spring to September	sun	10	30″	24″-30″	borders, cut flowers	white	Blooms June and July. Best grown as a biennial in well-drained location.
Delphinium elatum (delphinium)	spring to September	sun	20	24″	48″-60″	borders, background, cut flowers	blue, lavender, white, pink	Blooms in June. Best grown as a biennial. Needs dry location.
Dianthus deltoides (pink)	spring to September	sun	5	12″	12″	borders, rock garden, edging, cut flowers	pink	Blooms in May and June Best grown as a biennial. Needs dry location.
Foxglove (*Digitalis purpurea*)	spring to September	sun or partial shaded	20	12″	48″-60″	borders, cut flowers	pink, white, purple, rose	Blooms in June and July. Grown as a biennial. Shade summer plantings.
Gaillardia grandiflora (gaillardia)	early spring or late summer	sun	20	24″	12″-30″	borders, cut flowers	scarlet, yellow	Blooms from July until frost.
Gypsophila paniculata (baby's breath)	early spring to September	sun	10	48″	24″-36″	borders, cut flowers, drying	white, pink	Blooms early summer until early fall. Needs Needs lots of lime.
Hemerocallis (daylily)	late fall	sun or partial shade	15	24″-30″	12″-48″	borders, among shrubbery	pink, red,	Plant several varieties for longer blooming season.
Hibiscus moscheutos (Mallo Marvel)	spring or summer	sun or partial shade	15 or longer	24″	36″-96″	background, flower bed	white, pink, red, rose	Blooms July to September.

Fig. 25-11 Various Flowering Perennials (from USDA bulletin 114) (continued)

Plant	When to Plant Seed	Exposure	Germination time (days)	Spacing	Height	Best Use	Color	Remarks
Iris	bulbs or rhizomes fall	sun or partial shade	next spring	18″-24″	3″-30″	borders, cut flowers	blue, red, yellow, pink, bronze, wine	Blooms spring and summer if different varieties are used.
Liatris pycnostachya (gayfeather)	early spring or late fall	sun	20	18″	24″-60″	borders, cut flowers	rose-purple	Blooms summer to early fall. Easily started from seed.
Lupinus polyphyllus (lupine)	early spring or late fall (Soak before planting.)	sun	20	36″	36″	borders, cut flowers	white, yellow, pink, rose, red, blue, purple	Blooms most of summer. Needs excellent drainage. Does not transplant easily.
Peony (*Paeonia*)	Plant tubers in late fall 2″-3″ deep.	sun	variable	36″	24″-48″	borders, cut flowers, flower bed	pink, red, white, rose	Blooms late spring. Difficult to grow from seed.
Phlox paniculata (summer phlox)	late fall or early winter	sun	25 irregular	24″	36″	borders, cut flowers	red, pink, blue, white	Blooms early summer. Color of flower varies.
Phlox sublata (moss phlox)	grown from stolons	sun		8″	4″-5″	borders	blue, red, white, pink	Blooms in spring. drought resistant
Poppy, Iceland (*Papaver nudicaule*)	early spring	sun	10	24″	15″-18″	borders, cut flowers	white, pink, red	Blooms early summer. Does not transplant easily.

Fig. 25-11 Various Flowering Perennials (from USDA bulletin 114) (continued)

Plant	When to Plant Seed	Exposure	Germination time (days)	Spacing	Height	Best Use	Color	Remarks
Poppy, Oriental (*Papaver orientale*)	early spring	sun	10	24″	36″	borders, cut flowers	pink, red, rose, orange, white, salmon	Blooms early summer. Does not transplant easily.
Primrose (*Primula polyatha and P. veris*)	January, in a flat on surface. Allow to freeze; then bring in to germinate.	partial shade	25 irregular	12″	6″-9″	rock garden	white, yellow, pink, red, blue	Blooms April and May. May be seeded in fall.
Pyrethrum roseum (painted daisy)	spring to September	sun	20	18″	24″	borders, cut flowers	various, including gold, pink, and lavender	Blooms May and June. Prefers well-drained soil.
Rudbeckia (*Echinacea purpurea*) (cone flower)	spring to September	sun	20	30″	30″-36″	borders, flower bed, cut flowers	white, pink, red, rose	Blooms midsummer to fall. Shade summer plantings.
Salvia azurea grandiflora and S. farinacea)	spring	sun	15	18″-24″	36″-48″	borders	red	Blooms August until frost.
Sea lavender (*Limonium latifolia*)	early spring	sun	15	30″	24″-36″	flower bed, cut flowers, drying	pink, yellow, mauve	Blooms in July and August.
Stokesia cyanea (Stokes' aster)	early spring to Sept-	sun	20	18″	15″	borders, cut flowers	white, blue	Blooms in September if started early.

Fig. 25-11 Various Flowering Perennials (from USDA bulletin 114) (continued)

Plant	When to Plant Seed	Exposure	Germination time (days)	Spacing	Height	Best Use	Color	Remarks
Sweet pea (*Lathyrus latifolius*)	early spring	sun	20	24″	60″-72″	background	pink, white, purple, red	Blooms June to September. Easily grown as a vine on fence or trellis.
Sweet William (*Dianthus barbartus*)	spring to September	sun	5	12″	12″-18″ (Dwarf form also available.)	borders, edging, cut flowers	red, pink, white	Blooms May and June. Very hardy. Needs well-drained soil.
Veronica spicata (speedwell)	spring to September	sun	15	18″	18″	borders, rock garden, cut flowers	purple	Blooms June and July. Easy to grow.

Fig. 25-11 Various Flowering Perennials (from USDA bulletin 114)

SELF-EVALUATION

A. 1. Which of the following are not annual plants?

   a. asparagus and strawberry   c. marigold and ageratum
   b. petunia and geranium   d. morning glory and scarlet sage

2. Seed for most annuals should be

   a. started inside and transplanted outside later.
   b. directly seeded in the flower bed.
   c. soaked in water before planting.
   d. planted only in peat pots.

3. Which two annuals listed below are best used for edging?

   a. poppy and pink   c. stock and strawflower
   b. pink and sweet alyssum   d. sunflower and pink

4. Which two annuals listed below are best used for bedding?

    a. larkspur and morning glory   c. balsam and impatiens

    b. marigold and petunia   d. summer-cypress and sweet pea

5. Faded blossoms should be removed from annuals because

    a. it prevents the plants from setting seed and using strength for seed formation.

    b. it improves the appearance of the plants.

    c. new flowers develop faster when old ones are removed.

    d. all of the above

6. Annuals which are easiest to grow are

    a. marigold and zinnia   c. pansy and coleus

    b. ageratum and sweet pea   d. primrose and petunia

7. Most annuals should be transplanted outdoors

    a. as soon as the soil can be tilled in the spring.

    b. in early June.

    c. as soon as danger of frost has passed.

    d. after spring shrubs have bloomed.

8. When transplanting annual plants from the flat to the bed

    a. dip the roots in water to moisten them.

    b. shake any excess soil from the roots before planting.

    c. allow the roots to dry out before transplanting.

    d. keep as much soil on the roots as possible.

9. When planting annuals, organic matter is added to soil to

    a. give the soil better drainage.   c. lighten the soil.

    b. hold moisture in the soil.   d. mulch the plants.

10. When planting annuals, sand is added to the soil to

    a. give the soil better drainage.   c. make the soil heavier.

    b. hold moisture in the soil.   d. help warm the soil in the spring.

11. When preparing a flower bed for annuals, the best time to add sand, peat moss, and fertilizer is

    a. immediately before planting.
    b. two years before planting.
    c. in the fall before planting the following spring.
    d. in early spring for planting in May.

12. When testing soil for drainage, the water added to the hole on the second filling should drain in no more than

    a. two hours.   c. twenty hours.
    b. five hours.   d. ten hours.

13. If the soil is too acid, it may be necessary to add ground limestone to

    a. lower the soil pH.   c. prevent the pH from changing.
    b. raise the soil pH.   d. none of these

14. A perennial is a plant that lives

    a. from year to year without replanting.   c. for only two years.
    b. for only one year.   d. all of these

15. Soil that is to be planted with perennials should be well prepared because

    a. perennials are particular about soil conditions.
    b. perennials grow in one location for many years.
    c. all perennials require rich soil.
    d. none of the above

16. Important considerations when selecting perennials for planting are

    a. height and color.   c. soil fertility and drainage.
    b. time of bloom and color of flower. d. all of these

17. The best way to select flowering perennials for a particular area is

a. to observe plants that are growing there.
b. to judge according to what looks nicest in the seed catalog.
c. to purchase plants with colors that blend well together.
d. all of the above

18. Which of the following is *not* a part of soil preparation for flowering perennials?

a. adding organic matter
b. providing proper drainage
c. adding mulch
d. adding fertilizer and digging it into the soil

19. When starting flowering perennials from seed, how many weeks before the transplanting date should they be started?

a. two to three months
b. six to eight weeks
c. one to two weeks
d. three to four weeks

20. Plants are often seeded in peat pots. This is especially recommended for plants

a. that are difficult to transplant.
b. that are to be sold.
c. that require good drainage.
d. that have tap root systems.

B. 1. List the seven steps in maintaining a healthy annual planting.

2. List four uses of annuals in the landscape.

3. List five annuals that make attractive hanging baskets.

OBJECTIVE

To position and plant narrowleaf evergreens in the landscape.

COMPETENCY TO BE DEVELOPED

After studying this unit, the student will be able to

- differentiate between the two types of narrowleaf evergreen leaves.
- list four reasons for using narrowleaf evergreens in the landscape.
- describe the primary purpose of a lath house.
- describe the proper fertilizer practices for narrowleaf evergreens.
- demonstrate the proper procedure for planting narrowleaf evergreens.

Materials List

two different narrowleaf evergreen trees

several deciduous narrowleaf trees

fertilizer

shovel

watering can

---

unit 26
Narrowleaf Evergreens

Narrowleaf evergreens have long been popular landscape plants. This may be because they generally remain green year-round, thereby adding color to the landscape when most other plants are dormant. Another advantage is that they are adaptable to various soil types and weather conditions.

The leaves of narrowleaf evergreens may be needlelike or scalelike. The evergreens with needlelike leaves have leaves attached to the branch in bundles (known as clusters or *fascicles*). For example, the white pine, figure 26-1, has five needles per bundle or cluster. These evergreens carry their needles throughout the year, with old needles dropped in the fall and new needles produced in the spring.

Narrowleaf evergreens with scalelike leaves have leaves arranged so that each scale overlaps another to form a flat spray. An example of a narrowleaf evergreen with scalelike leaves is the juniper, figure 26-2.

Fig. 26-1 White pine - *Pinus strobus.*

Fig. 26-2 Pfitzer juniper — *Juniperus chinensis pfitzeriana.*

USES

The narrowleaf evergreen has long been a standard choice for foundation plantings because of its year-round green foliage. Figures 26-3 through 26-5 show several narrowleaf evergreens used as part of foundation plantings. Narrowleaf evergreens are also used extensively as boundary plants to accent borderlines of property, figure 26-6. The narrowleaf evergreen may also be used to screen a view of a less attractive area or to help control air currents. For example, in the winter months, a windbreak of narrowleaf evergreens can cause the temperature on the protected side of the windbreak to be 5 to 10 degrees warmer than on the unprotected side.

CARE

Narrowleaf evergreens are grown in nurseries. They are shipped three different ways: bare root (BR), balled and burlapped (BB), or as container stock (C).

Bare root narrowleaf evergreens are shipped while the plants are seedlings. When the seedlings arrive at their destination, the roots of the plants should be soaked in a bucket of water. The

Fig. 26-3 Mugho pine — *Pinus mugo mughus*.

Fig. 26-4 Dwarf Norway spruce (Bird's nest spruce) — *Picea excelsia*.

Fig. 26-6 Blue Colorado spruce — *Picea pungens 'Glauca'* ▶

Fig. 26-5 Hemlock — *Tsuga canadensis.*

roots must be prevented from drying out until they are planted in the ground.

Balled and burlapped narrowleaf evergreens may be moved at almost any time of the year except when the plant has shoots of new growth or when the ground is frozen. The new shoots must be allowed to harden before the plant is moved. If these plants are to be out of the ground any length of time, it is best to mulch them heavily with well-rotted sawdust, pine bark, peat moss, or hardwood bark to retain moisture around the roots.

For the production of healthy plants, narrowleaf evergreens should be watered regularly and stored in a lath house. A *lath house* gives protection from the sun by reducing the light available to the plant. This reduces the transpiration rate (water loss) of the plant.

Planting Balled and Burlapped Plants

The prepared hole should be one and one-half to two times as large as the ball of soil around the tree being planted. This allows room to set the tree in the hole and mix peat moss with the soil surrounding the hole. The ball of the tree is set 1 inch above the original soil line, since the tree will settle somewhat in the hole. Remove the burlap from the sides or roll it down to the bottom. It is not necessary to pull the burlap out from under the ball. Add prepared soil in and around the root ball until it is covered. When the hole is two-thirds filled with soil, fill it to the top with water and let it soak in. Then finish filling the hole with soil. With the remaining soil, form a saucerlike shape (known as a *berm*) around the tree to hold water that will be added later.

Planting Containerized Trees

Containerized trees are planted in the same way as balled and burlapped trees, except for the following differences.

(1) The container is removed from the tree roots. Care must be taken not to break the root ball apart.

(2) After the container is removed, roots which may be growing around the outside of the root ball in a circular fashion are straightened out before planting. If the roots cannot be straightened, they are cut off at the edge of the root ball. This stops circular root growth which tends to girdle root systems and restrict growth. In severe cases, trees can be killed if this growth pattern is allowed to continue.

Staking

Because evergreens sway in the wind, they must be staked. To stake them, wrap a piece of garden hose (about 1 foot long) around the trunk about one-half the height of the tree. Now slip a piece of number 9 wire through the garden hose, twisting it to hold the garden hose firm. Fasten the other end of the wire to the stake which is driven into the ground beside the root ball.

Fertilizing

The general recommendation for fertilizing a narrowleaf evergreen is three to six pounds of 5-10-10 fertilizer per 100 square feet placed in a circle under the drip line at the outer end of the branches. Fertilize before new growth starts in the spring. After applying the fertilizer, water it in with a garden hose sprinkler. Organic fertilizers such as cottonseed meal or soybean oil meal may also be used. Although there is an advantage to the use of organic fertilizers in that there is no danger of burning the plant, they are usually more expensive.

Watering

Some junipers will tolerate dry soil, while other narrowleaf evergreens may need to be watered every ten to fourteen days during the first year after planting. Thoroughly soak the soil to 6 inches or deeper when watering.

Pruning

There are two methods of pruning narrowleaf evergreens.

(1) **Pruning** individual branches gives the plant a natural, informal appearance. (Large branches may be removed from specific locations on the plant.) Spreading junipers and taxus should be pruned by removing the ends of long twigs. This gives an informal appearance to these plants.

(2) **Shearing** is done as when shaping a hedge. The ends of all small branches are clipped to shape the edges of the plant in straight lines, giving a more formal appearance. Hicks yew, columnar juniper, and globe arborvitae may be sheared in this manner.

NARROWLEAF EVERGREEN CHART

The following is a chart listing narrowleaf evergreens used in landscaping. The plants are listed in sequence according to their height. The general breakdown includes one group of those plants 3 feet or less in height; a second group of those plants ranging from 3 to 9 feet in height; and a third group of those plants greater than 10 feet in height.

The following information is given for each plant.

The hardiness of each plant is given. This number refers to one of the areas shown on the hardiness zone map on page 393. It indicates where the plant can be grown in the United States.

Foliage color is the color of the narrowleaf evergreen. Some types vary from a very light green to dark green.

Fig. 26-7 Black Japanese pine — *Pinus thunbergii.*

Period of interest is that time of the year during which the plant is most attractive. This may be when it flowers, bears fruit, or changes foliage color. All of these create interest during a particular season of the year.

Landscape use indicates how a particular plant is used in the landscape. The term *ground cover* refers to plants, other than grass, that are used to cover the ground. Ground covers also have special uses, such as preventing erosion by holding steep banks in place.

Foundation planting refers to plants which are used around buildings to help accent and tie the building into the landscape.

Rock garden refers to plants used in gardens which are planted in a rocky environment, whether natural or artificially reproduced.

Specimen plants refers to plants that are used alone for their own beauty or as an accent.

Screen plants and *hedges* refer to those plants used to confine certain areas, such as to reduce noise or an unappealing sight.

Other remarks are given to help the individual make a more educated selection of plants.

Narrowleaf Evergreens (3 feet or less in height)

Name	Hardiness	Foliage Color	Period of Interest	Landscape Use	Other Remarks
Erica carnea (Spring heath)	5	bright green	small, rosy red spikes in April	ground cover	attractive spring colors
Juniperus chinensis sargentii (Sargent juniper)	4	steel blue	blue berries in fall	ground cover	Excellent for planting along seashore. Does well on steep banks to prevent erosion.
Juniperus horizontalis Wilton (Blue rug)	2	steel blue	blue berries in fall	ground cover	excellent rock garden plant
Taxus baccata 'repandens' (Spreading English yew)	4	dark green	fall	foundation planting	excellent plant; low maintenance, good color all year
Taxus cuspidata aurencens (Dwarf Japanese yew)	4	light green	fall	foundation planting	compact plant which produces red berries

_____ **Narrowleaf Evergreens (3 to 9 feet in height)** _____

Name	Hardiness	Foliage Color	Period of Interest	Landscape Use	Other Remarks
Chamaecyparis obtusa 'nana'	3	deep dark green	all year	accent plant, specimen plant	compact form; pyramidal shape
Juniperus squamata Meyer (Meyer's juniper)	4	blue	all year	foundation planting	Needs good management and care in dry areas.
Juniperus virginiana tripartia	2	dark green	all year	foundation planting	good in dry areas
Pinus mugo mughus 'compacta' (Mugho pine)	2	dark green	all year	foundation planting	global shape; slow growing
Picea abies varietus (a) Dwarf Alberta spruce	2	light green	all year	foundation planting, specimen plant	fine-textured plant
(b) Bird's nest spruce	2	light green	all year	rock garden	Requires well-drained soil.
Juniperus chinensis pfitzeriana (Pfitzer juniper)	4	blue green	all year	foundation planting, screen plant	Plant in full sun. Requires well-drained soil. Control of bag worms necessary.
Juniperus chinensis Hetz	4	blue green	all year	screen plant, foundation planting	Plant in full sun; does well in dry areas.
Taxus media hicksii (Hicks yew)	4	dark green	all year; fall fruit	foundation planting, screen plant	Plant in full sun. excellent in formal gardens; columnar shape

_____ Narrowleaf Evergreens (over 10 feet in height) _____

Name	Hardiness	Foliage Color	Period of Interest	Landscape Use	Other Remarks
Cypressocyparis leylandi (Leyland cypress)	4	light green	all year	screen or hedge plant	fast grower, excellent windbreak; columnar shape
Taxus baccata (English yew)	6	dark green	all year	screen or hedge plant, foundation planting on large buildings	Female plant has berries. Will stand shade. good background shrub
Taxus cuspidata	4	dark green	all year	screen or hedge plant, foundation planting on large buildings	pyramidal form Tolerates shade. rapid grower Produces red fruit on female.

LEAF IDENTIFICATION CHART

The following chart, adapted from Extension Bulletin 238, University of Maryland, is presented to help the student in the identification of narrowleaf evergreens. Only by learning to identify specific plants can plants be properly matched to landscape requirements.

_____ KEY _____

Groups	Figure Numbers
I. Narrowleaf evergreens with single needles (This includes spruce, fir, and hemlock.)	1 to 7
II. Narrowleaf evergreens with two needles per cluster (This includes pines with two per cluster.)	8 to 11
III. Narrowleaf evergreens with three needles per cluster (This includes pines with three per cluster.)	12 to 13
IV. Narrowleaf evergreens with five needles per cluster (This includes pines with five per cluster.)	14 to 15
V. Narrowleaf evergreens with more than five needles per cluster (This includes deodar cedar.)	16
VI. Narrowleaf evergreens that are deciduous (This includes those evergreens that drop their needles in winter.)	17 to 19
VII. Narrowleaf evergreens with scalelike foliage (This includes arborvitae and red cedar.)	20 to 26

Group I. Narrowleaf evergreens with single needles

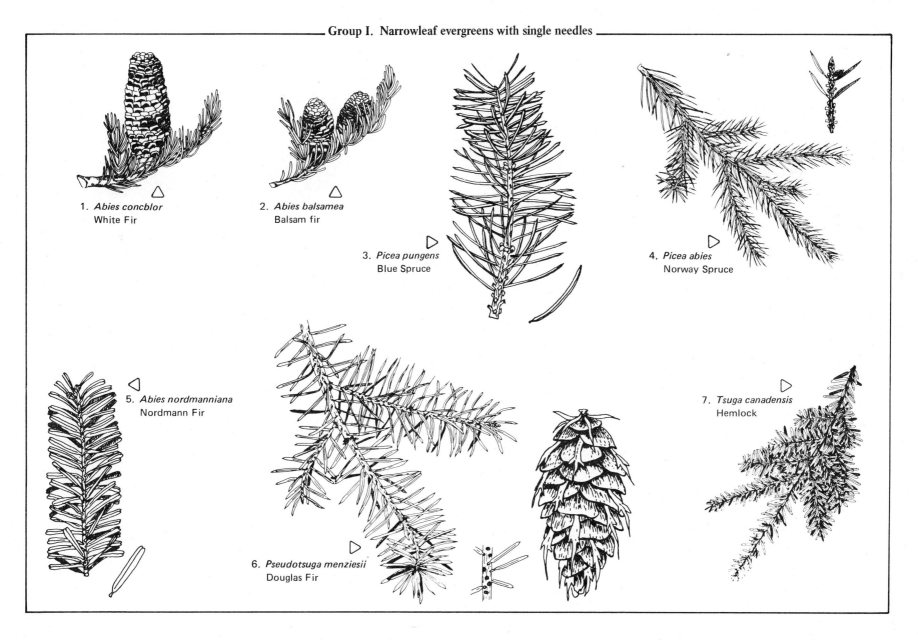

1. *Abies concblor*
White Fir

2. *Abies balsamea*
Balsam fir

3. *Picea pungens*
Blue Spruce

4. *Picea abies*
Norway Spruce

5. *Abies nordmanniana*
Nordmann Fir

6. *Pseudotsuga menziesii*
Douglas Fir

7. *Tsuga canadensis*
Hemlock

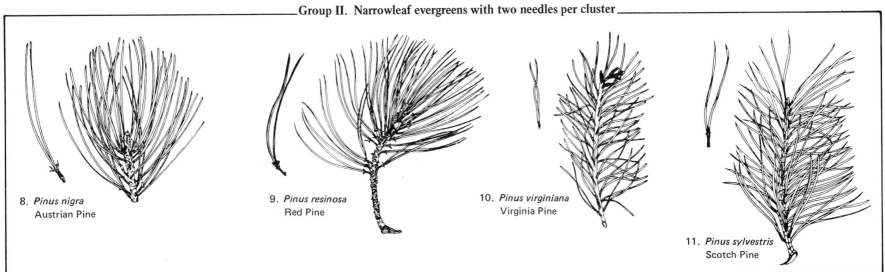

Group II. Narrowleaf evergreens with two needles per cluster

8. *Pinus nigra*
Austrian Pine

9. *Pinus resinosa*
Red Pine

10. *Pinus virginiana*
Virginia Pine

11. *Pinus sylvestris*
Scotch Pine

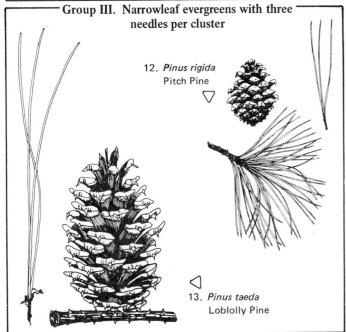

Group III. Narrowleaf evergreens with three needles per cluster

12. *Pinus rigida*
Pitch Pine

13. *Pinus taeda*
Loblolly Pine

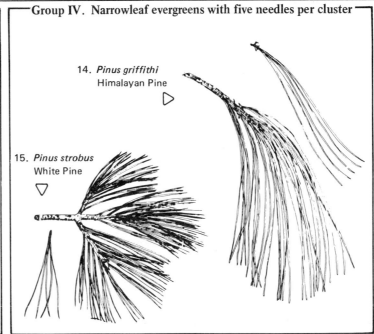

Group IV. Narrowleaf evergreens with five needles per cluster

14. *Pinus griffithi*
Himalayan Pine

15. *Pinus strobus*
White Pine

Group V. Narrowleaf evergreens with more than five needles per cluster

16. *Cedrus deodara*
Deodar Cedar

─── **Group VI. Narrowleaf evergreens that are deciduous** ───

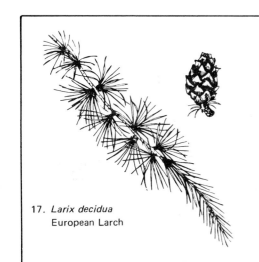

17. *Larix decidua*
European Larch

18. *Metasequoia glyptostroboides*
Dawn Redwood

19. *Taxodium distichum*
Bald Cypress

─── **Group VII. Narrowleaf evergreens with scalelike foliage** ───

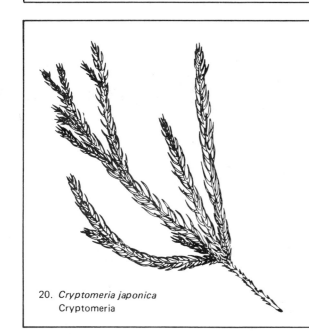

20. *Cryptomeria japonica*
Cryptomeria

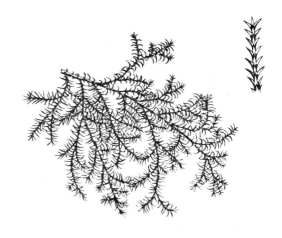

21. *Chamaecyparis pisifera 'Plumosa'*
Plume False-cypress

22. *Chamaecyparis psifera 'Squarrosa'*
Moss False-cypress

Group VII. Narrowleaf evergreens with scalelike foliage (continued)

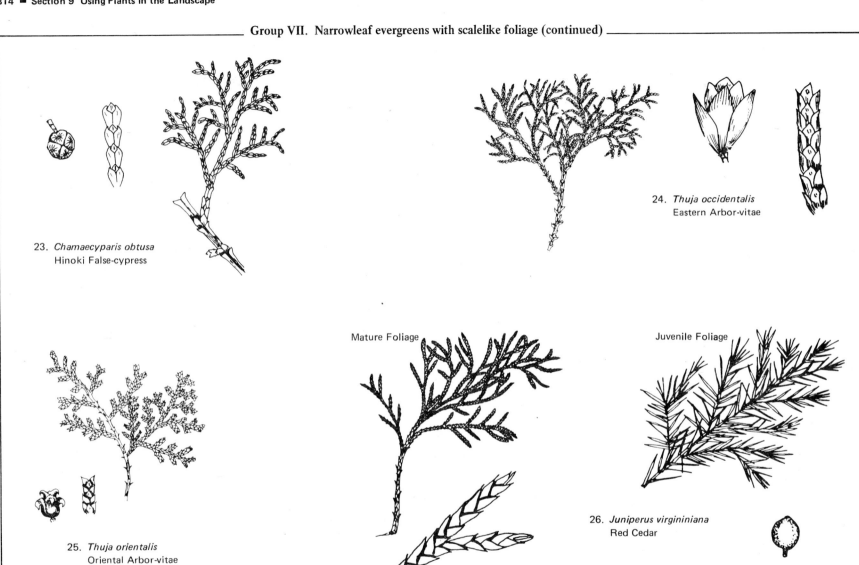

23. *Chamaecyparis obtusa*
    Hinoki False-cypress

24. *Thuja occidentalis*
    Eastern Arbor-vitae

25. *Thuja orientalis*
    Oriental Arbor-vitae

Mature Foliage

Juvenile Foliage

26. *Juniperus virgininiana*
    Red Cedar

STUDENT ACTIVITIES

1. Produce ten cuttings of selected narrowleaf evergreens and propagate them. When propagating, follow the procedure outlined in Section 6.

2. Transplant a bare root or balled and burlapped narrowleaf evergreen or explain the correct procedure for transplanting.

3. Demonstrate or explain the difference between pruning for an informal appearance and shearing.

4. Identify scalelike and needlelike narrowleaf evergreens from specimen twigs and branches.

SELF-EVALUATION

Select the best answer from the choices offered to complete each statement.

1. Narrowleaf evergreens are popular landscape plants because

   a. they are easy to transplant.
   b. they remain green year-round and are adaptable to a wide range of soil types and weather conditions.
   c. they are inexpensive.
   d. none of the above

2. The leaves of narrowleaf evergreens may be either

   a. needlelike or scalelike.   c. alternate or opposite.
   b. whorled or clustered.   d. simple or compound.

3. Narrowleaf evergreens are shipped bare root

   a. to reduce shipping weight when the plants are large.
   b. only when they are dormant.
   c. when the plants are still seedlings and quite small.
   d. when their foliage drops and they are dormant.

4. Narrowleaf evergreens are often chosen for

   a. their attractive flowers.
   b. their edible fruit which serves as food for wildlife.
   c. use as screen plantings.
   d. use as foundation plantings.

5. A windbreak of narrowleaf evergreens can result in temperatures from _____ to _____ degrees warmer on the protected side.

   a. 10 to 15
   b. 5 to 10
   c. 25 to 30
   d. 0 to 5

6. Balled and burlapped evergreens may be moved at any time of year except when

   a. new shoots are growing.
   b. they are dormant.
   c. the soil is frozen.
   d. the soil is too wet.

7. Burlap around the root ball should be removed

   a. entirely from the root ball.
   b. before the plant is set in the hole.
   c. by rolling it down the sides and leaving it under the ball.
   d. only from the top of the ball.

8. Balled and burlapped plants are set in the planting hole at a depth

   a. of one-half the depth of the root ball.
   b. equal to the depth at which it grew in the nursery.
   c. 3 inches above the surrounding soil line.
   d. 1 inch above the original soil line.

9. Three to six pounds of 5-10-10 fertilizer per 100 square feet should be applied annually

   a. in a circle at the drip line of the plant.
   b. in holes punched into the soil around the plant.
   c. under the mulch around the plant.
   d. on top of the mulch around the plant.

10. The best time of year to apply a 5-10-10 fertilizer is

   a. in early fall.
   b. in early spring before new growth starts.
   c. in midsummer to encourage a longer growth period.
   d. in late summer.

11. Informal pruning to remove individual branches is the technique usually used to prune

   a. Hicks yew.
   b. spreading juniper.
   c. columnar juniper.
   d. globe arborvitae.

12. Two narrowleaf evergreens that grow to a height of 3 feet or less are

   a. Sargent juniper and blue rug.
   b. dwarf Alberta spruce and Hicks yew.
   c. English yew and Japanese yew.
   d. all of these

unit 27
Broadleaf Evergreens

OBJECTIVE

To identify the cultural requirements, planting techniques, care, and uses of broadleaf evergreens in the landscape.

COMPETENCY TO BE DEVELOPED

After studying this unit, the student will be able to

- list five ways in which broadleaf evergreens are used in the landscape.

- list four cultural requirements of broadleaf evergreens.

- describe the soil and fertilizer requirements of broadleaf evergreens.

- explain the procedure for transplanting broadleaf evergreens.

- list three pests which attack broadleaf evergreens and one control for each.

Materials List

branches and leaves of broadleaf evergreens grown in the local area

one broadleaf evergreen properly balled and burlapped

The description of broadleaf evergreens is just as the name implies — *evergreen* plants (plants which hold their leaves all year) with broad leaves rather than the needlelike leaves of narrowleaf evergreens such as pines, yews, and junipers. Figure 27-1 illustrates some typical broadleaf evergreens.

Fig. 27-1 Leaves and twigs of three broadleaf evergreens: (A) American holly; (B) English holly; (C) Japanese holly.

Fig. 27-2 Rhododendron in full bloom, a beautiful specimen plant.

USES OF BROADLEAF EVERGREENS

Broadleaf evergreens are used in the landscape in several ways.

Around foundations. The smaller leaved broadleaf evergreens are commonly used as foundation plantings. Some of these are the Japanese holly, azalea, andromeda, dwarf English holly, and dwarf rhododendron.

Fig. 27-3 The large southern magnolia (Magnolia grandiflora), an impressive planting when covered with fragrant 8- to 10-inch blossoms (inset).

As specimen plants. A large rhododendron is beautiful as a singular specimen plant, figure 27-2. American holly, some of the larger varieties of English holly, and magnolia also make beautiful specimen plants, figure 27-3, page 319.

As hedges. Privet, a broadleaf evergreen, is often used as a hedge, figure 27-4. Japanese holly and boxwood are also used in hedges.

As corner plantings. Rhododendron, viburnum, and Japanese andromeda are often used in corner plantings, figure 27-5.

Along sidewalks. Broadleaf evergreens are frequently used to lead the way to the home entrance. A privet hedge, if cut short, and azalea also work well as plantings along sidewalks.

Figure 27-6 summarizes growing information and uses for some of the more common broadleaf evergreens.

Fig. 27-4 Privet used as a hedge for privacy from a highway. Notice that the hedge is being pruned at an angle so that the sun hits the total plant surface. (Richard Kreh, photographer)

Fig. 27-5 Japanese andromeda used as a corner plant in a foundation planting. (Richard Kreh, photographer)

Name	Size at Maturity	Soil and Soil pH Preference	Pruning Date	Blooming Date	Uses	Other Remarks
Andromeda (*Pieris japonica*)	8'-10'	loam pH 5.0-6.5	After blooming in spring	early spring	foundation, specimen, center of group	Gives very early white blooms. Excellent for corner of house. Hardy from Zones 6 to 9.
Azalea (*Rhododendron* varieties)	from creeping to 15' according to variety	loam pH 5.0-6.5	after blooming in spring	early spring through late spring depending on variety	foundation, beds, edging, woodland	Many variations in size, color, bloom date. Has a variety of uses. Should be sprayed for best results. Hardy Zone 6 to 8.
Barberry (*Berberis darvinii*)	8'	broad range	anytime	spring flowering; berries in fall	hedge, foundation, specimen	Few thorns, bright orange flowers, purple berries. Needs full sun. Hardy Zones 4 to 9.
(*Berberis wilsoniae*)	5'	broad range	anytime	spring flowering; berries in fall	hedge, foundation, specimen	Bright scarlet berries. Hardy Zones 4 to 9.
(*Berberis thunbergii*)	4'	broad range	anytime	spring flowering; berries in fall	hedge, foundation, specimen	Rich purple foliage turns scarlet in autumn.
(*Berberis verruculosa*)	4'	broad range	anytime	spring flowering; berries in fall	specimen, hedge	Yellow flowers in spring. Foliage white underneath. Hardy in Zones 4 to 9.
Boxwood, English (*Buxus sempervirens*) (*B.s. suffruticosa*)	slow growing up to 3'	heavy loam pH 5.0-7.0	early spring	inconspicuous	edging walkways, designs inside corners at foundation	Suffers windburn in strong winter winds. Grows in full sun or shade. Grows slowly (1 to 2 inches each year). Hardy in Zones 6 to 8.

Fig. 27-6 Common Broadleaf Evergreens

Name	Size at Maturity	Soil and Soil pH Preference	Pruning Date	Blooming Date	Uses	Other Remarks
Boxwood, Common (*Buxus sempervirens*) (*B.s. arborescens*)	fast growing up to 15'	heavy loam pH 6.0-7.0	early spring		hedge	Grows very fast. Full sun or shade.
Camellia (*C. sasaqua*)	8' to 15'	loam, high in organic matter pH 5.0-6.5	after flowering or early spring	autumn	specimen, border, hedge	White to pink single and double flowers. Will grow in full sun or partial shade. Hardy in Zones 7 to 9.
(*C. japonica*)	15' to 30'	loam, high in organic matter pH 5.0-6.5	after flowering	early spring	specimen	Dark green foliage; white to deep red single or double flowers. Needs sun to flower. Hardy in Zones 7 to 9.
Firethorn (*Pyracantha coccinea* 'Lalandi')	20'	loam wide range of pH (6.0-8.0)	early spring	June	specimen	Many varieties. Difficult to transplant (use container plants). Orange fruit.
Holly, American (*Ilex opaca*)	20' to 30'	sandy loam *pH 6.0-7.0*	winter holiday season	spring	specimen	Red berries in fall and winter; spiny leaves.
Chinese (*Ilex cornuta*) includes Burford holly	from 4' to 15' depending on variety	sandy loam pH 6.0-7.0	throughout the year except late summer	spring	foundation, hedge, specimen	Many varieties. Red or yellow berries. Full sun or shade. Needs heavy fertilization.
English (*Ilex aquifolium*)	2' to 10'	sandy loam pH 6.0-7.0 season	winter holiday season	spring	specimen	Glossy foliage. Yellow and red berries.
Japanese (*Ilex crenata*)	from 3' to 15' depending on variety	loam pH 5.0-6.0	throughout the year except late summer	early summer	hedge, topiary design, specimen, foundation	Small spineless leaves and black berries. Fertilizer necessary each year.

Fig. 27-6 Common Broadleaf Evergreens (continued)

Name	Size at Maturity	Soil and Soil pH Preference	Pruning Date	Blooming Date	Uses	Other Remarks
Magnolia (*Magnolia grandiflora*) Southern Magnolia	30' to 50'	loam pH 6.0-6.5	after flowering	midsummer	specimen	Native to the United States. 6" to 10" fragrant white flowers. Hardy in Zone 7 and in southern areas.
Sweet Bay Magnolia (*Magnolia virginiana*)	30' to 50'	loam	after flowering	early summer	background, specimen	Evergreen in south, Zones 7 to 10. Deciduous in Zones 5 and 6. Hardy in Zones 5 to 10.
Nandina (*Nandina domestica*)	6' to 8'	loam pH 5.0-6.5	after flowering	midsummer	shrub borders front of taller evergreens	White flowers; red berries in autumn and winter. Hardy in Zones 7 and 8.
Privet (*Ligustrum japonicum*)	9' to 18'	wide range	after flowering	June to July	hedge	Black berries. Hardy in Zones 6 to 7.
Glossy Privet (*Ligustrum lucidum*)	8' to 15'	wide range	after flowering	August-September	hedge, borders	Easy to grow. Hardy in Zones 7 to 10. Grows in shade.
Rhododendron varieties	3' to 10' for most commercial varieties	varies with variety pH 5.0-6.0	after flowering	spring	specimen, foundation	Requires mulch at all times and good drainage. High organic matter needed in soil. Hardy in Zones 4 to 8.
Viburnum (Many varieties; select for your area)	5' to 12' depending on variety	varies with variety	after flowering	spring	foundation, specimen	Fertilize each year. Attractive fruit, especially to birds.

Fig. 27-6 Common Broadleaf Evergreens (continued)

CULTURAL REQUIREMENTS

• **Ample moisture** is required by most broadleaf evergreens for best growth. This may be supplied by the addition of organic matter to the soil. Organic matter holds moisture well.

• **Good soil drainage** is also necessary. If drainage is a problem, the planting may be raised above soil level or sand may be added to the soil to improve internal drainage.

• **Mulching** is necessary for the proper supply of moisture. Since these plants have green leaves year-round, a ready supply of moisture must be available at all times to the roots to replace moisture lost through the leaves due to transpiration. If the soil freezes in winter, the roots cannot pick up moisture, causing the plants to dry out. A mulch prevents the soil from freezing as deeply and the roots are better able to supply lost moisture. In summer, a mulch keeps the soil cooler and saves moisture loss due to heat and drying winds.

• **A location which is sheltered from strong winds** is best for broadleaf evergreens. Cold winter winds do the most damage by removing moisture from the leaves faster than the roots can replenish it.

• **Enough sunlight** is needed to develop a full, compact plant. Plants tend to grow tall and leggy with too much shade. Full sunlight or partial shade is best for most of the broadleaf evergreens. For example, rhododendrons grow best in 40 percent shade. If plants do not flower well, they are probably receiving too much shade.

Soil and Fertilizer

Any good loam soil is satisfactory for most broadleaf evergreens. An acidic soil is usually the ideal choice. The rhododendron, azalea, and Japanese andromeda require a pH of from about 5.5 to 6.0 for best growth. A soil pH higher than this results in an iron deficiency.

Fertilizer should be used in small amounts. Too much fertilizer, especially nitrogen, causes soft, fast growth and results in open, straggly plants which are unattractive. A slower growing, compact plant makes a much better addition to the landscape picture. Also, if fertilizer is applied in smaller amounts, the plant will not outgrow its space in the landscape as quickly. If all other cultural conditions are right, short twig growth and small, pale green leaves may indicate a need for fertilizer.

Fertilizer is added in the spring. It is applied in a circular fashion around the plant at the drip line. (The *drip line* is an imaginary line directly below the outer edge of the branches.) A few handfuls carefully sprinkled around the plant are sufficient for small plants. For larger trees, use 1 to 1 1/2 pounds per inch of tree trunk. Any type of fertilizer that is not alkaline may be used for "acid-loving" broadleaf evergreens; cottonseed meal is the best and safest fertilizer for these plants.

The varieties of Chinese and Japanese holly require more fertilizer than most broadleaf evergreens. Two to three pounds of 10-5-5 fertilizer per 100 square feet should be used. On individual plants, use 1 1/2 pounds per 3 feet of height or spread of plant. Chinese holly grows best at a soil pH of 6.0-7.0

TRANSPLANTING

Since these plants are evergreen, the demand for moisture is always high. The roots must always be in close contact with moist soil. For this reason, broadleaf evergreens are always moved with the *root ball*. This means that the soil is left intact around the roots and moved with the plant, figure 27-7.

The hole for broadleaf evergreens should be wide enough so that 6 to 8 inches of space exists all the way around the root ball. The plants are always planted at the same depth as they were growing before; they are never planted deeper, as is done with many deciduous trees. No soil should be placed on top of the root ball, figure 27-8. Be sure that the soil in the bottom of the hole is firm so that the root ball does not sink after it is planted. Soil filled in around the plant should be 30 to 50 percent sphagnum peat moss for better moisture-holding capacity.

Firm the soil around and under the root ball with the hands and water with a garden hose at low water pressure. Do not tramp with feet or pack too tightly. Mulch around the plant with wood chips or tree bark to a depth of 2 to 3 inches and out beyond the branch ends.

Fig. 27-7 This rhododendron was dug from the ground with the root ball intact in preparation for transplanting. Careful handling is important so that the soil is not broken loose from the roots. (Richard Kreh, photographer)

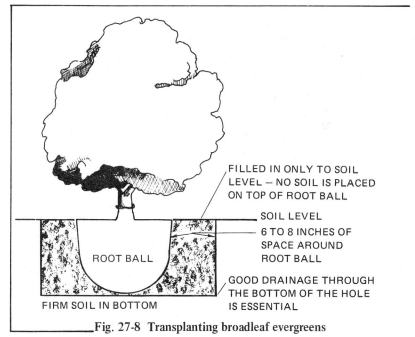

FILLED IN ONLY TO SOIL LEVEL — NO SOIL IS PLACED ON TOP OF ROOT BALL

SOIL LEVEL

6 TO 8 INCHES OF SPACE AROUND ROOT BALL

ROOT BALL

GOOD DRAINAGE THROUGH THE BOTTOM OF THE HOLE IS ESSENTIAL

FIRM SOIL IN BOTTOM

Fig. 27-8 Transplanting broadleaf evergreens

PRUNING

Broadleaf evergreens may be pruned for a special effect or cut back to thicken the plant. Pruning is most often done in early spring or immediately after blooming. Holly is often pruned during the winter holiday season and the prunings used for holiday decorations.

The rhododendron should be pruned as little as possible. The cut is made at the point where flush of growth stops and buds are more plentiful. This usually occurs at the point where branches extend from the plant.

INSECTS AND DISEASES

If varieties of broadleaf evergreens that are adapted to the area are selected, chances are good that few insect and disease problems will develop. Chewing insects may feed on leaves, causing some disfigurement. Except on small plants, this is seldom serious enough to require spraying. If a spray is necessary, use malathion and Sevin, or orthene.

Small sucking insects and mites, such as the lacewing fly, are the most serious pests that attack broadleaf evergreens. These insects attack the lower surface of the leaf and suck the chlorophyll from the leaf. An attack causes leaves to appear white or yellow. Afterwards, the plants do not grow as well. Small mites or insects can be seen on the undersurface of the plant's leaf with a 10-power microscope (one which magnifies the insects ten times their normal size). For control of these pests, spray with malathion if the pest can be hit with the spray. (Malathion will kill only if the insect comes in direct contact with the spray.) Cygon, or orthene, two systemics, kill without coming into contact with the pests and do a better job. However, they are more toxic to warm-blooded animals, including humans, and proper precautions must be taken.

Fungus diseases such as leaf spots, rusts, or dieback (caused by tips of branches being killed by a fungus) may be a problem. If these diseases are present, spray just after bloom and again in ten days with zineb, ferbam, captan, or benomyl.

Use a spreader sticker to apply all sprays to broadleaf evergreens. A *spreader sticker* is a chemical added to a pesticide which causes the spray to break into small droplets and spread evenly and thinly over the leaf surface. Shiny leaves are hard to get completely wet, and sprays tend to run off the plant without giving the proper protection. A spreader sticker helps to overcome this problem by causing the spray to stick to the plant more easily. Ordinary soap or detergent is acceptable as a spreader sticker if nothing else is available. Use only enough soap to produce light suds in the water.

Most broadleaf evergreens are propagated by semi-hardwood cuttings. See Unit 10 for details.

STUDENT ACTIVITIES

1. Draw up a chart such as the one shown in figure 27-6. Insert local broadleaf evergreens that are different from the ones listed in the text. Supply the information in each column for plants listed.

2. Tour the school grounds or part of the local community and identify various broadleaf evergreens sighted. Also note their use in the landscape.

SELF-EVALUATION

Select the best answer from the choices offered to complete the statement or answer the question.

1. Broadleaf evergreens are plants that

   a. hold their leaves all year round.
   b. have broad leaves rather than needles.
   c. require a constant supply of moisture.
   d. all of the above

2. Which broadleaf evergreen is often used as a hedge?

   a. privet
   b. rhododendron

   c. magnolia
   d. American holly

3. Which of the following are good choices as specimen plants in a landscape?

   a. Japanese holly, privet, magnolia
   b. magnolia, rhododendron, American holly

   c. boxwood, magnolia, English holly
   d. magnolia, privet, rhododendron

4. Broadleaf evergreens that work well in lining sidewalks and other entrance ways are

   a. English boxwood, azaleas, Japanese holly
   b. English boxwood, azaleas, magnolia
   c. rhododendron, magnolia, Chinese holly
   d. American holly, magnolia, Chinese holly

5. The four most important cultural requirements of broadleaf evergreens are

   a. ample moisture, good soil drainage, mulch, and a windbreak.
   b. dry soil, strong sunlight, heavily fertilized soil, and air circulation.
   c. ample moisture, dry soil, strong sunlight, and a windbreak.
   d. dry soil, good drainage, strong sunlight, and a windbreak.

6. Many of the broadleaf evergreens should be fertilized with only small amounts of nitrogen fertilizer because

   a. nitrogen makes the leaves turn yellow.
   b. nitrogen causes soft growth which is easily killed by frost.
   c. nitrogen causes long shoot growth and open, unattractive plants.
   d. nitrogen tends to dwarf the plants.

7. When dug from the ground, broadleaf evergreens should always be moved with soil around the roots because

   a. this makes it easier to plant correctly.
   b. the heavy root ball helps stabilize the plant.
   c. the roots will otherwise fall off the plant.
   d. the roots are less disturbed and are able to continue supplying moisture to the leaves.

8. Broadleaf evergreens require mulching for best growth. The two most important functions performed by the mulch are to (Select two answers.)

   a. make the plant look attractive.
   b. keep the soil cool in summer, thus conserving moisture.
   c. control weeds.
   d. prevent deep soil freezing in winter allowing the roots to obtain moisture easier.

9. Which of the following is *not* a reason for drying out of broadleaf evergreens due to lack of available moisture?

   a. freezing of soil around roots in winter.
   b. transplanting with a root ball.
   c. lack of mulch in summer.
   d. planting in areas having strong winds.

10. Two broadleaf evergreens that require relatively high amounts of fertilizer are

   a. Japanese and Chinese holly.
   b. the rhododendron and azalea.
   c. English and Chinese holly.
   d. the boxwood and andromeda.

11. A safe chemical spray used to control chewing insects on broadleaf evergreens is

    a. Sevin.   c. zineb.

    b. malathion.   d. captan.

12. A spray chemical used to control fungus diseases on broadleaf evergreens is

    a. Sevin.   c. captan.

    b. malathion.   d. Cygon.

13. A spray chemical used to control sucking insects on broadleaf evergreens is

    a. Sevin.   c. captan.

    b. zineb.   d. malathion or Cygon.

OBJECTIVE

To properly position and plant deciduous trees in the landscape.

COMPETENCY TO BE DEVELOPED

After studying this unit, the student will be able to

- list the six functions of trees in the landscape.
- select at least two specific trees to fulfill each of these six functions.
- demonstrate the proper planting technique for bare root, balled and burlap, and containerized trees.

Materials List

deciduous or flowering trees, at least one of each in a container, bare root, and balled and burlapped

shovels for digging holes

fertilizer

peat moss

mulching material (wood chips or bark)

tree stakes, tree wrap, #9 wire, small sections of garden hose

a bar for punching holes to fertilize an established tree

---

unit 28
Deciduous Trees

THE USE OF TREES IN THE LANDSCAPE

Trees are often positioned in the landscape to serve a specific purpose. The following are some of the functions of trees in the

Fig. 28-1

Fig. 28-2

landscape. (Figures 28-1 through 10 are from the University of Maryland Extension Bulletin 183 by Robert L. Baker and Carl N. Johnson.)

To Provide Shade. Shade trees keep temperatures inside houses cooler in summer as well as providing outdoor shade, figure 28-1. A well-placed tree can reduce the summer room temperature of a frame house as much as 20 degrees in certain areas. To serve this purpose, trees may be planted as close as 15 feet to the house. Larger spreading trees, such as oaks, must be given more generous spacing.

To Frame the House. Trees are sometimes used to make houses the center of interest in the view from a street, figure 28-2. Generally, the farther the house is set back from the street, the more effective the tree is as a framing device.

To Soften Lines. A tree placed at the back of a home is effective in softening the lines of the house, figure 28-3.

To Relieve Bare Spots. The end walls of many houses often have a bare look. This bareness can be relieved by planting a small or medium-sized tree near the side of the house, figure 28-4. This creates a softening effect that a planting of shrubs alone may not accomplish.

Fig. 28-3

Fig. 28-4

To Screen an Object. A single tree or a group of trees may be used to partially screen an object that would otherwise stand out too strongly in the landscape, figure 28-5.

To Frame a View. Tall trees with an underplanting of shrubs or small trees may be used to frame a view from a house or terrace, figure 28-6.

To Ensure Privacy. A carefully positioned tree can create privacy by screening a porch or terrace, figure 28-7.

To Accent the Landscape. Flowering trees or trees with graceful and interesting branching habits are often used as accents in the landscape, figure 28-8. They can be used to direct attention to certain areas or to act as a terminal planting for a hedge or wall.

To Break Monotony. Trees and shrubs can be used to break the monotony of an obtrusive architectural feature or enclosure, figure 28-9. The planting, however, should be placed off center so that the feature does not appear to have been cut in half. White brick walls, fencing, and other architectural means of enclosing property may be attractive in themselves, but can become monotonous unless softened by planting materials.

As Windbreaks. Tall evergreens are used as windbreaks, protecting the house from north and northwest winds in the winter, figure 28-10. These should be placed about 50 feet from the house to be effective.

TREE TYPES

Figure 28-11 lists deciduous trees that may be used in landscaping for shade and ornamental flowering. The trees in this chart are separated according to their mature height, with small deciduous trees classified as those up to 35 feet in height; medium

Fig. 28-5

Fig. 28-6

Fig. 28-7

Fig. 28-8

Fig. 28-9

Fig. 28-10

deciduous trees as those 36 feet to 75 feet; and large deciduous trees as those 75 feet and over in height.

The trees in the chart are classified according to various characteristics, under the headings *Form, Flowering, Color, Texture, How to Propagate, Height, Hardiness,* and *Ornamental Use.*

Form is the shape of the tree. This is important to consider when choosing a tree for a particular location. For example, a tall, columnar tree is used to fill a particular landscaping need and fits in a smaller space. A broad, spreading tree requires more room and has a different appearance and use in the landscape.

Flowering and Color. Whether or not the tree flowers, when it flowers, and the color of the blossoms are important factors in tree choice. Some trees have very large, attractive flowers while others have small, unnoticeable ones. Some trees vary in their leaf color from one season to the other, while others do not change at all. In some cases, a tree may be selected for its fall color.

Texture deals with the size of the leaves. Trees with large leaves appear to have a coarse texture while small leaves give an appearance of fine texture.

How to Propagate — Height describes the way in which the tree is reproduced. The next heading, **height**, is the average height to which that particular tree grows.

Hardiness is the ability of the tree to live in a particular climate zone. This is influenced by the temperature, rainfall, and the soil of that particular area. Some trees can tolerate lower temperatures than others. (In the chart, hardiness refers to U.S. Hardiness Zones). Hardiness zones give the average low temperature range for a particular area of the country.

Small Deciduous Trees (up to 35 feet in height)										
			Color			How to			Ornamental	
Name	Form	Flowering	Summer	Fall	Texture	Propagate	Height	Hardiness	use	Other Remarks
Acer palmatum (Japanese maple)	rounded moundlike		red to green	scarlet	fine	budding	20'	5	specimen	Many varieties are available. Needs well-drained soil.
Acer ginnola (Amur maple)	rounded upright	purplish	green	scarlet	fine	seed	20'	2	specimen screening	fragrant flowers
Cercis canadensis (Redbud or Judas tree)	rounded	purplish pink mid-May	green	yellow	coarse	seed	35'	4	border	Blooms same time as dogwood.
Chionanthus virginicus (White fringe tree)		white feathery late May	green	yellow	coarse	seed, softwood cuttings	30'	4	specimen	Does well in moist, shady site.
Cornus florida (Flowering Dogwood)	rounded	white mid-May	green	red	medium	seed	35'	4	specimen	Does well in shade or full sun. horizontal branching
Crataegus crus-galli (Cockspur hawthorne)		small white	green	scarlet	medium	seed	35'	4	border specimen	bright red fruit in fall Needs sandy soil.
Crataegus oxycantha pauli (Paul's scarlet hawthorne)	dense rounded	double scarlet late May	green	––	medium	seed	35'	4	border	thorny
Crataegus phaenopyrum	dense rounded	white mid-June	green	scarlet	medium	seed	30'	4	specimen accent	Bright red fruit in fall — good for wildlife.
Magnolia soulangeana (Saucer magnolia)	rounded open	white purple May	green	––	coarse open	seed cuttings grafting layering	25'	5	specimen border	
Magnolia stellata (Star magnolia)	rounded open	double white to mid-April	green	yellow	coarse open	seeds cuttings grafting layering	20'	5	specimen	Best color appears in direct sun.
Malus floribunda (Japanese flowering crabapple)	dense	white early May	green	––	fine	T-budded grafted	30'	4	specimen	Flowers yearly.

Fig. 28-11 Common Deciduous Trees

Name	Form	Flowering	Color Summer	Color Fall	Texture	How to Propagate	Height	Hardiness	Ornamental use	Other Remarks
Prunus americana 'thundercloud' (Thundercloud Plum)	rounded	white May	red	dark purple	medium	T-budded	20'	3	specimen	Has dark purplish fruit.
Prunus serrulata (Oriental cherry)	rounded	white to pink mid-May	green	––	coarse	T-budded	25'	5	specimen	Many varieties are available.
Prunus subhirtella pendula (Weeping cherry)	weeping	double pink early April	green	––	medium	grafted	25'	5	specimen	excellent as accent plant
Medium Deciduous Trees (36 to 75 feet in height)										
Aesculus carnea (Red horsechestnut)	pyramidal	pink to red spikes mid-May	green	brown	coarse	seed	75'	3	specimen	
Ailanthus altissima (Tree of Heaven)	rounded	small yellow late June	green	––	coarse	seed	60'	4	use on bank spoils	Grows well in dry areas.
Albizzia julibrissin (Silk tree)	rounded	pink summer	green	––	fine	seed	36'	7	specimen	Flowers all summer.
Betula pendula (Weeping Euorpean birch)	pyramidal	––	green	yellow	fine	seed	60'	2	specimen	white bark in winter Bronze birch borer can be a problem.
Franklinia alatumala (Franklinia tree)	rounded	white September	green	red	medium		36'	6	border specimen	late fall flower
Fraxinus pennsylvanica lanceolata (Green ash)	rounded		green	yellow	coarse	seed	60'	2	streets border	attractive form
Magnolia virginiana (Sweet bay)	rounded	white fragrant late May	green	––	coarse	seeds cuttings grafting layering	60'	5	specimen	

Fig. 28-11 Common Deciduous Trees (continued)

Name	Form	Flowering	Color Summer	Fall	Texture	How to Propagate	Height	Hardiness	Ornamental use	Other Remarks
Quercus virginiana (Live oak)	rounded wide-spreading		green		medium	seed	60'	7	specimen street	evergreen in southern United States
Oxydendrum arboreum (Sourwood or lily-of-the-valley tree)	rounded	white mid-July	green	red	medium	seed	75'	4	specimen	late summer flower; good fall color
Sorbus aucuparia (European Mountain ash)	pyramidal	clusters of white	green	reddish	medium		60'	5	specimen	Bright red berries appear in a cluster in fall.

Large Deciduous Trees (over 75 feet in height)

Name	Form	Flowering	Color Summer	Fall	Texture	How to Propagate	Height	Hardiness	Ornamental use	Other Remarks
Acer platanoides (Norway maple)	rounded	small yellow before leaves	green	yellow	coarse	seed	100'	3	shade	Provides dense shade. fast grower
Acer rubrum (Red maple)	rounded	small red early April	green	brilliant red	medium	seed	120'	3	border specimen	Several varieties are available
Acer saccharinum (Silver maple)	rounded	small green	green	— —	medium	seed	120'	3	wildlife area	fast growing; weak wooded
Acer saccharum (Sugar maple)	rounded	green	green	yellow	coarse	seed	125'	3	specimen shade	Used to produce maple syrup.
Fagus grandifolia (American beech)	pyramidal		green	bronze	medium	seed	90'	3	specimen	wildlife food
Fraxinus americana (White ash)	rounded		green	yellow	coarse	seed	120'	3	border	attractive form
Ginkgo biloba (Ginkgo)	pyramidal	inconspicuous	green	yellow	coarse	grafted	120'	4	street specimen border	Male form has fragrant fruit. fan-shaped leaf.
Gleditisia triacanthos 'inermis' (Thornless honeylocust)	rounded open	greenish pea-like June	green	yellow	fine	grafted	120'	4	lawn specimen shade	no thorns
Juglans nigra (Black walnut)	rounded open	catkins	green	— —	coarse	seed	150'	4	nut producing	Wood is used to make furniture.

Fig. 28-11 Common Deciduous Trees (continued)

Name	Form	Flowering	Color Summer	Color Fall	Texture	How to Propagate	Height	Hardiness	Ornamental use	Other Remarks
Liquidambar styraciflua (Sweet gum)	pyramidal	greenish clusters	green	red to scarlet	medium	seed leafy soft softwood cutting	125'	4	lawn specimen street shade	difficult to transplant good form corkey ridges on stem
Liriodendron tulipifera (Tulip tree)		greenish yellow tulip-shaped mid-June	green	yellow	coarse	seed	180'	4	specimen	timber tree
Magnolia acuminata (Cucumber tree)	pyramidal	greenish yellow early June	green		coarse	layering seed cuttings grafting	90'	4	specimen border	red to pink cucumber-shaped fruits
Nyssa sylvatica (Black gum)	rounded	incon-spicuous May-June	green	red	medium	seed	90'	4	specimen	Small blue berries appear in midsummer.
Platanus occidentalis (Sycamore)	pyramidal		green		coarse	seed	120'	4		Natives tend to defoliate.
Populus nigra italica (Lombardy poplar)	narrow tall		green		medium	seed hardwood cuttings	90'	2	quick temporary screen	short-lived tree
Quercus alba (White oak)	rounded		green	purple red	medium	seed	150'	4	specimen	excellent lumber tree slow grower
Quercus borealis (Red oak)	rounded		green	red	coarse	seed	76'	4	specimen street	transplants easily
Quercus coccinea (Scarlet oak)	rounded		green	bright scarlet	coarse	seed	100'	4	specimen street	difficult to transplant
Quercus palustris (Pin oak)	pyramidal		green	scarlet	coarse	seed	100'	4	specimen	very graceful excellent street tree
Tila americana (American linden)	rounded open	white June	green		medium	layering seed mound	100'	2	border	Wood is valuable for lumber.
Tila cordata (Little leaf linden)	rounded open	white to yellow mid-July	green		medium	layering seed mound	90'	3	shade	
Ulmus americana (American elm)	vase	yellow	green	yellow brown	medium	seed softwood cuttings	120'	2	street	possible problem with Dutch elm disease

Fig. 28-11 Common Deciduous Trees (continued)

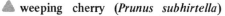

European white birch (*Betula pendula*)
▼

▲ weeping cherry (*Prunus subhirtella*)   ▲ pin oak (*Quercus palustris*)   sycamore (*Platanus occidentalis*) ▲

Fig. 28-12 Four deciduous trees (Richard Kreh, photographer)

Ornamental Use describes common uses of the tree in the landscape. Specimen trees are grown solely for special characteristics they possess which create interest in a landscape. Street trees are those trees which are able to withstand the special conditions occurring along roadways and streets.

Student Activity. Figure 28-12 illustrates four deciduous trees commonly used in landscapes. Examine the trees and identify their physical characteristics, such as shape and density.

PLANTING DECIDUOUS TREES

Ways Trees Can Be Purchased

Deciduous trees may be purchased in three different forms: bare root (BR), balled and burlapped (B&B), or containerized (C), figure 28-13.

Bare Root. Trees in this form should be purchased from a reliable nursery. The trees are dug in the nursery when the tree is dormant. No soil is left with the roots, hence the name bare root. Many nurseries purchase trees in this form for transplanting in the late fall or early spring. It is best to work with only small bare root trees, since they are easier to handle. Bare root stock is usually less expensive. The cost of shipping large quantities of these trees is smaller as compared with balled and burlapped and container grown stock.

Balled and Burlapped. Trees in this form have a ball of soil around the roots. Almost all ornamentals can be transplanted in this form. The soil is held around the roots by burlap which is secured by twine or nails. Evergreen trees should always be moved with a ball

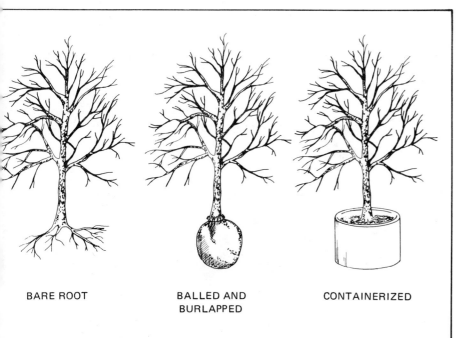

BARE ROOT   BALLED AND   CONTAINERIZED
BURLAPPED

Fig. 28-13 Ways in which deciduous trees are purchased

of soil attached to the roots, with the exception of tiny seedling evergreens, which may be moved bare root.

Container Stock. Trees in this form are planted in a container such as a basket, or plastic or tin can. Many trees are grown today in this fashion because they can be transplanted at any time of the year without disturbing the root system. Container stock becomes root bound in the container if allowed to grow there too long.

Planting Bare Root Trees

Bare root trees are planted only when they are dormant. Since the soil is removed from the roots, the tree cannot replenish moisture rapidly enough to support leaves. Only after the roots

have again made good soil contact can they supply the amount of water needed for a tree in leaf. The dormant tree does not require as much moisture and can survive transplanting.

The bare roots should never be allowed to dry out before planting. Keep them covered with wet peat moss, sawdust, burlap, or some material so that the roots remain moist and away from the wind and sun.

The hole should be large enough to receive the roots without crowding them. The roots should be spread in a natural manner without twisting the roots in the hole. Long roots should be cut off and not twisted around in the hole. Be sure to remove any broken roots with sharp pruning shears. After planting, one-third of the top growth of the original plant is removed by pruning, figure 28-14. This reduces the water loss from the plant.

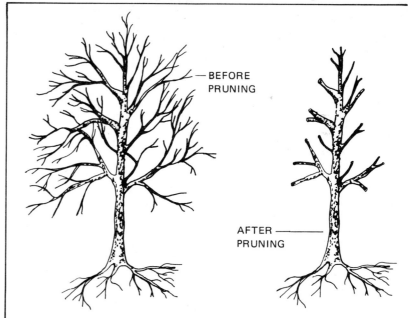

BEFORE
PRUNING

AFTER
PRUNING

Fig. 28-14 Pruning the bare root tree after planting

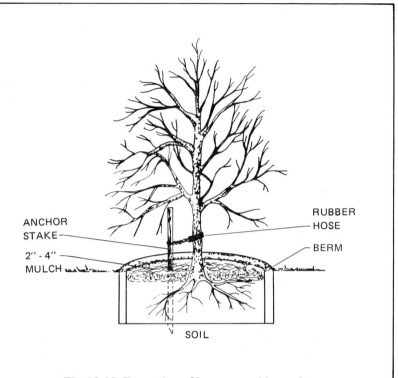

ANCHOR
STAKE

RUBBER
HOSE

2" - 4"
MULCH

BERM

SOIL

Fig. 28-15 Formation of berm around base of tree.

The hole should be large enough so that at least 4 to 5 inches of topsoil is under the roots. The tree is set at the same depth it was growing in its previous location, and the hole filled in with the remainder of topsoil. Be sure that all roots are covered and work the soil around the roots with the hands as the hole is being filled.

Watering. Use a garden hose (with the water running slowly) to settle the soil around the roots and to eliminate air pockets, which can cause the roots to dry unnecessarily. Water until the soil is saturated with water. Never stamp on the soil to firm it — this type of pressure will damage the root system.

To ensure a good supply of water, form a ring (berm) around the base of the tree about 3 inches high. Fill this ring with water, figure 28-15. The water gradually soaks in and settles the soil around the roots. The berm also catches rainwater and keeps it from running off. Drive a stake next to the tree and anchor the tree to the stake with a piece of rubber hose and wire. Mulch with bark or some other coarse material that will not blow away.

Planting Balled and Burlapped Trees

Select the area in which the tree is to be planted. Dig a hole one and one-half to two times larger across than the ball of the

Fig. 28-16 Transplanting a balled and burlapped tree. After the tree is placed in its new location, a prepared soil mixture of peat and soil is distributed around the base of the tree. (USDA photo)

tree to be planted. The hole should be dug so when the plant is set in the hole, the top of the ball sets slightly above the ground level. This allows for sinking of the tree caused by settling of the loose soil in the bottom of the prepared hole. Set the tree in the hole and loosen the burlap around the root ball, figure 28-16. Fill the area around the root ball with the prepared soil. Follow the same procedure for watering as for bare root stock.

Planting Containerized Trees

Containerized trees are planted in the same way as balled and burlapped trees except for the following differences.

1. The container is removed from the tree roots. It should be cut or otherwise removed carefully so that the soil is not broken from the tree.

2. After the container is removed, roots which may be growing around the outside of the root ball in a circular fashion are straightened out before planting. If the roots cannot be straightened, they are cut off at the edge of the root ball. This will stop the circular root growth which tends to girdle root systems and restrict growth. In severe cases, it can kill trees.

Wrapping

It is necessary to wrap the trunk of a transplanted tree to prevent sun scald, reduce water loss, and reduce chances of borer infestation. The best wrapping material is burlap or special tree wrap paper. On large trees, it is best to wrap from the ground line up to and including several lower branches. Each turn of wrap spiral should overlap about one-half the width of the previous wrap.

Staking

Staking the tree prevents the wind from swaying it and loosening the roots. It also helps to keep the tree standing straight.

When staking, use a piece of rubber garden hose and wrap it around the tree trunk above the first set of branches. Then slip a piece of number 9 wire through the rubber garden hose. Twist the wire around and fasten it to each of the stakes, figure 28-17. On large trees, the stakes should remain about two years. It is best to remove the plant ties after this time since the hose may girdle the

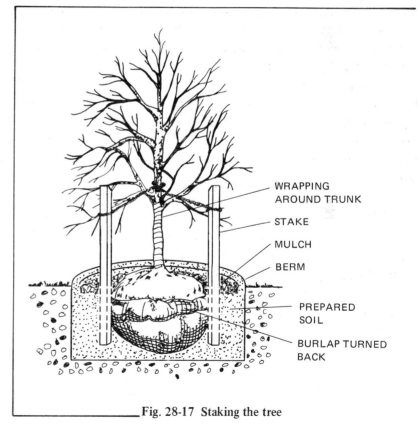

Fig. 28-17 Staking the tree

tree. As the tree grows older, it grows laterally and if the ties are not removed, the tree will grow over the hose, thus weakening the tree. In a bad storm, the top of the tree could break completely off at the weakened point.

Fertilizing

When planting trees, it is recommended that a well-prepared soil be used for backfilling. A slow-release fertilizer such as osmocote or Mag-Amp may be mixed in the soil at planting time according to manufacturer's recommendations.

Another way to fertilize is to apply 2 pounds of a 5-10-5 fertilizer for each 1 inch of diameter of the tree trunk at a height 3 feet above ground level. Use a soil auger, wrenching bar, digging iron, or post hole digger to form holes that measure 15 to 24 inches deep and about 2 feet apart around the drip line of the tree, figure 28-18. The fertilizer is distributed in equal amounts in the holes around the drip line of the tree. After the fertilizer has been applied, finish filling the holes in which the fertilizer was placed with soil.

Mulching

Mulching is necessary for moisture retention, weed control, supplying organic matter, and moderating the temperature for the roots. Mulching also improves the appearance of the area. Trees in the landscape may be mulched with very attractive materials such as hardwood bark, pine bark, coca hulls, and tanbark.

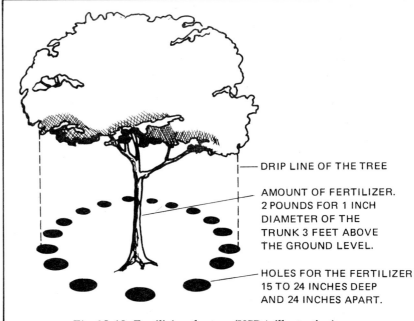

DRIP LINE OF THE TREE

AMOUNT OF FERTILIZER. 2 POUNDS FOR 1 INCH DIAMETER OF THE TRUNK 3 FEET ABOVE THE GROUND LEVEL.

HOLES FOR THE FERTILIZER 15 TO 24 INCHES DEEP AND 24 INCHES APART.

Fig. 28-18 Fertilizing the tree (USDA illustration)

Aftercare

The tree may require watering during dry conditions, especially during the first year after planting. Never add less than 1 inch of water at any one time. This is a sufficient amount for one week with no rain. Repeat every week from April through October if there is not sufficient natural rainfall to provide 1 inch of rain per week.

STUDENT ACTIVITIES

1. Visit a local park or other landscaped area and practice tree identification. Identify at least ten local trees.

2. Draw a simple landscape plan. Place specific trees in the plan in relation to buildings and other structures. List reasons for placing the trees in certain locations.

3. Demonstrate the techniques used in the planting of a bare root, balled and burlapped, and container grown tree.

SELF-EVALUATION

Select the best answer from the choices offered to complete each statement.

1. Deciduous trees may be purchased with their roots prepared in three different ways. These are
   a. bare root, root pruned, and containerized.
   b. balled, burlapped, and bare root.
   c. bare root, balled and burlapped, and containerized.
   d. root pruned, container grown, and root ball.

2. The form of a tree is a characteristic which describes
   a. the shape of the tree.
   b. the way the roots are treated for transplanting.
   c. whether it is evergreen or deciduous.
   d. how it is pruned.

3. Bare-rooted deciduous trees are planted only when they
   a. can be dug into the ground.   c. are beginning to leaf.
   b. are dormant.   d. are in full leaf.

4. Bare-rooted trees are dug when the trees are dormant and are shipped
   a. early in the spring.   c. for fall planting.
   b. with no soil on the roots.   d. with a layer of soil on the roots.

5. Balled and burlapped trees are shipped
   a. with a ball of soil around the roots wrapped in burlap.
   b. with bare roots.
   c. in early spring for immediate planting.
   d. in early fall for immediate planting.

6. Container grown trees become root bound if

   a. the roots are not pruned each year.   c. planted in the fall of the year.
   b. they are left in the same container too long.   d. all of these

7. In planting bare root trees, all broken and badly damaged roots are

   a. patched with tree wound dressing.   c. pruned with pruning shears.
   b. pulled aboveground to heal.   d. given extra fertilizer.

8. About one-third of the top is pruned off a newly planted tree to

   a. shape the new tree.
   b. thin the tree and let in more light.
   c. select and keep only the best branches.
   d. reduce water loss from the plant.

9. A bare root tree should be planted as deeply as

   a. necessary to keep it from blowing over.   c. possible for good anchorage.
   b. it was growing before being transplanted.   d. the level of the topsoil.

10. Trees must be watered well when transplanting to settle the soil and

    a. keep the roots moist so that they can grow.
    b. prevent the roots from drying out.
    c. close any air pockets around the roots.
    d. all of the above

11. A berm is built around the newly planted tree to

    a. help make watering the tree easier and to hold rainwater.
    b. keep rainwater from reaching the tree.
    c. prevent erosion.
    d. hold mulch in place.

12. When planting balled and burlapped trees, a hole is dug

   a. to fit the root ball.
   b. exactly as deep as the root ball.
   c. one and one-half to two times larger than the width of the root ball.
   d. with slanted sides.

13. Trees are staked and tied when planted to

   a. keep animals from running over them.
   b. mark the location so that they can be recognized.
   c. keep them from blowing over.
   d. hold them still and prevent movement which can cause root damage.

14. When applying a fertilizer such as a 5-10-5, a general rule of thumb is to apply about
   _____ pounds for each inch of trunk diameter.
   a. 4   c. 6
   b. 5   d. 2

15. Newly planted trees are generally fertilized by punching holes in the ground and placing the fertilizer in the hole. These holes are placed

   a. in a circle 3 feet from the tree with the holes 24 inches apart.
   b. at the drip line circling the tree with the holes 24 inches apart.
   c. at the drip line circling the tree with the holes 10 inches apart.
   d. close to the tree trunk where the feeding roots are located.

16. Watering newly planted trees is not necessary if natural rain supplies at least _____ of rain per week.
   a. 1/2 inch   c. 1 inch
   b. 2 inches   d. 1/4 inch

17. If watering the tree is necessary, no less than _____ of water should be applied at any one time.
   a. 1 inch   c. 2 inches
   b. 1/2 inch   d. 1/4 inch

OBJECTIVE

To select, use, and care for deciduous shrubs.

COMPETENCY TO BE DEVELOPED

After studying this unit, the student will be able to

- identify at least five deciduous shrubs.
- list at least two uses of deciduous shrubs in the landscape.
- explain orally or in writing the three ways deciduous shrubs are purchased.
- outline the planting procedure for deciduous shrubs.

Materials List

one bare root deciduous shrub

one balled and burlapped deciduous shrub

one container grown deciduous shrub

hardiness map

pick and shovel

mulch

fertilizer

pruners

---

unit 29
Deciduous Shrubs

USES IN THE LANDSCAPE

Deciduous shrubs are used widely in the landscape to form borders, screens, background plantings, and foundation plantings. They offer an especially attractive feature in that they change in color as seasons change. Deciduous shrubs are selected for the various effects that are created by their flower color, form, fruit they bear, color and shape of the stem, and color and height of the foliage. Through careful selection, an interesting and attractive landscape design can be developed with deciduous shrubs, either alone or in combination with other plant materials.

The following are some important factors to consider when selecting deciduous shrubs for the landscape.

Period of Interest

A *period of interest* is that time of year in which plant materials create the most interest in the landscape. For example, flower color or fragrance of the flower might be a notable feature determining the period of interest. One example is the mock orange (*Philadelphus*). When in bloom, the fragrance of the mock orange is often apparent before the plant is seen. The period of interest for many shrubs is during the fall months, when they bear colorful fruit. An example is the cranberry bush (*Viburnum opulus*). It produces attractive red fruit in the fall that also provides food for wildlife in the winter.

Foliage

The foliage of the plant has a definite influence on its selection for a particular role in the landscape. For example, some shrubs, such as the oakleaf hydrangea (*Hydrangea quercifolia*), have large, coarse foliage. Others, such as white baby's breath spirea (*Spiraea thunbergi*), are very fine in texture. These fine-textured plants are much more attractive as foundation plantings than are the coarse-textured plants.

Height

When selecting deciduous shrubs, it is important to consider the height of the mature plant. For example, tall growing shrubs should not be planted in foundation plantings where they may block windows.

Hardiness

The hardiness rating of a plant is based on the lowest minimum temperature that the plant can endure without noticeable damage to flower buds or wood. The adaptability of a plant to an area is further influenced by the amount of rainfall, wind velocity, exposure to sun, and soil type. Figure 29-1, page 348 gives necessary information for choosing shrubs to fill a particular role in the landscape.

PURCHASING PLANTS

After particular shrubs are identified for use in the landscape, a source must be located. It is always best to purchase plants from a reliable nursery. Deciduous shrubs may be purchased as bare root, balled and burlapped, or container grown.

Bare Root

A bare root shrub is transported and transplanted with no soil around the roots. Only small shrubs should be purchased bare root. After plants reach three or four years of age or older, the survival rate is better if the plant is moved and planted with a root ball. Bare-rooted plants are moved only when they are dormant. This allows enough time for the roots to establish themselves prior to leafing out (followed by a high demand for moisture).

Balled and Burlapped

Balled and burlapped plants are purchased with a ball of soil around the roots. The root ball is wrapped tightly in burlap to

	Date and Color of Flowering	Period of Interest	Hardiness	Exposure	Soil pH	Propagation	Remarks
Deciduous Shrubs Less Than 3 Feet Tall							
Name							
Chaenomeles japonica (Japanese quince)	orange to red	April	4	sun	6.0 to 7.0		border shrub Grows in partial shade.
Cotoneaster horizontalis (Cotoneaster)	small, pink mid-June	fall	5	southern slopes	6.0 to 7.0	cuttings	good for rock gardens low spreading branches
Daphne genkwa (Daphne, blue daphne)	blue April	April	5	sun	6.5 to 7.5	cuttings	difficult to transplant slow growing
Deutzia gracilis (Slender deutzia)	white May	May	4	sun		cuttings	showy, white
Forsythia viridissima (Dwarf forsythia)	yellow April	April	5	sun		layering cuttings	rock garden plant
Hypericum sp (St. Johnswort)	yellow summer	summer	5	sun		cuttings	many forms; yellow flowers throughout summer; dense
Potentilla fruiticosa (Cinquefoil)	yellow single rose mid-May	summer	2	sun		stem cuttings	fine textured good border plant
Spiraea bumalda 'Anthony watereri' (Anthony Waterer spirea)	pink	June	6	sun	well drained 6.0 to 7.0	cuttings	Blooms throughout summer compact and bushy
Deciduous Shrubs 3 to 6 Feet Tall							
Abelia grandiflora (Abelia) (See figure 29-2)	pink summer	summer	6	sun	6.0 to 8.0	cuttings	Flowers all summer. good border or foundation plant
Berberis thunbergi (Green leaf Japanese barberry)	yellow	fall winter	6	dry sun	6.0 to 7.5	cuttings	small hedge Grows well under all conditions. Green foliage in summer; red foliage in fall
Berberis thunbergi 'Atropurpurea' (Red leaf barberry)	yellow	fall winter	6	dry sun	6.0 to 7.5	cuttings	red all summer good specimen or hedge

Fig. 29-1

		Deciduous Shrubs 3 to 6 Feet Tall (continued)						
Name	Date and Color of Flowering	Period of Interest	Hardiness	Exposure	soil pH	Propagation	Remarks
Buddleia davidii (Butterfly bush)	white, pink, and blue spikes	summer	6	sun		cuttings	Cut back to ground in spring. Fast growing, arching plant
Euonymus alatus 'compactus' (Dwarfwinged euonymus)	greenish May	fall	4	sun	5.5 to 7.0	cuttings	brilliant red leaf in fall dense shrub hedge plant
Fothergilla monticola (Fothergilla)	white	fall May		sun	5.0 to 7.0	cuttings	
Hydrangea quercifolia (Oakleaf hydrangea)	white mid-July	summer fall	6	sun	moist soil 6.0 to 7.0	division cuttings	naturalistic effect when used to edge wooded lots
Kerria japonica (Japonica)	yellow	winter May		sun	6.0 to 7.0	cuttings division	green twigs in winter large perennial borders sometimes called *Mt. Vernon shrub*
Prunus glandulosa (Flowering almond)	double pink	early May	4	sun	6.0 to 7.0	cutting	double blossoms
Rhododendron calendulaceum (Flame azalea)	yellow to red	late May		sun	4.5 to 5.5	layering cuttings	acid soil open growth habit
Spiraea thunbergi (Baby's breath spirea)	white	April		sun	6.5 to 7.0	cuttings	informal hedge fine textured
Spiraea prunifolia (Bridalwreath spirea)	white	April	5	sun	6.0 to 7.0	cuttings	arching branches background plant
Spiraea vanhouttei (Vanhoutte spirea)	cascade of white, flat flowers	May	5	sun shade	6.0 to 7.0	cuttings	good hedge should not be sheared best to prune to base
Weigela sp (Weigela) (See figure 29-2)	pink to red	May	5	sun	6.0 to 7.0	cuttings	abundance of flowers used in boundary plantings

Fig. 29-1 (continued)

	Date and Color of Flowering	Period of Interest	Hardiness	Exposure	Soil pH	Propagation	Remarks
				Deciduous Shrubs 6 to 10 Feet Tall			
Name							
Aesculus parviflora (Bottlebrush buckeye)	white spikes	July	5	semishade	6.0 to 7.0	cuttings	green mass of foliage in summer
Elaeagnus umbellata (Autumn olive)		fall	3	sun	5.0 to 6.0	cuttings	Provides wildlife food. pink-red fruit silvery foliage used in border plantings
Forsythia intermedia spectabilis (Border forsythia)	yellow	spring	5	sun	6.0 to 7.5	cuttings	good in borders easy to transplant fast growing
Ilex verticillata (Winterberry)	inconspicuous	winter	4	sun	5.5 to 6.5		showy red fruit in winter Grows well in swamps.
Kolkwitzia amabilis (Beauty bush)	pink	June	5	sun	6.0 to 7.5	cuttings	boundary or specimen plant Requires sun.
Ligustrum ovalifolium (California privet)	white	May	4	shade sun	6.0 to 7.0	cuttings	good hedge material
Lonicera tatarica (Tatarian honeysuckle)	white to yellow	May	4	sun	6.5 to 8.0	cuttings	red berries in June or July
Philadelphus sp (Mock orange)	white	June	6	sun	6.0 to 8.0 well drained	cuttings	Many varieties are fragrant.
Syringa chinensis (Chinese lilac)	lavender	April	5	sun	6.0 to 7.5 moderately drained	cuttings	varieties available with white or red flowers
Syringa vulgaris (Common lilac) (See figure 29-2)	purple	April May	3	sun	6.0 to 7.5	cuttings	Powdery mildew is sometimes a problem. Suckers are pruned back to stimulate blooming.

Fig. 29-1 (continued)

Name	Date and Color of Flowering	Period of Interest	Hardiness	Exposure	Soil pH	Propagation	Remarks
Viburnum dentatum (Arrowwood)	flat white clusters	late May	3	wet to damp woods	6.0 to 7.5	cuttings	excellent wildlife food source and border plant
Viburnum macrocephalum (Chinese snowball)	large white flower heads	May	6	sun	6.0 to 7.5 rich soil	cuttings	largest flower heads of all snowballs
Viburnum opulus (Cranberry bush) (See figure 29-2)	white	May fall	3	sun	6.0 to 7.5	cuttings	park plantings red fruit purplish brown fall color
Viburnum plicatum tomentosum (Double file viburnum)	white	May fall	5	sun	6.5 to 7.5	cuttings	Flowers and fruit develop on upper side of branch.

Fig. 29-1

Deciduous Shrubs 6 to 10 Feet Tall (continued)

▲ COMMON LILAC

▲ ABELIA

EUROPEAN CRANBERRY BUSH ▲

▶ WEIGELA

Fig. 29-2 Four deciduous shrubs used in landscapes. (Richard Kreh, photographer)

hold the soil firmly around the roots and to prevent the soil from breaking apart and separating from the roots. The ball should be large enough to include most of the plant's root system, or as large a ball as can be lifted and moved. Balled and burlapped plants can be transplanted anytime during the year.

Container Grown

Container grown shrubs are cultivated in various containers, such as pots, baskets, and tubs. The root system is not damaged during transplanting, so these plants can be transplanted at any time during the year. The roots of root bound plants should be straightened after removal from the container. If twisted or curled roots cannot be straightened, they should be cut off at the edge of the ball to encourage the development of new roots as the plant continues to grow.

PLANTING TIME

The best time to plant deciduous shrubs is when they are in their dormant stage. This usually occurs in fall or early spring. In areas with mild winters, however, deciduous shrubs may be planted year round. This includes areas south of the Chesapeake Bay through the coastal plains of Texas and the coasts of Oregon and central California. Where winters are severe, spring is the best time to plant.

Planting Site

Most shrubs prefer a well-drained soil. A good garden loam is the best soil type for most shrubs. Soils may be improved upon by the addition of sand for drainage or organic matter to help hold moisture in dry soils. Soil pH preference and the effects of exposure to sun and wind vary for each shrub type. The selection

of the proper site is important since deciduous shrubs are relatively expensive and long living.

Planting Procedure

Note: Plant roots should never be allowed to dry prior to planting. Bare-rooted shrubs should be packed in moist peat moss or sawdust. Balled and burlapped and container grown plants should be kept properly watered until planting time.

1. Dig a hole large enough to allow the roots to spread in a normal manner without cramping or twisting. Dig deeply so that 2 to 4 inches of topsoil can be placed under the roots. The sides of the hole should be straight up and down and the bottom flat. This flat surface encourages straight growth of the root system.
2. Prune diseased and broken roots.
3. Plant the shrub at the same depth as that at which it grew before.
4. Work the soil in and around the roots.
5. Build up a slight berm around the outer edge of the hole to hold water.
6. Water well to settle the soil around the roots, eliminate any air pockets, and moisten the soil. Water regularly for the first year any time that the plant receives less than 1 inch of natural rainfall per week.
7. Mulch with chips or bark to a depth of 2 to 4 inches. Apply coarse material more heavily than fine material.
8. Prune the top to remove any broken branches and to reduce the size of the shrub by about one-third. This helps to balance the loss of roots in the transplanting process. Pruning should also be done to help shape the plant.

Holes for balled and burlapped and container grown plants should be at least 25 percent larger in diameter than the root balls to be planted. After the plant is placed in the hole, the container is cut away and removed or the burlap rolled down to expose the root ball as much as possible without loosening or breaking the ball apart. Soil is worked in around the root ball and the plant is watered and mulched.

CARE OF THE PLANT

Fertilizer

Slow-release fertilizer may be mixed into the soil that is placed around the plant, or regular fertilizer may be applied on top of the soil before mulching. A 10-10-10 fertilizer is a good choice as a top application for most shrubs.

Disease and Insect Control

Some deciduous shrubs require spraying for control of insects or diseases. Consult your local agriculture experiment station or extension service for details.

Pruning

Pruning is used to thin out old or dead wood, to shape plants, and to control plant size. The following are general rules that apply to pruning shrubs.

1. Shrubs that bloom on wood grown the previous season should be pruned immediately after flowering. (Examples are forsythia and spirea.)

2. Shrubs that bloom on the current year's growth should be pruned in fall or early spring. (One example is the rose.)

3. Some shrubs require annual pruning to thin out old wood or wood that is shaded out and killed. (Examples are hydrangea, privet, spirea.)

4. If shrubs send up shoots or suckers from the roots or at the base of the plant, at least some of them should be removed so that growth on the plant does not become too thick.

Deciduous shrubs are a long living investment in the landscape. Their uses are as many as are the rewards in year-round beauty.

STUDENT ACTIVITIES

1. Make ten cuttings of selected deciduous shrubs in the area and propagate them. Observe the growth habits of the resulting plants. Practice pruning, fertilizing, and insect and disease control. If possible, transplant the shrubs to school grounds.

2. Design and develop a small rock garden using shrubs discussed in this unit that are common to the local area.

3. Ask a representative from a local nursery to speak to the class on the uses of deciduous shrubs in the landscape.

SELF-EVALUATION

1. When used to describe a plant, the word *deciduous* means that
    a. the plant is perennial.
    b. the plant sheds its leaves each year.
    c. the plant is a shrub.
    d. all of the above

2. The period of interest for a deciduous shrub is
    a. the time of year during which it creates the most interest in the landscape.
    b. the time of year during which it sheds its leaves.
    c. pruning time.
    d. blooming time.

3. Pruning is used to
    a. shape plants.
    b. reduce the size of the top of the plant when transplanting.
    c. thin out dead wood.
    d. all of the above.

4. A bare-rooted plant is one that must be planted
    a. in the fall of the year.
    b. in the spring.
    c. when it is dormant.
    d. after severe pruning of the top.

5. One advantage in purchasing container grown plants is that
    a. they may be planted during any season of the year.
    b. they are less expensive than other forms.
    c. they are more readily available.
    d. they have better root formation.

6. If the roots of a container grown plant are twisted around in the pot,
    a. they are left undisturbed before planting.
    b. they are straightened out or pruned before planting.
    c. the plant is placed back in the container for growing.
    d. the root ball is broken apart.

7. Deciduous shrubs prefer a soil that is

    a. a well-drained loam.   c. fertile and well mulched.
    b. sandy and moist.   d. underlaid with a clay subsoil.

8. The planting hole for deciduous shrubs is formed with straight sides and a flat bottom

    a. to make it more attractive.
    b. to simplify packing soil around the plant roots.
    c. to encourage root growth through the sides of the hole.
    d. because this kind of hole is easier to dig.

9. The planting depth for shrubs is

    a. 2 inches deeper than the depth at which they previously grew.
    b. 4 inches deeper than the depth at which they previously grew.
    c. 2 inches above ground level.
    d. the same depth at which they were growing before.

10. Shrubs that bloom on previous season's wood are pruned

    a. immediately after blooming.   c. in the spring.
    b. in the fall.   d. in midsummer.

11. Shrubs that bloom on the current season's growth are pruned

    a. immediately after blooming.   c. in the spring.
    b. in the fall or early spring.   d. in the fall.

unit 30
Ground Covers

OBJECTIVE

To select, establish, and maintain ground covers in the landscape.

COMPETENCY TO BE DEVELOPED

After studying this unit, the student will be able to

- identify the three major types of ground covers.
- list three uses of ground covers.
- describe the cultural requirements of ground covers.
- identify five factors that must be considered when selecting ground covers.

Materials List

samples of different types of ground covers

5-10-5 fertilizer

peat moss

pine bark or hardwood mulch

Ground covers are low-growing plants that cover the ground in place of turf. When used functionally, ground covers fill in bare spots in the landscape, help prevent erosion of soil on steep banks, and fill in shady areas under trees where other plants have difficulty growing.

TYPES OF GROUND COVERS

There are three principal types of ground covers, including broadleaf evergreens, deciduous plants, and coniferous plants.

Name	Date & Color of Flowering	Height	Light	Foliage	Propagation	Hardiness	Remarks
Ajuga reptans (Bugleweed)	blue to purple April or May	4" to 8"	shade or sun	narrow leaves, 4" long	seed and division	4	Gives dense coverage.
Arabia alpina (Rock-cress)	white April to May	4" to 10"	full sun	evergreen	seed and division	3	Spreads by creeping rootstock; useful in small areas.
Asarum caudatum (Wild ginger)	brownish purple June	6" to 8"	shade	heart-shaped leaves with pungent taste	division	4	Grows best in moist soil which is high in organic matter.
Convallaria majalis (Lily-of-the-valley)	white spikes mid-May to mid-June	8"	shade	large oval leaves	division	2	Spreads by underground rootstock.
Cornus canadensis (Bunchberry)	small, yellow flowers May to June	9"	shade	evergreen	layering	2	Spreads by underground rootstock.
Coronilla varia (Crown vetch)	pinkish white June to September	1' to 2'	full sun	compound leaves	division or seed	3	excellent cover for steep slopes
Dianthus deltoides (Maiden pink)	red to pink May to June	2" to 6"	full sun	dense, glossy carpet	division or seed	2	Needs well-drained, acid soil.
Epimedium grandiflorum (Barrenwort)	white, yellow or lavender May to June	12" to 15"	semi-shade	dense	division	3	good in any soil
Hosta undulata (Plantain lily)	lavender August	2' to 3'	full sun to shade	dense, variegated	division	3	Dies to ground in fall.
Hypericum calucinum (Aaronsbeard, St. Johnswort)	bright yellow July to August	1'	semi-shade	dense after established	seed, division, or cuttings	6	Spreads freely by stolons.

Fig. 30-1 Common Ground Covers

Name	Date & Color of Flowering	Height	Light	Foliage	Propagation	Hardiness	Remarks
Iberis sempervirens (Evergreen candytuft)	white late May	12″	full sun	narrowleaf evergreen	division, cuttings, or seed (Roots easily.)	4	Needs moist soil.
Liriope muscari (Lilyturf)	lavender to purple August	10″ to 15″	sun or shade	glossy	division	4	Spreads underground stem forming a mat or sod.
Lysimachia nummularia (Moneywort)	yellow summer	2″ to 3″	sun or shade	leafy, soft stem	division	3	Spreads quickly. May become pest in lawns if not controlled.
Ophiopogon japonicus (Dwarf lilyturf)	violet July to August	10″	sun or shade	glossy	division	6	Forms a mat or sod.
Pachysandra terminalis (Japanese spurge)	small, white spikes early May	6″	shade	dark green, leafy	cuttings or division	4	excellent with evergreens
Sedum (Stonecrop)	yellow, white or pink summer	2″ to 8″	full sun	evergreen	division or cuttings	3 or 4	good for dry areas
Vinca minor (Periwinkle or myrtle)	blue violet or white late April	6″	partial to full shade	evergreen	division, root or stem cuttings	4	Needs fertile soil.
Arctostaphylos uva-ursi (Bearberry)	white to pink May	6″ to 12″	sun or partial shade	dark green	cuttings	2	hard to transplant unless pot grown
Calluna vulgaris (Heather)	white to red summer	4″ to 24″	sun	small needles	cuttings	4	Needs acid soil high in organic matter.

Fig. 30-1 Common Ground Covers (continued)

Name	Date & Color of Flowering	Height	Light	Foliage	Propagation	Hardiness	Remarks
Cotoneaster microphylla (Cotoneaster)	white June	6″ to 12″	sun	small, dark green leaves	cuttings	5	useful on banks
Erica carnea (Heath)	rosy red early April	6″ to 12″	shade	small needles	cuttings	5	Needs loose, well-drained soil.
Euonymus fortunei (Wintercreeper)	none	4″	shade or sun	dark green	cuttings or division	5	good on banks Holds leaves most of winter.
Gaultheria procumbens (Wintergreen)	white mid-May	3″	shade	small, shiny green leaves	cuttings division	3	Needs moist soil high in organic matter.
Hedera helix (English ivy)	small, greenish September	6″ to 8″	full or partial shade or full sun	dark green clinging vine	cuttings	5	many cultivars available excellent plant
Helianthemum nummularium (Sunrose)	yellow, pink or white June-July	6″ to 12″	full sun	evergreen	seed, cuttings, division	5	Does best in dry soil.
Mahonia repens (Creeping mahonia)	yellow early May	10″	sun	dull bluish green, leathery and spiny like holly	root cuttings or division	5	vigorous when established
Mitchella repens (Partridgeberry)	white	4″ to 6″	shade	dark green	division or seed	5	tolerant of dense shade and slow growing
Pachistima canbyi (Pachistima)	small reddish early May	12″	sun or shade	small, dark green, and spreading	cuttings, layering, division	5	Needs fertile, acid soil.
Sarcococca hookeriana humilis (Sarcococca)	fragrant white mid-spring	1′ to 2′	light to heavy shade	glossy green	seed, division, cuttings	5	Prune in early spring to keep plant compact.

Fig. 30-1 Common Ground Covers (continued)

Name	Height	Light	Foliage	Propagation	Hardiness	Remarks
Juniperus horizontalis 'Wiltoni' (Blue rug)	4″ to 6″	full sun	steel blue	cuttings	2	Does not flower.
Juniperus horizontalis 'Bar Harbor'	10″ to 12″	full sun	greenish blue	cuttings	2	Does not flower.
Juniperus horizontalis	12″ to 18″	full sun	summer — greenish blue winter — purple	cuttings	2	Does not flower.
Juniperus horizontalis 'Douglasi' (Waukegan)	12″ to 18″	full sun	summer — steel blue winter — tinge of purple	cuttings	2	Does not flower.
Taxus baccata 'Repandens' (Spreading English yew)	36″	full sun to partial shade	deep dark green	cuttings	2	Does not flower.

Fig. 30-1 Common Ground Covers (continued)

Most ground covers are perennial, but a few are annual plants. Figure 30-1 lists common ground covers and planting information for each.

Broadleaf evergreen ground covers retain their leaves year-round.

Deciduous ground covers lose their leaves during the fall and winter season.

Coniferous ground covers have needlelike or scalelike leaves. These plants retain their color throughout the year and thereby make excellent ground covers.

USES

Ground covers play an important role in landscaping because they can be used in many areas where the soil is not suitable for growing grass. These areas may be too steep, rocky, shaded, shallow soiled, or eroded for the proper growth of grass. However, ground covers are also used simply for the beauty they contribute to the landscape. Low-growing, dense ground covers that grow relatively slowly make attractive foreground plantings for shrubbery borders. They are also placed between plantings of broadleaf evergreens where they serve to keep the soil cool and shade the roots of certain plants, such as rhododendrons.

PERIWINKLE OR MYRTLE

▼ JAPANESE SPURGE

▼ VARIEGATED WINTER CREEPER

▲ SPREADING ENGLISH YEW

Fig. 30-2 Some commonly used ground covers (Richard Kreh, photographer)

SELECTION

Before selecting ground covers for use in the landscape, the following questions must be answered.

- In what type of soil does the plant grow best?
- Is it suitable for the locality?
- How is it propagated?
- How long will it take for the plants to cover the area in which they are planted?
- What is the mature height of the plant?
- Are diseases and insects a problem?
- How expensive are the plants?

There may be other questions which require answering, depending upon the particular landscape in question. Is the ground cover to serve a specific role in the landscape? If so, even more care must be taken when choosing the plant. If steep banks are to be covered for control of soil erosion, a plant that keeps its stem and plant top year-round should be selected. An evergreen would probably be best for this use.

Remember, also, that ground covers, as other plants, differ somewhat according to the soil type, moisture conditions, and amount of light they require. In a lightly shaded area, English ivy, pachysandra, and vinca minor would be good choices. For a sunny bank, a prostrate juniper might be chosen. For planting in an area with continuous shade on the north side of the building, English ivy, ajuga reptans, or lily-of-the-valley are wise choices. If the effect in the landscape is desired only for certain seasons of the year, select a plant that is most attractive during that season.

PLANTING GROUND COVERS

When preparing the soil for ground covers, consider any special problems in the area. On a steep slope, very little can be done except to add topsoil if needed. No tilling or digging in of organic matter is possible on a very steep slope. If tilling can be done, add organic matter in the form of peat moss or rotted manure and 2 to 5 pounds of a 5-10-5 fertilizer spread evenly over each 100 square feet. Dig the materials into the top 6 inches of soil before planting.

Ground cover plants are spaced according to their size, growth rate per year, and the time allowed for plants to spread and cover the area. On steep slopes, the area may need to be covered more quickly. In this case, place the plants closer together. On more level areas, the plants may be spaced farther apart. This reduces original plant costs. It will, of course, take longer for the ground cover to grow together and completely cover the soil surface. Local nursery employees may be able to give spacing directions for the area to be planted. Water well after planting to settle soil around the roots.

CARING FOR GROUND COVERS

Fertilizing

After the initial application of fertilizer at planting time, fertilizer should be applied only as needed to keep plants healthy. If fertilizing is necessary, it should be done in early spring. There are two ways in which it can be applied: by means of a dry, granular fertilizer which is scattered over the area and watered in, or in the form of a soluble fertilizer that is mixed with water and siphoned through a garden hose and sprinkler system.

Once ground covers are established, further soil cultivation is not necessary. However, it is good practice to mulch between newly positioned plants to aid in weed control and moisture retention until the ground cover has traveled over the entire soil surface. When mulching, cover the entire surface with the material before the plants are placed in the ground. This method is less time-consuming than if mulch were applied after the plants were installed. Push the mulch from the area where the hole is to be made and pull it back around the plant as soon as the plant is set in the soil. Pine bark, hardwood bark, pine needles, well-rotted sawdust, peat moss, and decomposed leaves are suitable mulching materials for ground covers.

Pruning

Pruning is only necessary to confine the planting to the area in which it is desired. This is usually accomplished by mowing around the edges or otherwise cutting back the outer perimeter.

Watering

Newly planted ground covers should be watered as needed during the first year. Any time the soil becomes dry or plants start to wilt, apply at least 1 inch of water to the entire planting area. If the proper plants are selected, watering should not be necessary after the first year.

Controlling Insects and Diseases

Insects and diseases are rarely a serious problem with ground covers. If control becomes necessary, identify the disease or insect and spray with a recommended pesticide. (Be sure to use protective clothing as required by the pesticide container label.)

STUDENT ACTIVITY

Draw a landscape plan depicting an area around the school or at home. Position ground covers where they could be used to best advantage. A class project might include actually planting and cultivating one type of ground cover.

SELF-EVALUATION

Select the best answer from the choices offered to complete each statement.

1. Ground covers selected for use should be

   a. adapted to the soil in which they are to be planted.
   b. adapted to the local hardiness zone.
   c. relatively free of disease and insect problems.
   d. all of the above

2. For steep banks with soil erosion problems, a ground cover should be selected that

   a. is an annual.
   b. is deciduous.
   c. keeps its stem and top cover aboveground year-round.

3. Before planting ground cover plants, _____ to _____ pounds of a 5-10-5 fertilizer should be applied to every 100 square feet of area.

   a. 5 to 10   c. 1 to 2
   b. 10 to 20   d. 2 to 5

4. The best time to apply fertilizer to an established ground cover is

   a. late fall.   c. early spring.
   b. midsummer.   d. midwinter.

5. The best time to apply mulch to an area planted to ground covers is

   a. before any plants are planted in the area.   c. as soon as weeds appear.
   b. after any plants are planted in the area.   d. every fall.

6. Pruning of ground covers is necessary to

   a. keep them from becoming too tall.
   b. confine the plants to the area set aside for them.
   c. renew growth.
   d. thin the plants.

7. A ground cover that is properly selected for the area planted should be watered

   a. every ten days.
   b. each year, in the heat of summer.
   c. every spring and fall.
   d. only during the first year after planting.

8. In the cultivation of ground covers, insects and diseases

   a. are almost always a problem and must be carefully controlled.
   b. never attack and therefore never require control.
   c. are generally not a serious problem, but sometimes require control.
   d. are a very limiting factor.

OBJECTIVE

To force bulbs indoors and to use bulbs in the landscape.

COMPETENCY TO BE DEVELOPED

After studying this unit, the student will be able to

- list four uses of bulbs in the landscape.
- describe the soil and fertilizer used in the flowering of bulbs.
- explain how planting depth and spacing of bulbs are determined.
- describe how to care for bulbs after they have flowered.
- list the steps in the forcing of bulbs.

Materials List

assorted bulbs and corms including tulip, daffodil, crocus, hyacinth, amaryllis, begonia, calla, ismene, and gladiolus

bulb planter and nursery spade

bag of 5-10-15 fertilizer

cold frame or cold treatment storage

bulb bed for demonstration

greenhouse space, if available

---

unit 31 Bulbs

Many people use the term *bulb* in a general way to refer to bulbs, corms, tubers, and rhizomes — all of which are structures containing an embryonic plant and the necessary stored food for plant growth. However, these structures are different in appearance and in their methods of propagation. Review Unit 12 for information on types of bulbs and descriptions of each.

USES OF BULBS IN THE LANDSCAPE

Many people purchase flowers which grow from bulbs, such as tulips, hyacinths, and daffodils, but are unsure about how to use them most effectively in the landscape. As a general rule, bulbs are most striking when they form a *massing* (grouping of color). This is usually more attractive than mixing various colors in the same bed.

In Wooded Areas

Bulbs grow well and produce beautiful color in natural wooded areas. Some bulbs grow well with evergreen ground covers in these areas. Growing taller than the ground covers, they give an excellent show of color. After they have flowered, the tops turn brown and die back. However, the ground cover remains to fill this area in the landscape until the next season's blooming of bulbs.

In Rock Gardens

Rock gardens are excellent places to use flowering bulbs. Bulbs that produce flowers that can grow low to the ground can be worked in very easily. To be most appealing, the flowering bulbs should be massed together. Using color masses throughout the garden accents the areas in which they are planted; this is an effective way to emphasize certain areas.

With Evergreen Shrubs

Many people use bulbs to add color around evergreen foundation shrubs. Bulbs planted in fall give early spring color to the landscape. Summer annuals can be planted to replace the color when the bulbs die back.

As Cut Flowers

Tulips and narcissus are used extensively by florists in January, February, and March to create an early touch of spring in arrangements. When bulbs are to be used in these months, they must be forced to bloom out of season, a process which will be discussed later in this unit.

If bulb flowers are to be used as a cutting garden, the beds for cutting should be located in a special area so that when flowers are cut, the appearance of the landscaped area is not spoiled.

SELECTING AND PREPARING SOILS

Bulbs grow well in a good garden loam which is well-drained. The organic matter content in the soil can be improved by adding sphagnum moss, leaf mold, garden compost, or well-rotted manure. To apply these materials, spread them over the ground in a layer of 3 to 4 inches thick and dig in with a Rototiller in the spring before planting in the fall.

Bulbs grow better if the soil is neutral, that is, neither acidic nor alkaline. If the pH value is below 6.0, ground limestone should be added at the rate of ten pounds per 100 square feet to raise the pH to 7.0.

PLANTING BULBS

Bulbs such as the crocus, narcissus, hyacinth, and grape hyacinth are planted in the fall. Others, such as the dahlia, amaryllis, and gladiolus, are planted in the spring. It is important to consider the height of various flowers when planting bulbs since they are most attractive when they grow in sequence of height. When planting bulbs, use a bulb planter, nursery spade, or hand trowel.

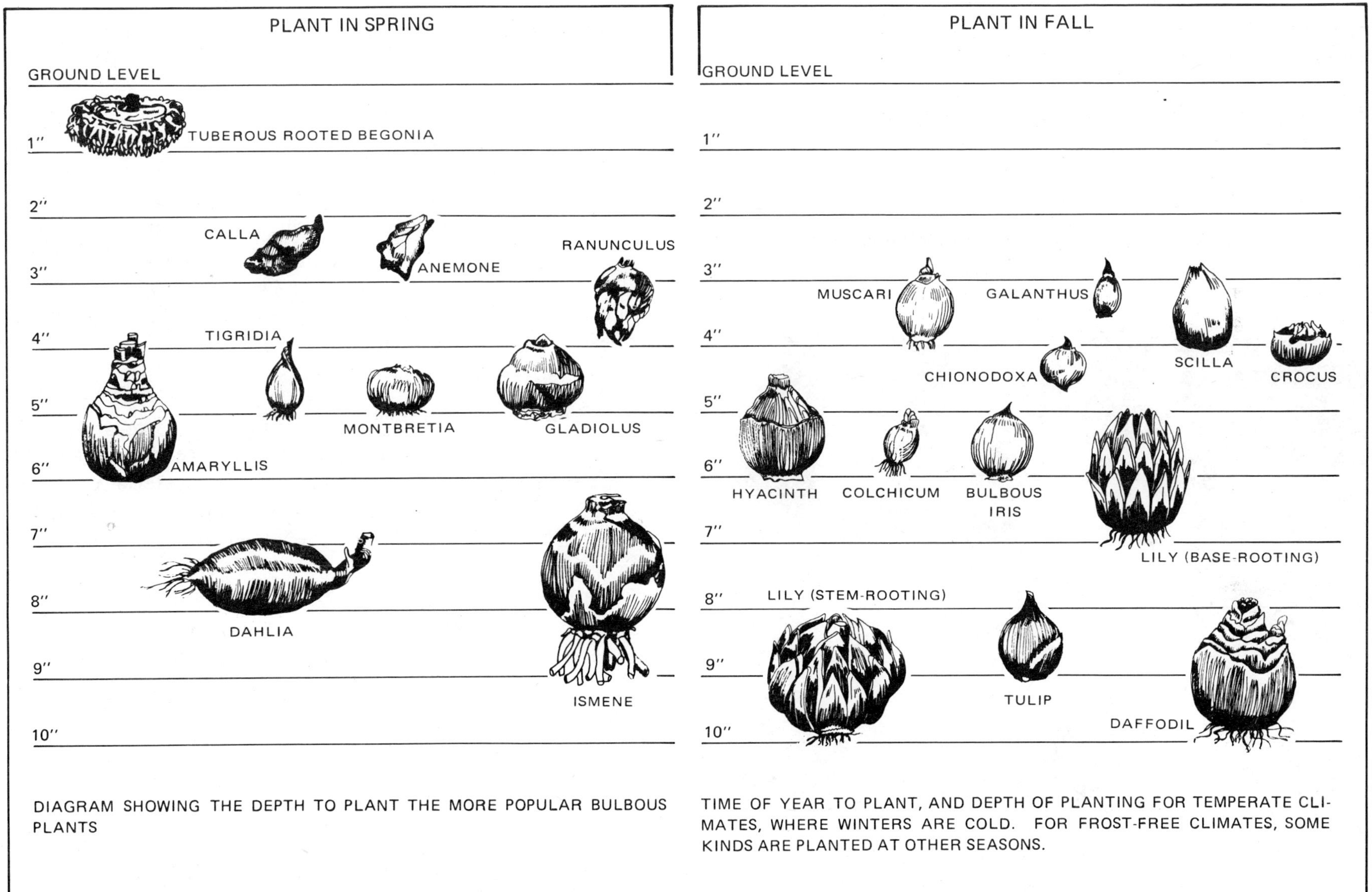

PLANT IN SPRING

GROUND LEVEL

1"   TUBEROUS ROOTED BEGONIA

2"

CALLA   ANEMONE   RANUNCULUS

3"

TIGRIDIA

4"

5"   MONTBRETIA   GLADIOLUS

6"   AMARYLLIS

7"

8"   DAHLIA

9"

10"   ISMENE

DIAGRAM SHOWING THE DEPTH TO PLANT THE MORE POPULAR BULBOUS PLANTS

PLANT IN FALL

GROUND LEVEL

1"

2"

3"   MUSCARI   GALANTHUS

4"   CHIONODOXA   SCILLA   CROCUS

5"

6"   HYACINTH   COLCHICUM   BULBOUS IRIS

7"   LILY (BASE-ROOTING)

8"   LILY (STEM-ROOTING)

9"   TULIP

10"   DAFFODIL

TIME OF YEAR TO PLANT, AND DEPTH OF PLANTING FOR TEMPERATE CLIMATES, WHERE WINTERS ARE COLD. FOR FROST-FREE CLIMATES, SOME KINDS ARE PLANTED AT OTHER SEASONS.

Fig. 31-1 Planting Depth of Bulbs (courtesy Brooklyn Botanic Garden)

Each type of bulb has a recommended planting depth and spacing, figure 31-1. Spacing of the plants may be varied to achieve different effects. If a bed is to be showy, the bulbs are spaced closer together, figure 31-2. As a general rule, bulbs should be spaced the same distance apart as their planting depth. This gives a more natural appearance to the landscape.

FERTILIZING BULBS

As stated before, bulbs have stored food which provides all nutrients for the current year's development of the flower. After flowering, the bulb builds reserve for the next year's flowers. This

Fig. 31-2 Tulip bulbs are planted at 6 inches center to center for maximum production of floral display. (USDA photo)

is the best time to fertilize bulbs for maximum production. Nutrients high in phosphorus and potassium are necessary for well-developed bulbs. **Example.** Five pounds of 5-10-15 fertilizer applied to every 100 square feet of area.

CARE AFTER FLOWERING

After the foliage dies, bulbs should be dug up and stored at 50°F (10°C) in dim light. Dusting with a fungicide prevents insects and rodents from attacking the bulbs while they are in storage. Remove any bulbs showing signs of disease or mechanical injury (those cut during digging); this prevents the spread of disease and rotting of other bulbs.

BULB PESTS AND DISEASES

Bulbs require special protection from certain diseases and insects. Many pests can be controlled by maintenance of good cultural conditions. This is accomplished by removing weeds and other rubbish which provide a natural home for insects and disease organisms.

Chemical control is also effective and usually necessary. Applying chemicals before placing bulbs in storage is a good practice. Before applying any chemical, check the label to be sure it will not injure the bulbs.

Consult local garden centers, nurseries, or hardware stores for information on pesticides available for use on bulbs.

Figure 31-3 lists the most common types of bulbs, pests and the damage they cause, and control methods which can be used.

COMMON BULB PESTS AND DISEASES AND THEIR CONTROL

HOST, Pest, Disease	DAMAGE	CONTROL	HOST, Pest, Disease	DAMAGE	CONTROL
AMARYLLIS			**LILY**		
Spotted Cutworm	Feeds on flowers at night.	Scatter cutworm bait or spray with Sevin.	Aphids (Lily, Bean, Melon, Peach, other species)	Curl leaves, transmit mosaic and other virus diseases.	Spray with malathion, being sure to cover underside of leaves.
Bulb Mites	Rotting bulbs. See Hyacinth.	Discard soft bulbs.	Botrytis Blight	Oval tan spots on leaves, which turn black, droop.	Spray with bordeaux mixture.
Narcissus Bulb Fly	Decaying bulbs. See Narcissus.	Discard soft bulbs.	Mosaic and other virus diseases	Plants mottled, stunted.	Rogue infected plants. Start lilies from seed in isolated portion of garden.
Leaf Scorch, Red Blotch	Reddish spots on flowers, leaves, bulb scales; stalks deformed.	Discard bulbs or remove diseased leaves. Avoid heavy watering.			
GLADIOLUS			**NARCISSUS**		
Thrips	Leaves silvered, flowers streaked, deformed.	Spray with lindane in spring. Dust corms before storing.	Narcissus Bulb Fly	Fly resembling bumblebee lays eggs on leaves near ground in early summer. Larva, fat, yellow maggot ½ to ¾ inch long, tunnels in rotting bulb.	Sprinkle naphthalene flakes around plants to prevent egg-laying. Before planting dust trench with 5% chlordane and dust over bulbs after setting.
Botrytis and other flower blights	Flowers, leaves, stalks spotted, then blighted.	Spray with zineb (Dithane Z 78 or Parzate).	Bulb Nematode	Dark rings in bulb.	Discard bulbs. Commercial growers treat with hot water, adding formalin to prevent rot.
Corm Rots, Scab	Lesions on corms, spots on leaves.	Dust with Arasan before planting.	Basal Rot	Chocolate-colored dry rot at base of bulbs.	Inspect bulbs before planting.
Yellows (due to a soil fungus)	Plants infected through roots, turn yellow and wilt.	Choose resistant varieties.	Smoulder (Botrytis Rot)	Plants stunted or missing; masses of black sclerotia on rotting leaves or bulbs.	Remove diseased plants. Put new bulbs in new location.
HYACINTH			Scorch	Yellow, red, or brown spots blight tips of leaves.	Spray or dust with zineb, maneb or copper.
Bulb Mites	Minute; less than 1/25 inch, white mites in rotting bulbs.	Discard infested bulbs.	**TULIP**		
Aphids, several species	Leaves are curled; virus diseases may be transmitted.	Spray with malathion, rotenone, or nicotine.	Tulip Bulb Aphid	Powdery white or grayish aphids common on stored bulbs.	Dust with 1% lindane before storing.
Bulb Nematode	Dark rings in bulbs.	Discard.	Green Peach, Tulip Leaf and other Aphids	Transmit viruses to growing plants.	Spray or dust with malathion or lindane.
Soft Rot	Vile-smelling bacterial disease. often after mites.	Discard.	Botrytis Blight, Fire	Plants stunted, buds blasted, white patches on leaves, dark spots on white petals, white spots on colored petals, gray mold, general blighting. Small, shiny black sclerotia formed on petals, foliage rotting into soil and on bulbs.	Discard all infected bulbs. Plant new tulips in new location. Spray with ferbam or zineb, starting early spring. Remove flowers as they fade, remove all tops as they turn yellow.
IRIS (Bulbous)					
Tulip Bulb Aphid	See Tulip.	See Tulip.	Cucumber Mosaic	Yellow streaking or flecking of foliage.	Do not grow near cucurbits or gladiolus.
Gladiolus, Iris Thrips	Leaves russeted or flecked, flowers speckled or distorted.	Spray or dust with malathion or lindane.	Lily Mottle Viruses	Cause broken flower colors, mottled foliage, in tulips.	Do not plant near lilies. Control aphids.
Leaf Spot	Light brown foliage spots with reddish borders.	Spray with zineb or bordeaux mixture; clean up old leaves.			

Fig. 31-3 (Courtesy Brooklyn Botanic Garden)

◄

Fig. 31-4 Bulbs used for forcing should be large. Here, bulbs are being graded for size. (USDA photo)

FORCING BULBS

Forcing bulbs is an interesting and challenging aspect of bulb production. The fragrance and color of flowering bulbs indoors during the late winter months can be very refreshing. When buying bulbs, be sure that they are the best quality and size by purchasing them from a local garden center or florist.

After bulbs are potted, they are placed outside in a cold frame and covered with straw. (A *cold frame* is a frame, usually glass covered, which is used to protect plants without artificial heat.) Roots will start to develop after the ground freezes. The pots should be left in the cold frame for eight weeks, then brought into a cool, partially lit room. This is done so that the bulbs will

Fig. 31-5 The bulbs are set in a 6-inch azalea pot or bulb pan. Media is added so that it almost covers the entire bulb. (USDA photo) ▶

develop a root system, break dormancy, and establish some growth before they are brought into a warmer room for flowering.

Procedure for Forcing Bulbs

1. Identify bulbs to use for forcing. Select Number 1 (large) bulbs, figure 31-4.

2. Select a potting media that is well-drained and high in organic matter: one-third soil, one-third sand, and one-third peat moss.

3. Place drainage material (a stone or piece of clay pot) in the bottom of the pot.

4. Place a layer of sphagnum moss and then a layer of the media in the pot. Set the bulb in the pot and fill in around the bulb with the growing media so that the top of the bulb is exposed, figure 31-5.

5. Water by setting the pot in a pan of water.

6. Be sure to correctly label the pot according to the variety of bulb.

7. Set the planted pot outside in a cold frame with a mulch of straw at a temperature of 50°F (10°C) from November 25 until January 15, figure 31-6. Bulbs may be protected against rodents and insects by treatment with a fungicide.

8. Remove the pot from the cold frame about January 15, figure 31-7, and place in the greenhouse. Check for root development. Water well. Maintain a temperature of 60°F (16°C).

9. Check the bulbs for bloom four weeks from the date of removal from the cold frame (January 15).

Fig. 31-6 After the bulbs are potted, they are set outside in a cold frame. A mulch of straw is added to retain moisture. (USDA photo)

Fig. 31-7 After eight weeks of cold treatment, the bulbs are brought inside to wait for blooming. (USDA photo)

STUDENT ACTIVITIES

1. Visit a local garden center and note what types of bulbs are available. If possible, select and purchase bulbs for home or school use.

2. Force several types of bulbs, including tulips, daffodils, hyacinths, and crocuses. Develop a bulb forcing schedule and produce the flowers for sale on a certain date.

3. Under the supervision of the instructor, build a cold frame for use in producing and forcing bulbs.

4. Design a flower bed using bulbs for a home or school garden.

5. Attend a local garden show and observe the use of bulbs in the landscape. If possible, discuss the design of the plantings with the landscaper.

6. Cut and care for bulbs used in arrangements.

7. Create a bulletin board showing different types of bulbs. (Bulbs may be placed in small plastic bags and fastened to the board.)

8. Choose one type of bulb and research further information on its culture and uses.

9. Compile a list of different bulbs and record their natural blooming dates. In this way, the sequence of their flowering can be determined.

SELF-EVALUATION

Select the best answer from the choices offered to complete each statement.

1. Flowers which are grown from bulbs include
   a. zinnias, hyacinths, and petunias.
   b. hyacinths, tulips, and daffodils.
   c. tulips, marigolds, and begonias.
   d. all of these

2. When landscaping with bulbs, the most striking effect is achieved by placing them in a
   a. single row.
   b. massing.
   c. grouping of mixed colors.
   d. all of these

3. Flowers grown from bulbs can be used
   a. as cut flowers.
   b. in rock gardens.
   c. in wooded areas.
   d. all of these

4. Bulbs planted in fall give color to the landscape in
   a. late summer.
   b. early spring.
   c. mid-fall.
   d. early summer.

5. The organic matter in soil used for bulb production can be increased by the addition of
   a. sand.
   b. 5-10-10 fertilizer.
   c. well-rotted manure.
   d. all of these

6. The best soil for bulbs has a pH of
   a. 6.0.
   b. 5.0.
   c. 7.0.
   d. 7.5.

7. Fertilizer used for bulbs should have an analysis of
   a. 10-10-5.
   b. 10-10-10.
   c. 5-10-15.
   d. 10-6-4.

8. Tulip bulbs should be planted
   a. 4 inches apart.
   b. 10 inches apart.
   c. 12 inches apart.
   d. 6 inches apart.

9. The best time to fertilize bulbs is
   a. one year after planting.
   b. after they have flowered.
   c. after it has rained.
   d. none of these

10. Bulbs that are dug and stored should be kept at a temperature of
    a. 50°F.
    b. 60°F.
    c. 70°F.
    d. 80°F.

11. To protect bulbs from insects, diseases, and rodents,
    a. place them in soil with additional lime.
    b. dust them with a pesticide.
    c. wrap them in paper.
    d. none of these

12. Begonia, calla, gladiolus, and dahlia bulbs should be planted in the
    a. summer.
    b. spring.
    c. fall.
    d. winter.

13. Colchicum, muscari, hyacinth, and crocus bulbs should be planted in the
    a. fall.
    b. summer.
    c. spring.
    d. winter.

14. The cold treatment for forcing tulip bulbs to bloom takes about
    a. eight weeks.
    b. twenty weeks.
    c. twenty-two weeks.
    d. four weeks.

15. Bulbs for forcing should be
    a. large.
    b. medium.
    c. small.
    d. none of these

16. Cold treatment for root formation can be accomplished by placing potted bulbs in a
    a. hot bed.
    b. greenhouse.
    c. nearing frame.
    d. cold frame.

To select proper pruning techniques for specific plants and demonstrate their use in the landscape.

COMPETENCY TO BE DEVELOPED

After studying this unit, the student will be able to

- list five reasons for pruning.

- describe four types of pruning.

- demonstrate pruning a stem at the proper angle.

- explain how the correct time to prune is determined.

- give three examples of each of the following types of plants and describe the pruning technique used for each type.

  — deciduous spring flowering shrubs
  — summer flowering shrubs
  — broadleaf evergreens
  — conifers

unit 32
Techniques of Pruning

Materials List

lopping shears	stepladder
anvil pruners	protective equipment
pruners	spring flowering shrubs
hand pruners	summer flowering shrubs
pruning saw	conifers
tree wound dressing	broadleaf evergreens

Plants are pruned to:

- remove dead, diseased, insect infested, or broken branches. This keeps the plant healthy by removing any parts which might hinder further growth.

- change the size or proportion of the plant. This is necessary when plants become overgrown in their landscape sites. This sometimes occurs when the mature height and growth rate of the plant are not considered before placing it in the landscape.

- offset root loss after transplanting. When plants are transplanted, the top of the plant is pruned to reduce transpiration from the plant. If the plant is pruned, it stands a better chance of surviving the transplant.

- develop a special form or shape. Hedge pruning and topiary and espalier work are examples of this type of pruning. *Topiary* is a practice used in formal gardens in which plant material is trimmed into different forms and shapes. *Espalier* is a pruning method of training plant material to grow flat against a trellis or wall.

- remove dead flowers and seed pods. Dead flowers give plants an unattractive appearance and should be removed. Seed pods use the plant's energy to develop seeds. Therefore, seed pods are removed to encourage further growth of the plant itself. For example, on rhododendrons and lilacs, seed pods develop rather than new shoots; the seed pods must be pruned to encourage new growth.

- rejuvenate a declining shrub. New growth is stimulated by pruning older wood. After pruning, the plant usually produces better quality flowers and develops a better form, figure 32-1.

Fig. 32-1 A rose bush before (left) and after (right) pruning. Rose bushes must be pruned every year to stimulate new growth. (Richard Kreh, photographer)

EQUIPMENT FOR PRUNING

There are a few basic pruning tools necessary for all pruning, figure 32-2. These tools must be kept clean and sharp so that when used they will give a good, clean cut.

Pruning Saw. This is a double-edged saw with very coarse teeth on one side for cutting green (live) wood. The other side has finer teeth to saw dead or dry wood.

Lopping Shears. These shears have long handles to give more leverage for cutting larger branches. The type shown in figure 32-2 is a scissor-action lopper.

Hand Shears. Hand shears are clippers consisting of a single blade that cuts against another piece of metal (as shown in figure 32-2), or two blades that work like scissors.

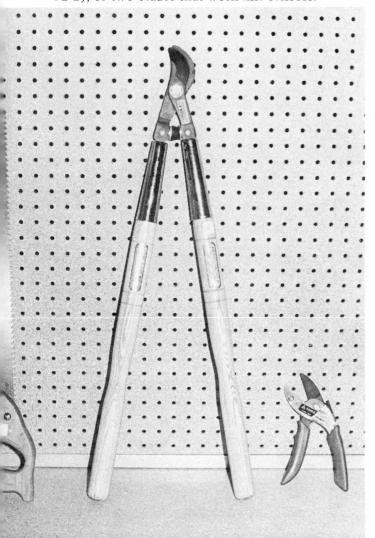

Fig. 32-2 Pruning tools. From left to right: pruning saw, lopping shears, and anvil pruners (hand shears). (Richard Kreh, Photographer)

Tree Wound Dressing. This is an asphalt-based wound dressing that prevents decay and disease organisms from entering the plant. All wounds with a diameter of 1 inch or more should be treated with wound dressing.

TYPES OF PRUNING

There are four basic methods of pruning. Before deciding which pruning method to use, consider the type of plant to be pruned and the desired finished effect.

Thinning

Thinning involves removing certain branches from a plant so that only branches with live buds remain. This method is usually used when the horticulturist wishes to keep the natural shape of the plant. Thinning allows more light to reach inner branches so that they can develop new growth more easily. The effect of thinning does not have to be noticeable, figure 32-3.

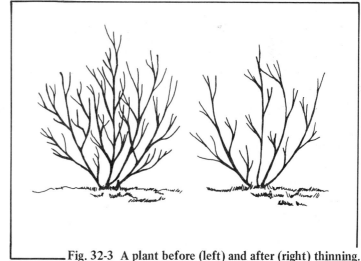

Fig. 32-3 A plant before (left) and after (right) thinning.

Fig. 32-4 As a result of pruning, the branches on this dwarf apple tree appear more evenly spaced. **(Richard Kreh, photographer)**

Fruit trees are a good example of plants that are commonly thinned. These trees are pruned while the plants are dormant (in the winter or dry season). All cuts are made as closely as possible to the remaining limb so that no stubs are left. It is important to leave about four to six lateral branches that are about 8 inches apart on the main trunk. The lowest branch should be about 2 feet from the ground. It is best to proceed in a spiral around the main trunk when pruning, keeping the lateral branches evenly spaced, figure 32-4.

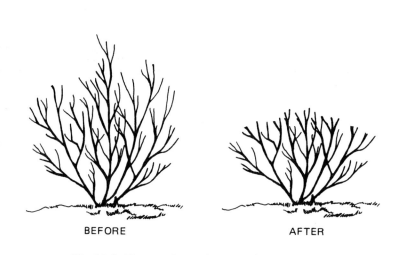

BEFORE   AFTER

Fig. 32-5 The pruning technique of heading back

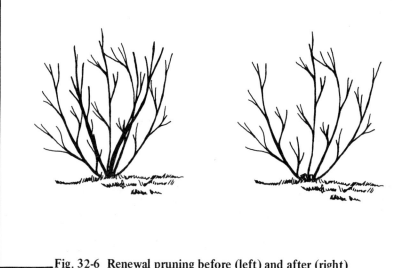

Fig. 32-6 Renewal pruning before (left) and after (right)

Heading Back

Heading back is the removal of the end section of the plant branches at the same height, figure 32-5. This causes new shoots to develop from dormant buds, therefore making the plant thicker. The plant will grow in the direction the bud is facing, so it is important to prune back to a live bud. The new growth will begin from that point. Heading back is usually considered the easiest of the four techniques discussed.

Renewal Pruning

In *renewal pruning*, old branches that are large and unproductive are removed. The oldest branches are removed by cutting them back to ground level, figure 32-6. This results in the development of new branches and better flowering on the new branches. Renewal pruning is usually effective on flowering shrubs.

Root Pruning

Root pruning, an important step in transplantation of plants, involves cutting off all the lateral roots of the plant with a sharp spade in a circle around the stem. In commercial nurseries, root pruners are mounted on tractors to allow for a more efficient operation. A general rule for root pruning is to cut in a circular motion around the plant so that 1 inch of stem diameter equals 10 inches of the circle diameter, figure 32-7. It is best to root prune one growing season prior to transplanting the plant. This results in the formation of many small, fibrous roots. Early fall is the best time to accomplish root pruning.

Fig. 32-7 A nursery worker indicating the point at which the root system of this tree will be pruned. (USDA photo by Dave Warren)

HOW AND WHEN TO PRUNE

Cutting at the Proper Angle

Cutting the stem at the correct angle is an important aspect in the pruning process. Figure 32-8 shows how pruning cuts should be made.

A cut that is made at too great an angle, as shown in the illustration below, leaves too great an exposed area for the plant to heal properly. It is also important that the cut be made the correct distance from the bud, figure 32-9. Cutting too far from the bud leaves too much stub above the cut, causing the branch tip to dry out and die. It may also provide a home for insects,

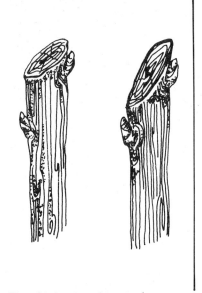

Fig. 32-8 Pruning at the proper angle of the stem. The stem on the left has been cut at the correct angle; the stem on the right has been cut at too great a slant.

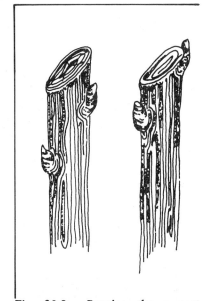

Fig. 32-9 Pruning the correct distance from the bud. On the left, the cut has been made too far from the bud; on the right, too close to the bud.

and results in an ugly appearance. Cutting too close to the bud may injure it.

When to Prune

The time schedule for pruning is usually dependent upon when the plant flowers or bears fruit. For example, if a plant flowers in the spring, it should be pruned after it flowers. This means that flower buds are developed during the previous season's growth. Remember that there must be flowering and fertilization of the flower buds before fruit will develop.

Shrubs which should be pruned after they flower are forsythia, deutzia, lilac, weigela, mock orange, butterfly bush, bridal wreath spirea, beauty bush, kerria, magnolia, sweet shrub and Juneberry.

Plants that flower in the summer or fall are usually pruned just before growth begins in the spring. In this case, the flower buds develop on the new growth (current season's growth). Some examples of plants which may be pruned in summer or fall are abelia, crape myrtle, Russian olive, bush honeysuckle, roses, spirea, five-leaf aralia, Norfolk Island pine, bamboo, aucuba, heather, boxwood, sweet shrub, bittersweet, croton, dogwood, winter hazel, viburnum, blueberry, hemlock, and bald cypress.

PLANTS WITH SPECIAL PRUNING NEEDS

Some plants require special attention at pruning time. Listed below are some examples and their pruning requirements.

Nandina. Prune the oldest canes to the ground to start new growth. Prune to thin out the plant.

Oleander. Prune the faded flower clusters. Remove the top branches to induce new growth and to control the size of the plant.

Privet. Prune to control shape and size.

Rhododendron. Pinch the flower truss after blooming to prevent the formation of seed pods. If they are permitted to set seed, additional energy is required from the plant to produce new growth. After pinching, the plant will produce new growth faster and set more flower buds for blooming the following year.

Hemlock. Prune in the early spring before new growth develops. This stimulates thicker growth and helps to control plant size and shape.

Spruce. Prune about one-half of the candle in August to result in a fuller tree. *Candles* are new growth on the ends of branches of pine species.

Fir. Prune the new growth in August to one-half the length of the candle. Growth may be retarded by removing the tip bud in the spring before the candle develops. Usually, the only pruning done is to shape the tree.

Yew. Prune after the new growth has hardened in the spring. Yews may be pruned in the summer if necessary to control plant growth.

Andromeda. Prune to remove dead faded flower clusters in late spring. The new shoots should be pinched back to shape the plant. This will cause the plant to develop new growth.

Aucuba. Pinching back results in a thicker plant with more shape. If some of the leaves have turned black from winter burn, remove them.

Arborvitea. Prune before new growth starts in the spring. Do not prune beyond the green leaves, since this will result in a permanent brown area.

Pine. Prune about one-half of each candle in June, figure 32-10A, B, and C. This promotes thicker plant growth and the production of new side shoots.

(A) Pine candles before they are pruned.

(B) To prune, hold the candle between the thumb and index finger and pinch.

(C) The pine candle after pruning.

Fig. 32-10 Pruning a pine tree. (Richard Kreh, photographer)

Azalea. Prune after flowering to promote the production of new growth. Pinching back new top growth stimulates heavier blooming.

Boxwood. Pruning boxwood is important to keep the plant compact and full. Prune in the fall by removing short branches with hand pruners. Remove diseased branches, but be sure to dip the hand pruners in a 70-percent alcohol solution between each cut to prevent the spread of the disease. Do not use hedge shears, since this causes the leaves to turn brown on the end where the leaf is cut.

Holly. Prune various types of holly in December so that the branches can be used for holiday decorating. Select and cut the branches to develop the plant's shape and form. Holly may also be pruned in early spring before new growth begins.

Mountain Laurel. Prune to encourage growth of new shoots. Oddly shaped, tall, or leggy plants may be pruned to the ground to start new growth. When pruning, be sure that the natural shape and form of the plant is maintained.

STUDENT ACTIVITIES

1. Collect various branches of trees and shrubs and bring them to class. Take turns demonstrating the proper cutting technique to the rest of the class.

2. Under the supervision of your instructor, prune shrubs on the school grounds in accordance with procedures specified in the text. If possible, visit a local orchard before pruning to observe professional techniques.

SELF-EVALUATION

1. Pruning is done to
   a. remove dead wood.
   b. remove diseased wood or create special tree forms.
   c. remove insect infested or broken wood.
   d. all of the above

2. The pruning method in which all the terminal ends of the plant branches at the same height are removed is called
   a. thinning.
   b. heading back.
   c. root pruning.
   d. renewal pruning.

3. In the root pruning process, lateral roots are cut, resulting in the formation of
   a. small fibrous roots.
   b. large branches.
   c. better quality flowers.
   d. none of these

4. The pruning procedure that removes old and unproductive branches is called
   a. heading back.
   b. root pruning.
   c. renewal pruning.
   d. none of these

5. Pruning plants to result in a special shape or form is called
   a. heading back.
   b. thinning.
   c. root pruning.
   d. topiary pruning.

6. Pruning plant branches on an angle promotes better healing. Branches are pruned at an angle of
   a. 42 degrees.
   b. 36 degrees.
   c. 50 degrees.
   d. 45 degrees.

7. Deciduous plants that flower in spring are pruned
   a. before flowering.
   b. after flowering.
   c. in late spring.
   d. none of these

8. New growth on pines is called
   a. needles.
   b. branches.
   c. candles.
   d. none of these

9. The spruce and fir are pruned after the new growth has hardened off in
   a. late summer.
   b. early spring.
   c. fall.
   d. early winter.

10. Yews are pruned in the spring after the new growth has
    a. fully developed.
    b. hardened off.
    c. become green.
    d. none of these

OBJECTIVE

To apply the principles of landscaping to an actual setting for better understanding of the goals of landscape professionals.

COMPETENCY TO BE DEVELOPED

After studying this unit, the student will be able to

- describe the three major career fields within the landscape profession.
- state the main objectives of good residential landscaping.
- list the five principles of landscape design and examples of an application of each principle.

Materials List

    photographs or drawings of well-landscaped property

    felt, construction paper, scissors

---

unit 33
Principles of Landscaping

THE LANDSCAPING PROFESSION

Landscaping is the development of the outdoors to serve the needs and desires of people. Some view landscaping as an art; others see it as a craft. It may correctly be regarded as both an art and a craft.

The art of landscaping is centered within the career field known as *landscape design* or *landscape architecture.* It is in the mind's eye of the landscape designer or architect that the landscape first takes shape. The successful designer must be able to look at undeveloped, scarred, and ugly land, and recognize its potential for becoming more attractive and/or functional for human use.

Environmental Design For:
SCHOHARIE BOCES CENTER
Schoharie, New York

Fig. 33-1 A completed landscape plan (Walter Ressler, photographer)

Many states set high standards and legal requirements for those individuals calling themselves landscape designers or landscape architects. Often, four to five years of college training are required. State licensing may also be necessary. Only through training can the designer's full imagination be developed. Only through on-the-job work experience can the designer channel creativity and imagination in the development of practical design solutions.

The language of the landscape designer is *graphic art*. Through this art form (graphics), designers are able to reduce the actual dimensions of the area to be landscaped to a size that can be drawn on paper. The designer's ideas can then be illustrated to the client in the form of a reduced drawing. The landscape plan is a collection of graphic symbols, which represent trees, shrubs, flowers, buildings, and other constructed materials necessary to the proposed landscape, figure 33-1. When the land dimensions and the symbols are reduced in the same proportion, the drawing is said to be done to *scale*. Two examples of commonly used landscape scales are 1 inch on the drawing equal to 10 feet of real land (expressed as 1″ = 10′), and 1 inch on the drawing equal to 20 feet of real land (expressed as 1″ = 20′).

By means of landscape plans, designers are able to show their ideas to clients in such a form that they can be understood and discussed, figure 33-2.

Fig. 33-2 A landscape designer and client discussing plans. (Mark Bosworth, photographer)

Once the design has been approved by the client, it must be installed and brought into reality. *Landscape contracting* is the career field which deals primarily with the installation of landscapes. To be successful, landscape contractors must have knowledge of plant materials and proper planting techniques and engineering and building skills. It is as important that they know how to lay a patio or build a brick wall as how to plant a tree or shrub. The landscape contractor must be able to read and follow the scaled plans prepared by the landscape designer. If the landscape designer/architect is responsible to the client for the actual development of the landscape, he or she may be the one who hires the landscape contractor to build the landscape. This practice is known as *subcontracting*. A sizeable landscape may involve numerous subcontractors, all responsible to the landscape architect. The landscape architect is, in turn, accountable to the client.

The care of the landscape after it has been installed is the job of the *maintenance landscaper*. Professionals in this career field are sometimes called gardeners or caretakers. Some maintenance landscapers spend their time totally on a single landscape, such as a large estate, shopping center, cemetery, or park, while others serve numerous landscapes accomplishing all or part of the maintenance tasks required by the landscape. Typical landscape maintenance tasks include lawn care (fertilizing, weeding, mowing); tree and shrub care (fertilizing, pruning, mulching); flower care; repair of walls, fences, walks, and drives; painting; and snow plowing.

The simplicity or complexity of the landscape design will make the maintenance of landscape easy or difficult. It is vital that both the designer and the contractor keep maintenance requirements of the landscape in mind while developing the landscape. It is a weak design that contains so many high maintenance features that the cost of upkeep is prohibitive.

Likewise, the maintenance landscaper must know how to read a landscape plan. As the landscape grows over a period of years, certain plants may need to be replaced. Others may require special pruning techniques. Flower plantings often require special care to obtain the desired color patterns year after year. A poorly trained or careless maintenance person can cause the best landscapes to appear shabby.

While there are different levels of involvement by various professionals in the landscaping business, all are inter-related. Neither the designer, the contractor, nor the maintenance landscaper can work without an appreciation for the role of the others if landscapes that both serve and satisfy the client are to be created.

THE OBJECTIVES OF RESIDENTIAL LANDSCAPING

The objectives of residential landscaping are evident in the definition of the term *landscaping*: that is, to serve the needs and desires of people in development of the outdoors. Specifically, the goals of residential landscaping are:

- to determine the exact needs and desires of the homeowners for their landscapes.

- to determine the capabilities of the land (site) to fulfill those needs and desires.

- to develop the outdoor living areas of the landscape in a manner similar to the way indoor living areas are developed.

- to design the landscape in such a way that maintenance practices do not exceed that which the homeowner is willing to do.

- to keep costs within the budget of the homeowner.

The needs and desires of the homeowner are best determined by an interview. A direct conversation between the designer and the client promotes trust and confidence between the two parties. This helps the designer to recognize the client's actual desires much more easily. During the interview, questions such as the following should be answered:

a) How many family members are there? What are their ages?

b) How much does the family use the outdoor areas around the home?

c) Does the family entertain frequently? Large groups or small?

d) How much privacy from the neighbors and passing cars do they desire?

e) How much maintenance are they willing to do in the upkeep of the landscape?

f) Are there certain plants they are fond of or dislike?

g) What service needs will the landscape be expected to accommodate? (Examples: clothesline, trash cans, pets, vegetable garden, compost pile, garden tool storage)

h) Will the family be using the garden after dark?

i) How much does the family want to spend on the total development of the landscape? Do they want to spread the cost over several years?

j) Is the family willing to wait several years for the plants to reach maturity or do they want large plants installed for an immediate effect?

To determine the capabilities of the site requires a thorough *site analysis* by the landscape designer. Becoming familiar with the site helps the designer to determine how easily and how many of the client's needs and desires can be met. The site analysis usually involves several visits to the property by the designer. A good site analysis often suggests additional possibilities for development that had not occurred to the homeowner, but which might be of interest once identified.

The following are some of the things to look for and take note of during a site analysis.

- dimensions of the property

- topography of the site (how flat or rolling it is)

- quality of the topsoil and subsoil

- condition of the lawn areas

- types and condition of existing plants

- location of utility lines, meters, and utility easements

- good and bad views from the site

- locations of glass areas in the house and where they open onto in the landscape

- architectural style of the neighborhood

- environmental setting of the site and the neighborhood

- existing natural features such as streams, rock outcroppings, specimen plants, and wildlife habitat areas

The Concept of the Outdoor Room

When developing the design for a landscape, it is helpful to visualize the outdoors in the same manner as the indoors. The home is a series of connecting living units called rooms. When the landscape designer sees the outdoor areas as rooms, the landscape becomes more familiar and easier to work with. Outdoor rooms

Fig. 33-3 An outdoor room. Notice the materials which act as the walls, floor, and ceiling. (Jack Ingels, photographer)

have walls, ceilings, and floors, just as indoor rooms do, figure 33-3. The main difference between indoor and outdoor rooms is in the materials used to construct them. Outdoor walls may be developed using shrubs, fences, brick or stone, exterior walls of buildings, or trellises. Outdoor floors may be the natural earth, sand, crushed stone, poured concrete, brick, decking, turf grass, or many other similar surfacing materials. Outdoor ceilings can be developed with trees, awnings, canopies, or other overhead structures.

Trees are excellent materials for use in developing the outdoor ceiling. If an intimate patio setting is desired, a small tree branching overhead can create the low-ceiling effect. Also important to consider is the shade that the tree provides on a hot summer day. The southwest corner of a house will benefit greatly from the cooling effect of a medium to large tree planted nearby. Deciduous trees will allow the sun to pass through and warm the house with solar energy during the cold winter months.

The material selected for outdoor walls is usually determined by how much privacy and security is needed in the landscape. When total privacy or security is desired, the outdoor wall should be at least 6 feet tall and solid. When less privacy is needed or security is not a factor, the outdoor wall can be lower and more open.

Fig. 33-4 An outdoor wall constructed of natural and man-made materials (Jack Ingels, photographer)

The main function of the outdoor wall is to define the shape and limits of the outdoor room. Plants or constructed materials used alone may accomplish this; however, the combination of natural and constructed materials often creates the most attractive effect of all, figure 33-4.

There are several groups of materials from which to select for the outdoor floor. *Hard paving* includes concrete, flagstone, tile, decking, and brick, among other materials. Hard paving is expensive to install but inexpensive to maintain. *Soft paving* includes crushed stone, wood chips, marble chips, and other loose materials. Soft paving has a moderate cost for both installation and maintenance. It requires periodic replacement of worn areas. Turf grass is the most popular of all surfacings. Turf has a low installation cost but a high maintenance cost, since it must be cared for regularly. *Ground covers* are good for surfacing areas where no one will be walking, such as on slopes and directly under trees. Figure 33-5 describes the various outdoor room surfacings.

TYPE OF SURFACING	COST OF INSTALLATION	COST OF MAINTENANCE	DURABILITY
Hard paving	high	low	high
Soft paving	moderate	moderate	moderate
Turf grass	low	high	moderate
Ground covers	moderate	high at first	low

Fig. 33-5 Comparison chart of various outdoor room surfacings

The type of material chosen for the outdoor floor is certainly affected by what the client can afford. However, there are other needs of the individual client to consider:

Does the material absorb noise and/or dust?

How well does it blend in with the rest of the landscape?

Does it produce a glare from the sun?

Does it track easily?

Does it become slippery when wet?

THE PRINCIPLES OF LANDSCAPE DESIGN

There are five basic principles which guide the landscape architect or designer in planning the landscape's outdoor rooms. These principles are:

1) simplicity
2) balance
3) focalization of interest
4) rhythm and line
5) scale and proportion

The principle of simplicity is very important to the overall unity of the design. When this principle is correctly applied, the landscape is understood and appreciated by the viewer. The principle of simplicity is accomplished by repeating specific plants throughout the design, and by massing plant types or colors into groups rather than spacing them so that each plant or color is seen separately. The fewer different objects there are for the eye to focus upon, the more simple the design will seem. Finally, straight-lined or gently curving bedlines around shrub plantings, rather than fussy, scalloped bedlines, add to the design's simplicity, figure 33-6, page 390.

The principle of balance is applied by imagining the area of landscape placed on a seesaw. If properly balanced, the left side of the landscape should have no more visual weight than the right side. Balance may be either symmetrical or asymmetrical. *Symmetrical* balance is attained when one side of the landscape is an exact duplicate of the other side, figure 33-7, page 390. This is a

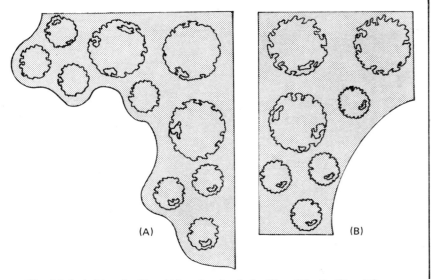

Fig. 33-6 A fussy bedline (A) and a simple bedline (B). Bedline (B) was more thoughtfully planned and would be much easier to maintain.

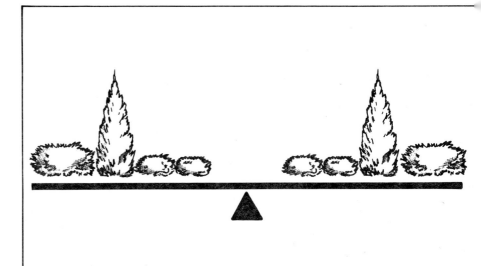

Fig. 33-7 Symmetrical balance. Each side of the landscape duplicates the other.

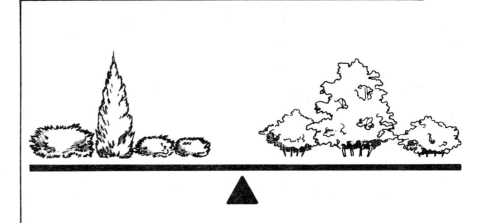

Fig. 33-8 Asymmetrical balance. One side of the landscape provides as much interest as the other, but does not duplicate it exactly.

form of balance common in formal designs. It is sometimes applied to modern residential design, but more frequently asymmetrical balance is used. With *asymmetrical* balance, one side of the landscape has the same visual weight as the other side, but they are not duplicates, figure 33-8.

The principle of focalization of interest recognizes that the viewer's eye wants to see only one feature as being most important within any given view. All other elements complement that important feature (the focal point) but do not compete with it for attention. When looking at a house from the street, the viewer's eye should go quickly to the front door. When sitting on the patio looking out across the backyard, the viewer may have no focal point at which to look unless one is created by the designer. Focal points may be created using especially attractive plants (*specimen*

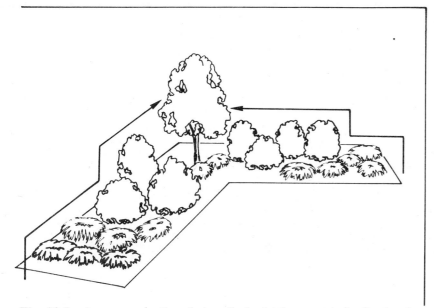

Fig. 33-9 A corner planting that easily leads the eye to the focal point.

plants), statues, fountains, pools, and flower masses. Once created, all bedlines and plant arrangements should be designed to lead the eye of the viewer to the focal point, figure 33-9.

The principle of rhythm and line also contributes to the overall unity of the landscape design. This principle is responsible for the sense of continuity between different areas of the landscape. One way in which this continuity can be developed is by extending planting beds from one area to another. For example, shrub beds developed around the entrance to the house can be continued around the sides and into the backyard. Such an arrangement helps to tie the front and rear areas of the property together. Another means by which rhythm is given to a design is to repeat shapes, angles, or lines between various areas and elements of the design.

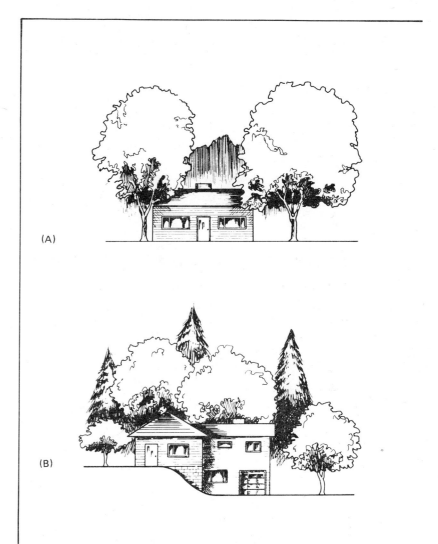

(A)

(B)

Fig. 33-10 The trees in (A) overpower the structure to such an extent that it is dehumanizing and uncomfortable to view. The landscape in (B) is more carefully planned, with plantings in proportion with the structure.

The principle of scale and proportion helps to keep all elements of the landscape in the correct size relationship with not tower over the building when fully grown, figure 33-10. Plants selected for the landscape should add to human com-fort in the setting. For example, plants and other materials used around a childrens' play area should be small so that the children can relate to them. In a world of giant adults, it is nice to feel as "tall as a tree," even when child and tree are only 3 1/2 feet tall.

STUDENT ACTIVITIES

1. Use cutouts of symbols on a felt board to practice arranging different types of plantings. Prepare the symbols (to represent various plantings and structures) beforehand. Try some arrangements which will (a) draw attention to a corner; (b) give partial screening; (c) give total screening; and (d) provide an interesting silhouette for the viewer's eye.

2. Invite someone who is in a landscaping profession (architect, contractor, or maintenance person) to the class to discuss his or her occupation.

3. As a class, install a planting around the school or in a nearby park. Discuss the function of the planting before the actual work is begun.

SELF-EVALUATION

A. List the five principles of landscape design.

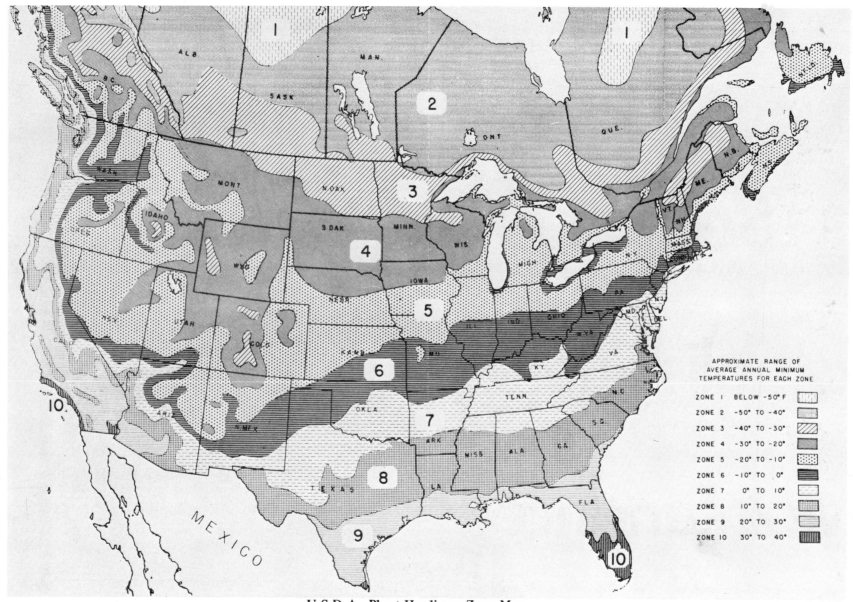

U.S.D.A. Plant Hardiness Zone Map

APPROXIMATE RANGE OF
AVERAGE ANNUAL MINIMUM
TEMPERATURES FOR EACH ZONE

ZONE 1   BELOW -50° F

ZONE 2   -50° TO -40°

ZONE 3   -40° TO -30°

ZONE 4   -30° TO -20°

ZONE 5   -20° TO -10°

ZONE 6   -10° TO 0°

ZONE 7   0° TO 10°

ZONE 8   10° TO 20°

ZONE 9   20° TO 30°

ZONE 10   30° TO 40°

Section 10
Lawn
Establishment
and
Maintenance

unit 34
Establishing
New Lawns

OBJECTIVE

To start a lawn by seeding, sodding, and plug or strip planting.

COMPETENCY TO BE DEVELOPED

After studying this unit, the student will be able to

- list three reasons for establishing and maintaining a lawn.
- describe two methods of establishing proper drainage in a lawn.
- list three materials that are used to increase the organic matter in a new lawn.
- name eight items that must be included on a seed label.
- describe three ways turf grasses are started in the United States.
- demonstrate the five steps in seeding a lawn.

Materials List

bag of lawn seed

bag of lawn fertilizer and lime and spreader

sample of drainage tile

samples of peat moss, well-rotted sawdust, and weed-free manure

samples of commercial fertilizer and lime

samples of cool-season and warm-season turf grasses

several steel rakes

Rototiller

straw

lawn roller

Lawns are a major part of most home landscapes. Basically, lawns are established for three reasons.

- They add beauty to the landscape. A well-kept lawn is very appealing and inviting.

- They are used as play areas for sports such as baseball, football, basketball, or lacrosse. Since these types of sports are tough on lawns, it is important to select a lawn grass that can take wear.

- They provide an excellent cover to help control soil erosion while allowing the movement of air and water to roots of trees and shrubs in the soil below.

Fig. 34-1 A Rototiller is used to distribute and mix 6 inches of topsoil over the surface of the ground. (Richard Kreh, photographer)

SOIL AND GRADING

The first thing to consider in establishing a new lawn is the present condition of the soil. Is this an area in which the builder has graded off all the topsoil? Is the slope too steep to establish a lawn and mow it safely? Is the drainage adequate? These are all questions that must be answered before establishing a lawn.

The builder usually establishes the rough grade. The lot is graded so that the land slopes away from the foundation of the house, to help prevent any water from entering the basement. After the rough grade is established, topsoil which the builder has set aside may be spread over the subsoil which was kept in the rough grading. Six inches of topsoil should be spread evenly over the surface and tilled to loosen and break up clods, figure 34-1.

The general slope for the lawn after the topsoil is spread should not exceed 15 percent; that is, no more than a 15-foot drop for every 100 feet of lawn area. Slopes greater than 15 percent are unsafe to mow. If a slope steeper than 15 percent cannot be avoided, it should be covered with plants that do not require mowing.

DRAINAGE

Good drainage ensures a balance between the air and water in the soil. This, in turn, encourages proper root growth. Grasses can endure relatively wet soil, but not bog or swamp conditions. There are two ways of establishing proper drainage in lawns. One method is to install drainage tile about 3 feet below the surface of the soil to drain the subsoil. Another method is to make use of the slope of the land to drain surface water away.

PREPARATION OF THE SOIL

As mentioned earlier, 6 inches of topsoil should be spread over the rough graded subsoil. A good garden loam is the best medium for most grasses. If quality topsoil is not available, the organic matter content of the soil should be increased by adding well-rotted sawdust, weed-free, well-rotted manure, or commercial peat moss at the rate of 6 cubic feet per thousand square feet of land. Work the material well into the soil with a Rototiller. It may be necessary to remove stones or dirt clods. Use a hand rake or a stone rake mounted on a tractor to remove stones. A good seedbed should have a firm and smooth, but not powder fine, surface texture.

FERTILIZER

A complete fertilizer with a high phosphorus content is recommended for use in establishing new lawns. Some companies manufacture a special fertilizer known as a *starter fertilizer* which is especially high in phosphorus.

Before applying fertilizer, the soil should be tested to determine the correct amount of fertilizer and lime to add. This can be done in one of two ways: with a special portable soil test kit, or by sending a sample of the soil away to a land grant university for testing. When test results are returned from these schools, they also contain a recommendation regarding the amount and analysis of fertilizer to apply.

The results of soil tests sometimes indicate a need for the addition of lime to the soil. Lime reduces the acidity of soil and encourages root development. There are different forms of lime available for use on lawns. They include ground limestone (calcium carbonate), burned lime (calcium oxide), and hydrated lime (calcium hydroxide). When choosing a lime for use in a lawn, pick the one which gives the most calcium for the money.

Most lawn grasses grow best in a well-limed soil with a pH level from 6.0 to 6.5. If the soil test calls for lime, add the recommended amount evenly over the entire soil surface. Work it into the ground 4 to 6 inches before seeding.

Spreading the Fertilizer

Fertilizer should be spread in two different directions on new lawns. Spread one-half the recommended amount in a north-south direction and the rest in an east-west direction. This gives a better distribution of the fertilizer.

Fertilizers are usually applied most efficiently by use of a spreader. It is important that the fertilizer be applied uniformly. There are two basic types of spreaders available. Figure 34-2 shows a rotary spreader, which spins the fertilizer out over a relatively wide path. A typical turf fertilizer is shown with the

Fig. 34-2 Rotary spreader with a typical lawn and turf fertilizer. (Richard Kreh, photographer)

Fig. 34-3 Conventional spreader (Richard Kreh, photographer)

spreader. Figure 34-3 shows a conventional spreader which spreads fertilizer in a strip as wide as the spreader. Both spreaders can be adjusted so that the proper amount of fertilizer is spread. The fertilizer is then worked into the soil surface with a hand rake, figure 34-4.

STARTING THE LAWN

Lawns are started in one of three ways: (1) seeding, (2) sodding, or (3) plug or strip planting.

Selecting the Seed

It is very important to purchase the best lawn seed. All lawn grass seed is required by law to have a label with the following information.

Fig. 34-4 The fertilizer is worked into the topsoil with a hand rake. (USDA photo)

Name. The law requires that the seed package give the correct name of all the seeds included in the package. Information on where this type of seed grows best — in a shaded area, partial shade, or full sun — is also provided.

Purity. These figures give the makeup of the seed by percent of the different grasses included in the package.

Percent of Germination. This figure indicates the percentage of seeds in the package that will sprout when planted. The higher this percentage is, the more seed there is that will germinate. (A germination percentage of 85 means that 85 seeds out of 100 will germinate.)

Other Crop. The percentage of other crop seed is important to consider. Crop seeds such as wheat, barley, orchard grass, and timothy are undesirable seeds for lawns.

Inert Matter. This is material that will not grow. It includes pieces of seed, sand particles, and chaff from the seed. All of these things add weight to the package, but they do not germinate. This type of material is undesirable.

Weed Seed. A weed is any plant that is growing where it is not wanted. Weed seeds are not desirable in a lawn grass mixture. Some weed seeds are very small and difficult to remove from the desirable seed. Most packages of seed contain some weed seed, but the manufacturer is required by law to list the percentage present in each container.

Noxious Weeds. Certain weeds have been declared by each state as noxious weeds because they are particularly difficult to control. In some cases, these weeds may be harmful. This material is listed on the package by a specific number; for example, one listing might read, *four garlic per ounce.*

Year Tested. Commercial seed must be tested every year for the correct germination percentage. The month and year that the seed was tested are sometimes given also.

Company Name. The name of the company selling the seed and its correct address must be given on the label. This is done so that anyone having problems with the seed can contact the grower.

Seeding

Sowing Seed. Seed may be planted by hand or with a mechanical seeder such as the seeder shown in figure 34-2. To obtain uniform distribution, the seed is mixed with small amounts of a *carrier*, such as sand. The mixed material is divided into two equal parts; one part is sown in one direction, and the other part crosswise to the first sowing. Figure 34-5 illustrates hand sowing.

Covering the Seed. The seed is lightly covered by hand raking. Large seeds are covered with 1/4 to 3/8 inch of soil and small

Fig. 34-5 Lawn seed is sown by hand or by means of a mechanical seeder. (Richard Kreh, photographer)

Fig. 34-6 A hand rake is used to cover the seed. (Richard Kreh, photographer)

seeds with 1/8 to 1/4 inch of soil. It is important that all seed is covered by and in close contact with the soil, figure 34-6.

Mulching. Mulching with a light covering of weed-free straw or hay helps to hold moisture and prevent the seed from washing away during watering or rainfall. Straw also helps to hide the seed from birds. One 60- to 80-pound bale of straw or hay mulch will cover 1,000 square feet of area. Mulches applied evenly and lightly may be left in place and the grass allowed to grow through. Peat moss or other fine material does not make satisfactory mulch. These materials become packed too tightly, resulting in the seed being planted too deep.

On terraced areas or sloping banks, cheesecloth, burlap, or commercial mulching cloth helps to hold in moisture and keep the seeds in place. Grass is able to grow through these mulching materials, which may be left to rot.

Firming the Seed. The seeded area is firmed by rolling it with a light roller or cultipacker, figure 34-7.

Watering. New seedlings should be kept moist until they are well established, figure 34-8. Once seeds have begun to germinate, they must not be allowed to dry out, or they will die. Avoid saturating the soil, however; excessive moisture is favorable for the development of *damping off*, a fungus disease.

Vegetative Planting

There are some grasses for which seed is not available, or the seed that is available does not produce plants that are true to type. These grasses must be planted by one of several vegetative methods, such as spot or plug sodding, strip sodding, sprigging, or stolonizing. Grasses planted by vegetative methods include zoysia, improved strains of Bermuda grass, St. Augustine grass, centipede grass, creeping bentgrass, and velvet bentgrass.

Whether plugged, strip sodded, sprigged, or stolonized, the planted material must be kept moist until well established. During the first year, light applications of a nitrogenous fertilizer every two to four weeks during the growing season helps speed the spread of the grass.

Sodding

Sod consists of grass and grass roots in a thin layer of soil which is removed from the area in strips. It is rolled and transported to the area to be sodded. Unless good quality sod is available and complete coverage is needed immediately, the expense of sodding is only justified on steep slopes or terraces where erosion may be a serious problem.

Sod should not be cut more than 1 inch thick. Sod which is cut 3/4 inch thick will knit to the underlying soil faster than

Fig. 34-7 Rolling firms the seedbed and helps to hold the mulch in place. In this photo, straw covers the newly seeded area. (Richard Kreh, photographer)

thicker sod. The sod pieces are laid in the same way as bricks are and fitted together as tightly as possible. After the first strip is laid, a broad board is placed on the sodded strip. The board is knelt on and moved forward as the job progresses. This eliminates tramping on the prepared seedbed.

The planting area is prepared and fertilized in the same manner for sodding as for seeding. A roller is used to firm the seedbed after final hand raking and leveling. After the sod is laid it is tramped or rolled lightly and top-dressed with a small amount

Fig. 34-8 The newly seeded lawn should be given at least 1 inch of water. The cans shown are used to measure the amount of water applied to the lawn. (Richard Kreh, photographer)

of topsoil to fill in any cracks between the pieces of sod. The back of a rake or a straw broom may be used for this job. It is important that the sod be kept moist until the roots have grown well into the soil.

Spot Sodding or Plugging. *Spot sodding* or *plugging* is the planting of small plugs or blocks of sod at measured intervals. The plugs are spaced from 8 inches to 1 foot apart depending on how rapidly the area is to be completely covered. The closer the plugs

Fig. 34-9 A turf plugger is sometimes used to make the holes into which plugs of grass are planted. (Richard Kreh, photographer)

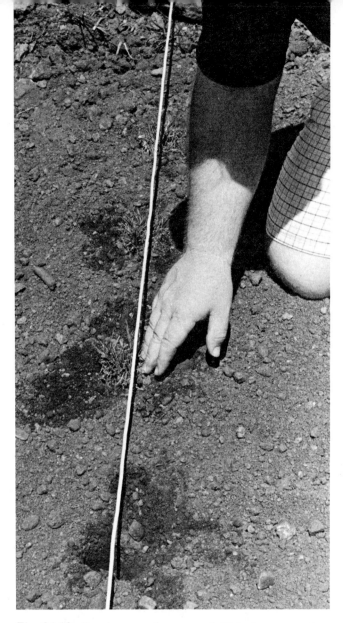

Fig. 34-10 Zoysia grass plugs spaced 12 inches on center. A line is being used to keep the rows straight. (Richard Kreh, photographer)

are set, the faster the lawn will be completely covered with grass. The plugs are fit tightly into prepared holes, figure 34-9. They are then tamped firmly into place. Zoysia grass is started in this way, figure 34-10.

Strip Sodding. *Strip sodding* is the planting of strips of sod end to end in rows that are 1 foot apart. The sod strips should be 2 to 4 inches wide. Firm contact with surrounding soil is necessary.

Sprigging. *Sprigging* is the planting of individual plants, runners, cuttings, or stolons at certain spaced intervals. The *runners* (sprigs of grass) are obtained by shredding solid pieces of sod. The spacing of the runners is determined by how fast that particular grass grows, how fast the area is to be covered, and the amount of plant material available. Runners may also be planted end to end in rows.

Stolonizing. In *stolonizing*, shredded stolons are spread over the area with mechanized equipment. The spreading is followed by disking or rolling the planted area and topdressing with fertilizer. (*Topdressing* is the spreading of fertilizer on top of soil.) Stolonizing is usually done only when large areas are to be planted or when the area to be planted is highly specialized, such as golf course putting greens.

Large areas of Bermuda grass may be established by spreading shredded stolons with a manure spreader and disking lightly to firm into the soil. (*Disking* is the mixing of soil with a disc harrow.) This method requires 90 to 120 bushels of stolons per acre. Creeping bentgrass and velvet bentgrass are stolonized by spreading shredded stolons at the rate of 10 bushels per 1,000 square feet, topdressing with topsoil to a depth of 1/4 inch, and rolling to firm the stolons into the topdressing.

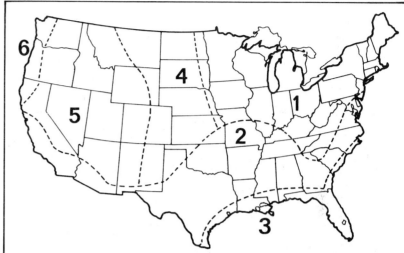

Climatic regions of the U.S. in which the following grasses are suitable for lawns: Region 1. Common Kentucky bluegrass, Merion Kentucky bluegrass, red fescue, and Colonial bentgrass. Tall fescue, bermudagrass, and zoysiagrass in southern portion of the region. Region 2. Bermudagrass and zoysiagrass. Centipedegrass, carpetgrass, and St. Augustinegrass in southern portion of the region with tall fescue and Kentucky bluegrass in some northern areas. Region 3. St. Augustinegrass, bermudagrass, zoysiagrass, carpetgrass, and bahiagrass. Region 4. Nonirrigated areas: Crested wheatgrass, buffalograss, and blue gramagrass. Irrigated areas: Kentucky bluegrass and red fescue. Region 5. Nonirrigated areas: Crested wheatgrass. Irrigated areas: Kentucky bluegrass and red fescue. Region 6. Colonial bentgrass and Kentucky bluegrass.

Fig. 34-11 Choosing a grass type for a certain area. (from USDA bulletin 51)

CHOOSING THE GRASS TYPE

When deciding upon which grass to plant in a certain area, the most important factors to consider are the climate conditions, such as the temperature of the area and the moisture available. Figure 34-11 shows a map of the continental United States which is divided into areas suitable for the planting of certain grass types. Before planting, also check other sources, such as your instructor and local extension service, for varieties that are recommended for your locality.

STUDENT ACTIVITIES

1. Create a display showing labeled samples of lawn seed.

2. Establish small yard-square plots of various lawn grasses. Practice leveling the area with a steel rake to establish a seedbed.

3. Practice using a lawn spreader to obtain uniform coverage of areas to be seeded. If the seed and the necessary ground are not available, sand can be used in place of seed, and applied to marked areas of a parking lot. Sweep up and measure the sand to determine the correct amount of planting material that would be applied to the area.

SELF-EVALUATION

Select the best answer from the choices offered to complete each statement.

1. The finished grading around a home is done so that rainwater will
   a. stay in place and soak into the soil.
   b. gradually drain away from the house.
   c. drain toward the house.
   d. quickly run off the lawn area.

2. For easiest and safest mowing, the grade or slope of the lawn should not exceed
   a. 5 percent.
   b. 10 percent.
   c. 25 percent.
   d. 15 percent.

3. After the rough grade of a lawn is established, topsoil is added to a depth of
   a. 6 inches.
   b. 2 inches.
   c. 3 inches.
   d. 12 inches.

4. A complete fertilizer which is high in _____ is best for seeding grass to start new lawns.
   a. phosphorus
   b. nitrogen
   c. potash
   d. lime

5. A good seedbed for sowing grass consists of topsoil prepared so that it is
   a. loose and porous.
   b. coarse and open.
   c. powdery fine.
   d. firm, smooth, and relatively free of rocks.

6. When establishing new lawns, the fertilizer is spread evenly over the soil surface and

   a. watered in.
   b. seeded immediately.
   c. worked into the soil surface with a rake.
   d. dug deeply into the soil.

7. The percentage of germination listed on seed container labels indicates

   a. the purity of the seed.
   b. the percentage of seeds that will sprout and grow.
   c. the percentage of weed seeds in the mixture.
   d. the amount of inert matter in the seed.

8. Grass seed should be spread evenly over the soil surface to reduce the chance of missing areas. To accomplish this,

   a. sow half the recommended amount of seed in one direction, and the other half at a 90-degree angle to the first seeding.
   b. always sow by hand.
   c. always sow using a rotary seeder.
   d. always rake the seed into the soil.

9. Lawn seed should be carefully covered with soil to a depth of

   a. 1/4 inch.   c. 1/2 inch.
   b. 1/8 to 3/8 inch, depending on seed size.   d. 1 inch.

10. A newly seeded area is often mulched with clean, weed-free straw or other material to

    a. help shade out weeds.
    b. make it possible to walk on the area immediately after seeding.
    c. hold moisture in the soil and prevent seed from washing away.
    d. eliminate the appearance of bare soil.

11. Once grass seed germinates and new seedlings begin to grow, it should

    a. never be allowed to dry out.
    b. be mowed regularly.
    c. be fertilized regularly.
    d. be rolled to firm the roots against the soil.

12. Planting grass by use of a solid covering of sod is an expensive method of establishing a lawn and is generally used

    a. only by wealthy people.
    b. on steep slopes and terraces where erosion is a serious problem.
    c. by contractors.
    d. in the off-season, when seed would not grow otherwise.

13. In establishment of lawns, spot sodding is a process in which

    a. small plugs or blocks of sod are placed at measured intervals.
    b. sod is used to patch bare spots in an old lawn.
    c. sod is used on steep slopes.
    d. sod is used on terraces.

14. Sprigging is

    a. cutting sprigs off grass to keep it short.
    b. sowing a new lawn.
    c. chopping up a lawn to provide better drainage.
    d. planting individual plants, runners, cuttings, or stolons at spaced intervals to establish a new lawn.

15. Stolonizing is

    a. spreading shredded stolons of grass to establish a lawn.
    b. rolling a seedbed to make it firm.
    c. removing stolons from grass to thicken the lawn.
    d. all of the above

OBJECTIVE

To use proper lawn maintenance techniques.

COMPETENCY TO BE DEVELOPED

After studying this unit, the student will be able to

- list the six factors of good lawn maintenance and explain each orally.

- describe the analysis of a good turf fertilizer.

- determine the best time to apply fertilizer to a lawn.

- demonstrate how to set the mowing height of a rotary mower.

- list the three causes of fungus disease in lawns.

Materials List

lime and fertilizer

tape measure and enough string to lay out 1,000 feet of lawn area

one dull and one sharp mower blade

knapsack sprayer for applying weed killer

crabgrass weed killer (in combination with fertilizer)

2,4-D for broadleaf weed control

rotary mower

---

unit 35
Maintaining
the Lawn

After the lawn has been seeded or planted vegetatively, care must be taken to keep it healthy. Proper lawn maintenance is dependent upon six factors:

- planting the proper variety or species.

- applying fertilizer and lime at the proper time and in the proper form and amount.

- mowing to the proper height at the correct time.

- watering properly.

- using chemicals for weed, insect, and disease control if necessary.

- using the lawn in such a way that the traffic is not too heavy.

PLANTING THE PROPER VARIETY OR SPECIES

Different varieties or species of grasses grow best under different temperature, light, and moisture conditions. If planting is to be done in a cool climate, for example, cool-season grasses should be selected. Some grasses, such as the creeping red fescues, grow better in shade than in full sun. They also grow better in dry areas. No grass does well in dense shade, however. Bluegrass grows well in cool climate areas receiving full sun and adequate moisture. Bermuda grass grows well in hot climates.

Another factor to consider when choosing a variety of grass is the way in which the lawn is to be used and the amount of traffic to which the lawn will be exposed. The more a lawn is used as a recreation or work area, the more stress there is on the grass. If heavy use is anticipated, select a hardy grass that can withstand more traffic. (The proper placement of walks and rest or play areas reduces traffic on grass plants.)

> **Check with a local agricultural experiment station for the varieties of grass that grow best in your area.**

APPLYING LIME AND FERTILIZER

Lime

Lime should be used whenever necessary to keep the soil pH in the proper range. A pH of 6.0 to 6.5 is a good pH range for most grasses. Finely ground limestone is generally the best and least expensive form of lime to use. Late fall and winter are generally the best times to apply lime.

Lime moves down through the soil very slowly. For this reason, it is usually necessary to apply it to the soil before planting so that it mixes with the top 4 to 6 inches of soil by the time planting begins. Additional applications are spread on top of the soil as needed. This lime gradually works its way down into the soil at a rate of 1/2 to 1 inch a year.

Fertilizer

If a healthy lawn is to be maintained, annual applications of a nitrogen fertilizer are needed. Nitrogen is leached from soil and must be replaced regularly.

Lawns require a fertilizer high in nitrogen (indicated by the first number on the fertilizer label). Of the three analysis numbers on fertilizer labels, the first number is always highest on high quality maintenance turf fertilizers.

Many turf fertilizers supply nitrogen in an organic form. This form of nitrogen is released slowly and thus does not burn the grass. It also supplies nutrients over a longer period of time. The urea form of nitrogen is often used as a slow-release fertilizer for turf grasses. If an inorganic form of nitrogen which is released rapidly (such as nitrate of soda) is used, it must not be applied to wet, actively growing grass. Such an application would burn the grass badly.

Applying Fertilizer

Fertilizer may be applied by hand if uniform coverage can be obtained. However, it is better to use a spreader.

> **Proper fertilizer use is one of the most important factors in maintaining a good lawn.**

Fertilizer should be applied just prior to the active growing season. For cool-season grasses, this is in the fall or very early spring. These grasses grow in spring and fall and are either dormant or very slow growers in the hot, dry summer months. To fertilize them in summer would only encourage weeds that grow in hot weather and could use the fertilizer for growth.

Care must be taken not to skip spaces or overlap with the fertilizer spreader. Difference in green shades and growth rate of the grass will result, ruining the appearance of the lawn. Fertilizer rates depend upon such things as soil type and the amount of rainfall available in the area.

> **Check local recommendations by contacting your local extension service for a lawn care calendar which gives the proper dates for fertilizing as well as spraying and seeding.**

MOWING THE LAWN

There are two types of mowers used for mowing lawns, the reel mower and the rotary mower. The type of mower that is used generally depends upon individual preference; either does a good job. Close mowing of 1/2 to 1 inch is done most efficiently with a reel mower.

Most lawns are cut too short because the homeowner believes that the grass looks better when short. A very short cutting reduces the leaf area of the grass to such an extent that it cannot manufacture enough food. Also, grasses that are cut too short encourage weed growth since the grass plants cannot kill weed seeds by "shading them out." Cool-season grasses should not be cut shorter than 2 to 3 inches, figure 35-1. Warm-season grasses are cut shorter, from 1/2 to 1 1/4 inches, depending upon the grass variety. Since warm-season grasses grow rapidly during warm weather, they are better able to compete with weeds growing during the same season. Bermuda grass should be clipped to 1/2 to 1 inch and zoysia 3/4 to 1 1/4 inches.

Lawns should be mowed often enough so that no more than one-third of the top, or 1 inch of growth, is cut off in any one mowing. For example, if grass is kept at a height of 2 or 3 inches, only 1 inch should be cut off. This means that when the grass reaches 3 or 4 inches in height, it is time to mow again. This may require mowing two or three times a week when the grass is in

Fig. 35-1 This lawn of creeping red fescue grass has been mowed to a height of about 2 1/2 inches, ideal for this cool-season grass. When it reaches a height of 3 1/2 inches, it will be ready for another mowing. Thus, no more than one-third of the leaf area is cut at any one mowing. (Richard Kreh, photographer)

periods of most active growth. This is a very important point to remember, since if grass is allowed to grow to a height of 5 to 6 inches and is cut back to 3 inches in one mowing, most of the food-producing leaf blade is removed. This causes the lawn to appear yellow, an indication that it cannot manufacture enough food until more new leaves are formed. This constantly drains food reserves from the roots, weakening the root system. Tall weeds also have a better chance to establish a root system if the lawn is not mowed often enough.

Cutting the lawn to the proper height requires adjustment of the mowing height of the mower. This is done by setting

Fig. 35-2 **Adjusting the mower height to the height required by the grass. This blade will cut the grass to a height of about 2 inches. (Richard Kreh, photographer)**

the mower on a flat, smooth surface and measuring from the surface up to the blade with a yardstick, figure 35-2. Adjust the height of the mower according to the height of the grass being mowed.

Caution: **Remove the spark plug wire before turning the blade to measure it.**

The mower should be kept sharp at all times so that grass blades are cut and not torn off. Jagged, split ends cause the lawn to appear uneven and provide open wounds for the attack of fungus diseases.

Mowing Technique

The lawn should be mowed in such a pattern that it is cut in one direction at one mowing and at right angles to that direction the next mowing. This helps eliminate soil compaction and gives the lawn a more even appearance. It may also help reduce thatch buildup in the lawn. A slight overlap is necessary on each pass over the lawn. This overlap prevents missed areas and picks up any grass that the mower wheel may have only pushed down on the previous trip.

If the lawn becomes too tall and it is necessary to cut off a large part of the grass blades, the grass should be collected in a catch bag attached to the mower or raked up and removed. Heavy accumulation of grass clippings shades out light and kills grass. It also mats down the grass and causes a buildup of thatch on the soil surface. (*Thatch* is a layer of dead grass.) Thatch reduces soil aeration, thus damaging roots and also providing a breeding place for insects and diseases.

WATERING THE LAWN

A great deal of water is required to give lawns the moisture they need. Unless an adequate supply is available, the lawn should not be watered at all. Shallow watering does more harm than good. This is because shallow watering causes the grass roots to move to the surface to absorb water. Here, they are more easily dried out and require still more frequent watering. Surface roots are more easily torn loose from the soil during winter freezing and thawing. Frequent watering also encourages the growth of fungus disease on the grass blades.

At least 1 inch of water should be applied at each application. An amount less than this does not penetrate the soil deeply enough. To determine when 1 inch has been applied by a sprinkler, set a rain gauge or any container with straight sides on the lawn on the area to be watered. When the container has 1 inch of water in it, 1 inch has been applied to the lawn.

Do not apply water faster than it can soak into the soil surface; runoff is wasted water. Do not water until the grass wilts and needs water, and then water only if you have enough water to do the job well.

A healthy lawn can become dormant and withstand a great deal of dry weather without being permanently damaged. The grass will become green and active again with the first good rain. The appearance may not be as pleasing when the grass turns yellow or brown, but little actual damage is being done.

SOLVING PROBLEMS IN THE LAWN

Controlling Weeds

If a lawn is heavily infested with weeds, chemicals should be used to eliminate the problem. However, development of a thick, healthy turf is the best way to guard against a serious weed problem.

Weeds in a lawn are usually an indication of an unhealthy lawn caused by poor maintenance practices. When weeds are spotted, first check maintenance practices and correct them if necessary. If weeds are still a problem, consider chemical control.

Two types of chemical weed killers are used on lawns. One is a preemergence weed killer, which kills germinating seeds and very tiny weed seedlings before the weeds become established. This is the type of weed killer which is applied in the spring to cool-season grasses for the control of crabgrass. It is often mixed with a fertilizer before application. Since crabgrass is a grass, the mature plant cannot be controlled without killing the lawn grass. However, the tender seedlings can be killed with small amounts of weed killer which are not strong enough to kill the lawn grass.

Postemergence weed killers are applied after the weeds sprout and begin to grow. A sprayer is generally used to apply these types of weed killers, figure 35-3, page 412. Notice that the sprayer is labeled to avoid using the container for any other purpose. Since many weed killers are selective and only kill certain types of plants, it is possible to select a chemical that does not kill grass, but does kill broadleaf weeds. One chemical of this type is 2,4-D. It is used to kill many broadleaf weeds in lawns, such as dandelions and plantain.

> For specific weed identification and control, contact your local extension service to obtain bulletins written for the local area.

Lawn Diseases and Their Control

Most turf diseases are caused by parasitic plants called fungi (plural of *fungus*). These parasites live in and on dead grass and in the soil where they attack the green grass and rob the soil of

Fig. 35-3 Small pressurized sprayer used to apply 2,4-D weed killer to broad-leaf weeds in a lawn. Notice that the sprayer is marked WEED KILLER to guard against its use on ornamental plants. (Richard Kreh, photographer)

nutrients. This weakens and can kill the lawn grass plant. Fungus diseases are spread easily by mowing or simply walking through the lawn. The tiny seedlike spores spread rapidly.

For fungus diseases to cause a serious problem, there must be, in the area,

- grass plants on which the fungus can live.
- fungus spores and a means of spreading them to the grass.
- temperature and moisture conditions favorable for the growth of fungi.

Disease Prevention

The best control measure for fungus disease is prevention. To reduce the chance of attack by fungus spores on lawns,

1. do not overuse nitrogen fertilizer.

2. add lime as needed to maintain a pH of 6.0 to 6.5. (This helps to control thatch buildup.)

3. avoid thatch buildup by collecting clippings or by raking and removing thatch.

4. water only when necessary, and then water deeply. Do not repeat watering for one week.

5. mow frequently, and remove only one-third of the top growth at any one mowing. Cut grass to the proper height and keep it within 1 inch of that height throughout the entire growing season.

Insect Control

Insects can cause serious damage to lawns. There are insects that attack the top of the plant, such as the chinch bug, sod web-worm, and the flea beetle. Serious damage is also done by insects that attack grasses below the soil level by eating roots and stems. Some below-the-surface feeders are the grubs (larvae) of the Japanese beetle and many other beetles and cutworms.

When insect attacks are serious, spraying or application of granular insecticides may be necessary. Consult local extension service bulletins for pictures to assist in identification of the pests and control recommendations.

Thatch Control

Thatch is a layer of dead stems, leaves, and roots of grass which builds up on the soil surface and under the green leaf area of grass. A thick layer of thatch may prevent water from penetrating the soil, prevent proper soil aeration, and provide a breeding area for insects and disease. Heavy thatch buildup and reduced aeration of the soil result in shallow root systems.

The amount of thatch buildup varies somewhat according to the variety of grass that is planted. For example, Merrion Kentucky bluegrass builds up thatch faster than common bluegrass. Bentgrass and some red fescues also tend to build up thatch.

Other factors that may cause thatch to build up are:

- not adjusting the mower blade properly, resulting in too much of the grass blades being cut.

- returning clippings to the lawn.

- heavy fertilization.

- heavy clayey soil.

- acidic soil (apply lime).

On small areas, thatch may be removed with a hand rake, figure 35-4. However, this is a very difficult job. A better method is use of a power driven thatch machine; it is well worth the rental fee. Early spring is usually the best time to remove thatch from the lawn. Another solution to the problem of thatch is a light application of topsoil (1/8 inch). This buries the thatch, causing it to rot. However, the best thatch control measures are preventive. Proper mowing, fertilizing, and liming and removal of clippings usually provide adequate control.

Fig. 35-4 This pile of thatch was raked from an area measuring about 1 square foot. Raking is not the most effective way of collecting thatch. (Richard Kreh, photographer)

Loosened Grass

Rollers are used to press down grass that may have been loosened, pushed, or heaved out of the ground by freezing and thawing during the winter. Rolling should be done in early spring after the last freezing and thawing. The roller should be only heavy enough to push the grass firmly against the soil. Too much weight may make the soil too compact and reduce aeration in the soil. This is especially true for heavy clayey soils.

Moss

Moss sometimes grows in shady, wet areas of lawns. It is not a weed. Moss grows only where grass has died for some reason. There are four problem areas which can result in the growth of moss in the lawn.

Problem	Solution
too much shade	Reduce shade or plant a variety of grass that is shade tolerant.
poor growth of grass	Fertilize properly. (A thick turf prevents the growth of moss.)
acidic soil conditions	Test soil and add lime.
a wet area (Moss will grow in soil too wet for grass.)	Fill low, wet areas with topsoil or drain the area.

STUDENT ACTIVITIES

The following activities are to be accomplished on an area of lawn assigned by the instructor.

1. Examine the area and make a list of problems found in the lawn. Write down possible reasons for the problem areas and report to the class. Topics included in the report might be weed control, fertilizer needs, and insect and disease problems.

2. Develop a lawn maintenance and/or renovation schedule for the lawn area. Have the schedule approved by the horticulture teacher.

3. Locate a good source of lime and a turf fertilizer. Apply each to the lawn area at the proper time and rate using both a rotary and conventional fertilizer spreader. Calculate how much fertilizer to spread on 1,000 square feet and set the spreader for that application. Check your accuracy by seeing how much fertilizer remains after the job or if you run out of fertilizer before the job is finished.

4. Adjust a rotary mower blade to the best mowing height for the grass to be mowed. Repeat with a reel mower.

5. Examine a sharp and a dull mower blade. Sharpen or observe the proper sharpening technique of the dull blade as demonstrated by the instructor. **Caution: Be careful of the sharp cutting edge.** Balance the blade after sharpening.

6. Identify weeds in the lawn area. Determine if a weed killer is needed for control and, if so, which type. Apply the chemical. Observe and record the results.

7. Demonstrate a proper mowing pattern on a lawn area.

8. Visit a local golf course. Have the greens keeper discuss the type of grass used on the greens, the height of the grass and adjustment of the blade, and maintenance schedule used to keep the greens in order. The student can expect this schedule to be more time consuming than that used on the home lawn.

SELF-EVALUATION

A. List the five management practices that reduce or help to control turf grass diseases.

B. Select the best answer from the choices offered to complete the statement or answer the question.

1. A cool-season grass that thrives in partial shade is
   a. creeping red fescue.
   b. Kentucky bluegrass.
   c. Bermuda grass.
   d. centipede grass.

2. Lime moves slowly through the soil to correct soil acidity. When applied to the soil surface, lime moves down through the average loam soil at a rate of about
   a. 2 inches a year.
   b. 3 inches a year.
   c. 1/2 to 1 inch a year.
   d. 6 inches a year.

3. Grasses feed heavily on the fertilizer element
   a. phosphorus.
   b. nitrogen.
   c. potash.
   d. calcium.

4. What form of nitrogen in fertilizer is released slowly?

   a. organic

   b. inorganic

   c. chemical

   d. soluble

5. Fertilizer should be applied to a lawn grass

   a. when it is available at the lowest price.

   b. in midsummer for cool-season grasses.

   c. when the grass begins the dormant rest cycle.

   d. just before the beginning of the active growth cycle of the grass.

6. Cool-season grasses should be cut to a height of

   a. 3 to 4 inches.

   b. 1/2 to 1 inch.

   c. 2 to 3 inches.

   d. 1/2 to 3/4 inches.

7. Warm-season grasses such as Bermuda grass should be cut to a height of

   a. 3 to 4 inches.

   b. 1/2 to 1 inch.

   c. 2 to 3 inches.

   d. 4 to 5 inches.

8. Frequent shallow watering of the lawn

   a. helps to keep the lawn looking fresh and green.

   b. helps control insects in the lawn.

   c. causes shallow root development of the grass.

   d. helps control fungus diseases in the lawn.

9. Preemergence weed killers are applied before weeds become established. This type of weed killer is used to control

   a. dandelions.

   b. broadleaf weeds.

   c. all lawn weeds.

   d. crabgrass.

10. Postemergence weed killers are used after weeds are present in the lawn. The weed killer 2,4-D is a selective weed killer used to kill

    a. annual grasses in the lawn.   c. broadleaf weeds in the lawn.
    b. crabgrass in the lawn.   d. all lawn weeds.

11. Most lawn diseases are caused by parasitic plants called

    a. toadstools or mushrooms.   c. fungi.
    b. molds.   d. slime.

12. Insects that attack grass below the soil level and eat the grass roots are

    a. chinch bugs.   c. flea beetles.
    b. beetle larvae and cutworms.   d. sod webworms.

13. Thatch buildup in a lawn causes all of the following except

    a. faster growing grass that requires more frequent mowing.
    b. slow penetration of water.
    c. poor soil aeration.
    d. development of a shallow root system in the lawn.

14. Rolling the lawn each spring is necessary to

    a. level the lawn.
    b. push grass roots which have heaved loose because of frost into close contact with the soil.
    c. provide a firm surface for mowing.
    d. roll down any big clumps of grass that might be present.

15. A light application of topsoil (about 1/8 inch) on a lawn is often used to

    a. help control thatch.   c. fill in the small cracks in the lawn.
    b. help to level the lawn.   d. add depth to the topsoil.

16. When mowing the lawn, no more than _____ of the top of the grass plants is cut at any one time.

   a. one-half   c. seven-eighths
   b. three-quarters   d. one-third

17. The best weed control for any lawn is

   a. a good, healthy thick turf.
   b. a combination of preemergence and postemergence weed killers.
   c. a weed killer/fertilizer combination.
   d. none of the above

18. Moss generally grows in a lawn when (select the *incorrect* or *wrong* answer)

   a. there is too much shade.
   b. inadequate fertilizer results in poor growth of grass.
   c. acid, wet soil conditions exist.
   d. too much lime is used.

OBJECTIVE

To select and use the proper renovation technique for a specific problem in the lawn.

COMPETENCY TO BE DEVELOPED

After studying this unit, the student will be able to

- inspect a lawn area and determine if it requires renovation.
- determine which of the four methods of renovation should be used on a lawn.
- use the step-by-step approach to renovate a lawn.

Materials List

grass seed

fertilizer and lime, if needed

weed killer and sprayer or other applicator

garden rake

thatch rake or thatching machine

lawn mower

shovel or sod-lifting tool

straw or burlap for mulching seeded areas

---

unit 36
Renovating the Lawn

When good maintenance practices are not carried out or weather conditions have prevented the growth of healthy grass plants, it is time to consider lawn renovation.

WHY DO LAWNS FAIL?

The wrong species or variety of grass was planted. Be sure to check the type of grass best suited for the area. It is especially important to consider soil and light conditions. Before purchasing seed, check the label for percentages of the different varieties of seed it contains. Good seed is expensive, but it saves money over a period of time since the lawn will require less reseeding. Seed cost is a small part of the total cost of establishing a lawn.

Improper mowing. The lawn may have been cut too short. Some people prefer a lawn that is cut very short, but cool-season grasses cannot survive if mowed to heights of 1/2 to 1 1/2 inches. Grass should not be allowed to grow to be 3 inches high and be cut back to 1 inch. Only one-third of the top should be removed at any one mowing. Remember that close mowing does not reduce the number of times the lawn needs to be mowed.

Improper fertilizing — The most common mistakes in fertilizing are applying the fertilizer too late in spring for cool-season grasses or too late in fall for warm-season grasses. Fertilizer should be applied a few weeks before the grass begins its active growing stage. Use a slow-release fertilizer that is high in nitrogen. Apply fertilizer at the proper rate.

Improper watering — The most common mistake in watering is failure to apply enough water to soak into the soil to a depth of from 4 to 6 inches. This requires at least 1 inch of water at each watering. As a general rule, lawns should be watered no more than once a week, and then only when the soil is dry.

Heavy traffic — Too much traffic on lawns, especially on those planted with a variety not known for its hardiness, can result in bare spots. To solve this problem, reduce traffic or change the grass variety.

Excessive shade — Dense shade and competing shallow roots of trees such as maples kill lawn grasses. In these cases, the tree, and thus the source of the shade must be removed, or a shade-tolerant grass or other ground cover must be planted under the tree.

The soil was not prepared properly for planting — A good topsoil, proper fertilizer and lime application, and proper planting techniques are essential for a healthy lawn. (Test the soil to determine fertilizer and lime needs.)

Infestation of weeds, diseases, and insects — Weed, disease, and insect problems are usually the result of one or more of the above problems. A thick, healthy lawn turf is seldom overcome by insects, diseases, or weeds.

Thatch — When old grass leaves, roots, and stems accumulate on the soil surface faster than they rot, the buildup is called thatch. Thatch buildup on the soil surface prevents air from circulating in the soil — a very necessary condition for root growth. Thatch also causes water to run off rather than soaking into the soil, and harbors insects and disease organisms. To help speed up thatch rotting, increase the soil pH and apply more fertilizer. Another solution is to remove the thatch with a thatch removing machine or a special thatching rake.

RENOVATION TECHNIQUES

If the lawn still fails to prosper, renovation, or rebuilding of the lawn, must be considered. Lawns should be renovated at the time of year best suited for starting a new lawn; that is, just prior to the most active growing season for the type of grass being planted in the renovation. The entire lawn may be reseeded, resodded, or planted by other vegetative means such as strip planting or with plugs, or a technique known as spot patching may be used.

Weed-Infested Lawns

When more than one-third of the lawn grass plants are dead and large bare spots exist, it may be better to kill the existing grass with a weed killer such as paraquat and reseed the entire area, figure 36-1. More than one application may be necessary to kill some grasses. After the grass is killed, the entire lawn area is dug up, the old sod is removed, and the lawn is reseeded. The procedure after the soil is cleared of all old grass is the same as for seeding a new lawn. (Refer to Unit 34 for details.)

Thinly Covered Lawns

If the lawn grasses cover the entire lawn area but do not produce a thick turf, the lawn may be salvaged as follows, providing good topsoil is present and drainage is not a problem.

1. If a heavy layer of thatch exists, remove it. A thatch-removing machine should be used to dig up the thatch and vertically slice the root area for improved aeration.

2. Add seed on the condition that it will come in close contact with the soil surface. If seed cannot rest firmly on the soil surface, it will not grow. A roller helps bring the seed into contact with the soil.

3. Apply weed killer to control broadleaf weeds if they are present. Do not apply weed killer if new seed was sown as directed in step 2. Weed killer should not be used for six months after a new seeding is made.

A decision must be made as to whether step 2 or step 3 should be followed; both cannot be accomplished on the same lawn area.

4. Fertilize the lawn with a good turf fertilizer (a slow-release, high nitrogen fertilizer), figure 36-2, page 422.

5. Mow the lawn properly. Remove no more than one-third of the top with any one mowing. Mow at the proper height for

Fig. 36-1 A section of lawn with less than one-third of the area producing healthy lawn grass. Notice the distinct difference in the appearance of the healthy grass (top right corner) as compared with the rest of the area, which is infested with various weeds. If the entire lawn suffers from such severe infestation, it is best to kill all vegetation and reseed. (Richard Kreh, photographer)

Fig. 36-2 Although the lawn in this area is thin, the bare spots are very small and the plants that are present are healthy ones. There are few weeds present and drainage is good. The best treatment for such an area is an application of fertilizer. Lime may be added if a soil test indicates a need. (Richard Kreh, photographer)

the grass species planted. (See number 2 under "Why Lawns Fail" in this unit.)

6. Follow all good maintenance practices.

Spot Seeding

When there are isolated spots of dead or weak grass, the lawn may be renovated by a "patchup" of these spots. Any spot that measures 1 foot and over should be treated.

The same procedure is used to renovate these spots as when seeding a new lawn. The only difference is that the process is used on a much smaller area.

1. Remove any dead grass or weeds from the area to be reseeded. Dig up the spot with a steel rake and remove any old sod or weeds. Loosen the soil well to prepare a shallow seedbed 1 or 2 inches deep.

2. Add fertilizer as needed. Use a fertilizer with extra phosphorus, such as a 5-10-5. Dig the fertilizer into the top 1 or 2 inches of the soil with the rake.

Fig. 36-3 A small, weedy area in which most of the grass is dead. The weedy area is dug out and pieces of sod (one of which is shown) are placed in the hole. (Richard Kreh, photographer)

3. Level the soil and firm the seedbed. Remove any rocks or large clods of soil.

4. Spread the seed uniformly over the surface by hand.

5. Lightly rake the seed into the soil surface. Cover with 1/8 to 1/4 inch of soil.

6. Lightly cover with straw or burlap mulch.

7. Tamp with the feet or roller to bring the seed in close contact with the soil.

8. Practice proper maintenance techniques.

Spot-Patching

When good sod can be taken from inconspicuous areas, small spots in the lawn may be patched up by digging out the weak area about 1 inch deep and replacing it with a piece of sod cut to fit the spot, figure 36-3. For large spots, more than one piece of sod may be fitted into the hole. Tamp the sod so that it rests tightly against the loosened soil in the bottom of the hole. Fertilize and water well to stimulate new root growth. It is important that the sod knit well and quickly to the new soil area, and that it not be allowed to dry out until its roots have grown into the soil, figure 36-4.

Fig. 36-4 Sod piece in place. The sod is tamped in place so that it makes good contact with the soil. Notice that there are still single weeds scattered throughout the lawn area — a weed killer sprayed over the area will control these. (Richard Kreh, photographer)

STUDENT ACTIVITIES

1. Take a field trip to observe lawns in the local area. Note lawns that are in need of renovation and ones that appear to be healthy. Inquire about the present condition of the lawns and their maintenance schedules. Determine why the lawns are in varying conditions and report to the class.

2. Establish a 1,000-square foot plot for use in practicing lawn renovation. Keep notes of the problems found, treatment, and results of treatment.

3. Invite a nurseryman, golf course superintendent, or other lawn specialist to the class to discuss the most common lawn problems in the local area and the methods used to correct them.

4. Note any soil drainage, shade, or nutrient deficiencies that exist in the local area.

SELF-EVALUATION

Select the best answer for each question below.

1. The most common mistake in mowing cool-season grasses is to

   a. mow them too short.
   b. mow too often.

   c. use a dull mower blade.
   d. mow them too high.

2. The renovation technique for weak, bare lawn areas under dense shade is

   a. to fertilize more heavily.
   b. to let more sun in and/or plant a shade-tolerant grass or other ground cover plant.
   c. to lime the area heavily.
   d. to mow more often.

3. To renovate a lawn means to

   a. dig the lawn completely up.
   b. fertilize the lawn.
   c. rebuild or restore a strong stand of grass to the lawn.
   d. mow the lawn properly.

4. When over one-third of the lawn grass is dead and large bare spots exist, the best renovation practice is to

   a. kill all existing plants with weed killer and reseed the lawn.
   b. spray with 2,4-D and fertilize to stimulate plant growth.
   c. patch up the dead areas and fertilize the entire lawn.
   d. none of the above

5. Where lawn grasses are thin but cover the entire lawn area and few weeds are present, the best renovation practice is to

    a. remove the thatch layer if it is too thick.
    b. either add seed or use weed killer — whichever is needed most.
    c. fertilize the lawn and establish a good mowing schedule.
    d. all of the above

6. When mowing the renovated lawn, no more than _____ of the top of the grass plant should be removed at any one mowing.

    a. 1/2 inch   c. 2 inches
    b. one-third   d. 3 inches

7. When patching spots in the lawn, it is best to treat spots 1 foot across and larger by

    a. reseeding, sodding, or sprigging, as if starting a new lawn.
    b. applying fertilizer so that the grass will spread to these spots.
    c. sodding.
    d. mowing the lawn properly until the spots are again covered with grass.

8. Sod is an expensive method of starting a lawn and is usually used in renovation work to

    a. resod all weak and weedy areas, no matter how extensive the work is.
    b. patch up small spots in the turf.
    c. lay out areas for gardens.
    d. all of the above

9. Lawns are often mowed too short because

    a. it results in mowing the lawn fewer times.
    b. it is difficult to adjust lawn mowers to higher cutting heights.
    c. insect problems are better controlled.
    d. some people prefer a close-cut lawn.

Section 11
The Vegetable Garden

unit 37
Planning and Preparing the Vegetable Garden Site

OBJECTIVE

To select and prepare a vegetable garden site.

COMPETENCY TO BE DEVELOPED

After studying this unit, the student will be able to

- list four items to consider when choosing the location of a vegetable garden.

- draw to scale a garden plan that includes at least four vegetables. Also include plans for successive plantings of two vegetables which are planted early and harvested early and two vegetables that are planted after earlier crops and harvested in the fall.

- explain the difference between preparing a heavy, clayey garden soil and a sandy garden soil.

- take a soil sample and have it tested. Write a fertilizer program for the garden plan developed in objective 2 with this type of soil.

Materials List

graph paper, ruler, and pencil

seed catalogs

No matter where you live, there is probably space for at least a small vegetable garden. Gardening is a good hobby which provides exercise, the satisfaction of growing some of your own food, and a savings on grocery bills. For the beginner, it is probably best to start with a small garden and work up in size as skill increases. The best techniques of planning and planting a garden are useless unless proper weed control, watering, and insect and disease control are practiced. Many of the skills developed in growing a small garden can be transferred to large-scale commercial vegetable production.

LOCATION

If irrigation of the garden is necessary, the vegetable garden must be located near a supply of water. More importantly, the location should have a healthy, loamy, well-drained soil and plenty of sunshine. Some vegetables will grow in partial shade, but most need full sun. A location in the shade of large trees is especially bad, since the tree roots may extend into the garden

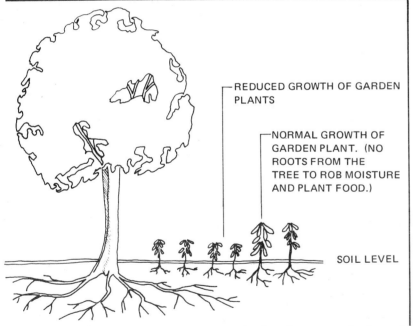

REDUCED GROWTH OF GARDEN PLANTS

NORMAL GROWTH OF GARDEN PLANT. (NO ROOTS FROM THE TREE TO ROB MOISTURE AND PLANT FOOD.)

SOIL LEVEL

Fig. 37-1 The effect of large trees on growth of garden crops. As shown in the illustration, tree roots usually extend farther than the limbs.

and rob garden crops of necessary plant food and moisture, figure 37-1. If some of the garden plants must be grown in shade, leafy green vegetables and pumpkins suffer less from shade than other vegetables.

The garden should not be located on a steep hill where rainwater will run off rapidly, since the runoff water is lost for use by the plants and also causes soil erosion. If the garden is located on a slope, form the rows along or around the hill instead of up and down the hill. This helps prevent soil erosion and slows down water runoff, allowing it to soak into the soil.

THE GARDEN PLAN

Deciding what to plant and how much to plant are important steps in vegetable gardening. First, list the vegetables you and your family like to eat. Decide if you want to plant enough to freeze or can for winter use, or only grow enough to eat during the summer. A garden plan saves time and work. Draw a sketch of the plot to scale, preferably on graph paper. Keep in mind that a small, well-kept garden gives better returns than a large, weedy one.

The following are some points to consider when planning the garden.

- Plant perennials such as asparagus, rhubarb, horseradish, and sorrel together on one side of the garden where they will not interfere with the plowing and working of the rest of the garden.

- Group quickly maturing crops together, or plant them between rows of crops that mature later. When they are harvested, the area can either be replanted or left to provide more room for the later maturing crops.

- Plan the distance between rows according to the type of cultivation methods that will be used. Hoeing and mulching by hand allows the rows to be closer together.

- Crops such as snap beans and sweet corn can be planted at intervals of every two weeks so that they can be harvested at different times during the season.

- Replant areas where early crops, such as peas and lettuce, are harvested with fall crops, such as kale or turnips.

Beets, carrots, leaf lettuce, onions, early peas, radishes, and spinach can be planted in rows a foot apart (although 18 inches is preferable), but other vegetables must have more room. Rows planted 18 inches apart with plants 2 feet apart is a good rule of thumb for most other vegetables except staked tomatoes, which must be planted about 2 feet apart in rows 3 feet apart. Sweet corn should either be in hills 3 feet apart each way, or in rows 2 1/2 feet apart with plants spaced 1 foot apart. Vine crops (cucumbers, melons, and squash) require 2 1/2 to 5 feet of space in all directions. For space savers, remember that cucumbers can be grown on a fence and bush varieties of squash require less room than other vegetables.

All these spacing rules are for gardens that are hand cultivated. If a tractor or cultivator is to be used, more room must be allowed between rows. In a location with a considerable slope, make the rows across the grade to help prevent erosion during heavy rains. Plant tall growing vegetables on the north or west side of the plot to prevent shading of lower growing varieties.

If your garden is tiny, a good garden plan would be a half row each of radishes and leaf lettuce (to be followed by a succession planting of snap beans); a row of snap beans planted at the regular time; and a few tomato plants. With more space, plan to include snap beans, lima beans, beets, broccoli, carrots, cabbage, a hill or two of cucumbers, greens such as spinach, Swiss chard and tampala, leaf and head lettuce, a few early peas, and tomatoes. If there is plenty of room, include brussels sprouts, cauliflower, sweet corn, eggplant, melons, onions, peas, peppers, squash, and any other varieties of vegetables you especially like.

A garden guide from the W. Atlee Burpee Company is given in figure 37-2 to help in the planning of a vegetable garden. The guide is designed to provide vegetables for four people.

Definition of Terms in Guide

Hardiness is a measure of the crop's resistance to frost. See notes under growing suggestions.

Days to germinate is the number of days from planting of the seed to the date on which the plant surfaces.

Quantity to grow is how much to plant for four. Increase or reduce to fit personal needs.

Days to harvest is the number of days from planting of the seeds or plants in the garden to the date on which the crops are gathered.

pH range is the soil acidity at which the plants grow best. Check soil with a soil test. Add lime as recommended.

Uses gives the best use for each vegetable.

THE GARDEN LAYOUT

Figure 37-3 shows a garden plan for spring and summer planting. Figure 37-4 shows a garden plan for the same garden later in the summer. Notice that the early maturing vegetables have been harvested and the same areas replanted with late summer

VEGETABLES AND TYPES	HARDINESS	DAYS TO GERMINATE	GROWING SUGGESTIONS	QUANTITY TO GROW	DAYS TO HARVEST	SATISFACTORY PH RANGE	USES
BEAN, snap bush and pole green, yellow	T	7-14	Sow bush types every 2 weeks until mid-summer. Support pole types.	50 ft. bush; 8 hills pole	50-70	5.5-6.7	Fresh, frozen, canned. Vitamins A,B,C.
BEAN, bush shell red, white, green	T	7-14	Fava or English Broad Bean hardier than other types. Sow as early in spring as soil can be worked.	50 ft.	65-103	5.5-6.7	Fresh shell beans, or use dried for baking, soup, or Spanish or Mexican dishes. Vitamins A,B,C.
BEAN, lima bush and pole	T	7-14	Wait until ground is thoroughly warm before planting. Bush types mature earlier. Support pole varieties.	70 ft. bush; 8 hills pole	65-92	6.0-6.7	Fresh, frozen, canned, dried for baking. Vitamins A,B,C.
BEETS, red, golden, white	HH	10-21	For continuous harvest, make successive sowings until early summer. Do not transplant. This may cause forked or split roots.	25 ft.	55-80	6.0-7.5	Fresh, pickled, canned. Cook "thinnings" first, and tops later on for delicious greens. Vitamins A,B,C.
BROCCOLI	H	10-21	Plant again in midsummer for fall harvest. Grows best in cool weather.	25-40 plants	60-85*	5.5-6.7	Fresh, frozen. Vitamins A,B,C.

Fig. 37-2 Burpee Vegetable Garden Guide for Four. See key to chart on p. 435. (Courtesy W. Atlee Burpee Company)

VEGETABLES AND TYPES	HARDINESS	DAYS TO GERMINATE	GROWING SUGGESTIONS	QUANTITY TO GROW	DAYS TO HARVEST	SATISFACTORY PH RANGE	USES
BRUSSELS SPROUTS	H	10-21	Pick lowest "sprouts" on stem each time; break off accompanying leaves but do not remove foliage.	25-40 plants	80-90*	5.5-6.7	Light frost improves flavor. Sprouts delicious fresh or frozen. Vitamins A,B,C.
CABBAGE, early, late, red, green	HH	10-21	Do not plant where any of the cabbage family grew the previous year.	25-40 plants	60-110*	5.5-6.7	Fresh, salads, coleslaw, sauerkraut. Winter storage. Vitamins A, B, C.
CARROT, long, short	HH	7-14	Short root types best for shallow or heavy soil. Plant again in midsummer for fall harvest.	25-30 ft.	65-75	5.2-6.7	Salads, relish, juice. Stews, soup. Vitamins A, B, C.
CAULIFLOWER, white, purple	HH	10-21	Tie leaves over heads to whiten.	16-24 plants	50-85*	6.0-6.7	Fresh, frozen; salad, relish. Vitamins A,B,C.
CELERY	HH	10-21	To whiten, mound soil up around mature stalks.	30-36	115-135*	5.5-6.7	Raw in salads and as relish. Cooked and creamed, soups; Vitamin A.
CHARD red, white stalked	HH	7-14	Pick frequently to encourage fresh leaves. Stands summer heat.	20-30 ft.	60	6.0-6.7	Cook leaves for greens; midribs and stalks like asparagus. Vitamins A,B,C.
COLLARDS	H	7-14	Easily grown, nonheading, cabbagelike leaves.	20-30 ft.	80	5.5-6.7	Cook leaves for greens. Popular in southern states. Vitamins A,B,C.

Fig. 37-2 Burpee Vegetable Garden Guide for Four. (Courtesy W. Atlee Burpee Company) (continued)

VEGETABLES AND TYPES	HARDINESS	DAYS TO GERMINATE	GROWING SUGGESTIONS	QUANTITY TO GROW	DAYS TO HARVEST	SATISFACTORY PH RANGE	USES
CRESS, garden and water	H	3-14	Sow garden or curly-cress every 2 weeks for continuous supply. Also grows well on sunny window-sill. Grow watercress in moist, shady spots or along a shallow stream.	20-30 ft.	10-50	6.0-7.0	Salads, sandwiches, garnish, seasoning. Vitamins A,B,C.
CUCUMBERS, slicing, pickling	T	7-14	Grow on fence to save space. Keep picking to encourage new fruit.	8-12 hills	53-65	5.5-6.7	Salad, relish, pickles. Vitamin A.
EGGPLANT	T	10-21	Needs warm tempera-ture — 70° to 75°F for good germination. Pick fruits when skin has high gloss.	8-12 plants	62-75*	5.5-6.7	Delicious fried, sauteed, or in casseroles. Vitamin A.
ENDIVE	H	7-14	Grows best in cool weather.	20-30 ft.	90	6.0-7.0	Salad, greens. Hearts can be cooked and served with cream sauce or grated cheese. Vitamins A,B,C.
KALE	H	14-21	Mature plants take cold fall and winter weather. Frost improves flavor.	25-30 ft.	55-65	5.5-7.0	Chop young leaves for salads and sandwiches. Cook for greens. Vita-mins A,B,C.
KOHLRABI	HH	14-21	Grow for spring or fall crop; thrives in cool weather.	16-20 ft.	55-60	5.5-6.7	Fresh, frozen; cooked like turnips. Vitamins A,B,C.

Fig. 37-2 Burpee Vegetable Garden Guide for Four. (Courtesy W. Atlee Burpee Company) (continued)

VEGETABLES AND TYPES	HARDINESS	DAYS TO GERMINATE	GROWING SUGGESTIONS	QUANTITY TO GROW	DAYS TO HARVEST	SATISFACTORY PH RANGE	USES
LEEK	H	14-21	Whiten and improve flavor by mounding soil around mature plants.	25-40 ft.	130	5.5-6.7	Fresh in salads. Cooked in soups, stews, or creamed.
LETTUCE, leaf	H	7-14	Make successive sowings in spring and another in late summer. Keep seedbed moist to get good germination for a fall crop.	25-40 ft.	40-47	6.0-7.0	Salad, sandwiches, garnish. Vitamins A,B,C.
LETTUCE, head	H	7-14	Needs cool weather in spring or fall to head well.	25-30 ft.	65-90	6.0-7.0	Salad, sandwiches, garnish. Vitamins A,B,C.
MUSTARD GREENS, fringed, smooth leaves	H	7-14	Grow as fall, winter and spring crop in mild winter areas; spring and fall in north.	25-30 ft.	35-40	5.5-6.5	Greens. Vitamins A,B,C.
MELONS, cantaloupe, crenshaw, casaba, honeydew, watermelon	T	7-14	Very sensitive to frost. Black plastic mulch speeds maturity. Needs warm sunny weather when ripening for good flavor.	12-20 hills	75-120	6.0-6.7	Fresh, frozen. Ripe cantaloupes slip easily from stems. Ripe watermelons sound dull and hollow when tapped. Vitamins A,B,C.
OKRA	T	7-14	Needs hot weather to mature well. Pick pods young.	16-20 ft.	52-56	6.0-7.0	Soups, stews. Vitamins A,B,C.

Fig. 37-2 Burpee Vegetable Garden Guide for Four. (Courtesy W. Atlee Burpee Company) (continued)

VEGETABLES AND TYPES	HARDINESS	DAYS TO GERMINATE	GROWING SUGGESTIONS	QUANTITY TO GROW	DAYS TO HARVEST	SATISFACTORY PH RANGE	USES
ONIONS, yellow, white	H	10-21	Grow best in fine, well-drained sandy loam soil.	50-100 ft.	95-120	5.5-6.7	Fresh, salads, pickling. Vitamins B,C.
PARSLEY, curled or plain leaves	H	14-28	Attractive edging for flower garden; pot herb on sunny window-sill in winter.	10-20 ft.	72-90	6.0-7.5	Salad, garnish, seasoning. Dries or freezes well. Vitamins A,B,C.
PEA, dwarf, tall	H	7-14	Plant as early as ground can be worked.	40-100 ft.	55-79	5.5-6.7	Fresh, frozen, canned, dried. Vitamins A,B,C.
PEPPER, sweet, hot	T	10-21	Needs warm tempera-ture – 70° to 80°F for good germination.	8-10 plants	60-77*	5.5-6.5	Salad, stuffed, relish, seasoning. Vitamins A,B,C.
PUMPKIN, large, small bush, vine	T	7-14	For huge "contest" pumpkins, let only 1 or 2 grow per plant.	12-20 hills	95-120	5.5-6.5	Fresh, canned, frozen. Vitamins A,B,C.
RADISH, red, white, black	H	7-14	Make successive sowings until early summer; again a month before fall frost.	15-30 ft.	22-60	5.2-6.7	Relish, salad. Vitamins B,C.
RUTABAGA	H	14-21	Grows best in cool weather.	20-30 ft.	90	5.2-6.7	Fresh. Winter storage. Vitamins B,C.
SPINACH, crinkled, smooth	H	7-14	New Zealand and Malabar take hot weather; other varieties cool.	20-40 ft.	42-70	6.0-6.7	Greens, frozen, canned. Vitamins A,B,C.
SQUASH, summer, bush, vine	T	7-14	Keep fruits picked so plants produce more.	8-12 hills	48-60	5.5-6.5	Fresh, frozen. Vitamin A.

Fig. 37-2 Burpee Vegetable Garden Guide for Four. (Courtesy W. Atlee Burpee Company) (continued)

VEGETABLE AND TYPES	HARDINESS	DAYS TO GERMINATE	GROWING SUGGESTIONS	QUANTITY TO GROW	DAYS TO HARVEST	SATISFACTORY PH RANGE	USES
SQUASH, winter, bush, vine	T	7-14	Black plastic mulch speeds maturity.	8-12 hills	80-120	5.5-6.5	Fresh, frozen, canned. Winter storage. Vitamin A.
SUNFLOWER	T	7-14	Use for screen plant. Protect maturing heads with bags to prevent bird damage.	25-50 ft.	80	6.0-7.5	Bird, poultry seed.
SWEET CORN, white, yellow	T	7-14	Plant in blocks of short rows for good pollination and well-filled ears.	50-100 ft.	63-90	5.2-6.7	Fresh, frozen, canned. Vitamins A,B,C.
TOMATO, red, pink, yellow	T	7-14	Hybrids especially need warm temperature — 70° to 80°F for good germination.	16-20 plants	52-68*	5.2-6.7	Fresh, salad, canned, juice, pickles. Vitamins A,B,C.
TURNIPS, white, yellow	H	14-21	Grow best in cool weather.	20-30 ft.	35-60	5.2-6.7	Fresh, raw or cooked. Leaves of some types for greens.

KEY TO CHART

H — HARDY VARIETIES OF VEGETABLES AND FLOWERS — Tolerate cool weather and frost. Plant fall to early spring in zones G,H,I,J. Two to four weeks before last killing spring frost in all other zones.

HH — HALF HARDY VARIETIES OF VEGETABLES AND FLOWERS — Tolerate cool weather and very light frost. For earlier maturity, varieties that transplant well can be started inside or in a cold frame 6-10 weeks before last expected light frost.

T — TENDER VARIETIES OF VEGETABLES AND FLOWERS — Cannot stand frost; plant in spring after last frost date.

* — Time from when plants are set into garden.

Fig. 37-2 Burpee Vegetable Garden Guide for Four. (continued) (Courtesy W. Atlee Burpee Company)

NORTH

	EARLY SWEET	MIDSEASON SWEET CORN	LATE SWEET
16½'	CORN	(Block Planting)	CORN

3' ——— TOMATOES (8 STAKED PLANTS) ———————————

1½' · · · BROCCOLI ·

1½' ——— TOMATOES (8 STAKED PLANTS) ———————————

1½ · · · LEAF LETTUCE — HEAD LETTUCE · · · · · · · · · ·

1½' ——— CUCUMBERS (5 HILLS) ———————————————

1½' · · · RADISHES ·

1½' ——— SQUASH (4 ACORN BUSH, 4 SUMMER BUSH) ———————

1½' · · · EARLY CABBAGE PLANTS · · · · · · · · · · · · · · · ·

1½' ——— 6 PEPPER PLANTS AND 4 EGGPLANTS ——————————

2' ——— BUSH LIMA BEANS ———————————————————

2' ——— BUSH LIMA BEANS ———————————————————

1½' ——— SNAP BEANS — FOLLOW WITH SPINACH ————————

1½' ——— SNAP BEANS — FOLLOW WITH KALE AND LETTUCE ———

1½' ——— SNAP BEANS — FOLLOW WITH CABBAGE AND CAULIFLOWER—

1½' · · · BUSH PEAS ·

1½' ——— BUSH PEAS — FOLLOW WITH SNAP BEANS ——————

1½' · · · SPINACH ·

1½' ——— BEETS — FOLLOW WITH BRUSSELS SPROUTS AND ENDIVE——

1½' ——— BEETS ———————————————————————

1½' ——— CARROTS (SECOND SOWING) ———————————————

1½' ——— CARROTS ——————————————————————

1½' ——— SWISS CHARD, PARSLEY ———————————————

1½' ——— ONION SETS —————————————————————

1½' ——— RADISHES — FOLLOW WITH BEETS ————————

SOUTH

*Dotted rows indicate varieties which mature quickly and are removed to make room for long season vegetables.

Fig. 37-3 Suggested Plan for Vegetable Garden, 20 feet x 55 feet, Spring and Summer.

NORTH

	EARLY SWEET	MIDSEASON SWEET CORN	LATE SWEET
16½'	CORN	(Block Planting)	CORN

3' ——— TOMATOES (8 STAKED PLANTS) ———————————

3' ——— TOMATOES (8 STAKED PLANTS) ———————————

3' ——— CUCUMBERS (5 HILLS) ———————————————

3' ——— SQUASH (4 ACORN BUSH, 4 SUMMER BUSH) ——————

3' ——— 6 PEPPER PLANTS AND 4 EGGPLANTS ——————————

2' ——— BUSH LIMA BEANS ———————————————————

2' ——— BUSH LIMA BEANS ———————————————————

1½' ——— SPINACH ——————————————————————

1½' ——— KALE AND LETTUCE ———————————————————

1½' ——— CABBAGE AND CAULIFLOWER ———————————————

3' ——— SNAP BEANS ————————————————————

3' ——— BRUSSELS SPROUTS AND ENDIVE ——————————

1½' ——— BEETS ———————————————————————

1½' ——— CARROTS ——————————————————————

1½' ——— CARROTS ——————————————————————

1½' ——— SWISS CHARD, PARSLEY ———————————————

1½' ——— ONION SETS —————————————————————

1½' ——— BEETS ———————————————————————

SOUTH

Fig. 37-4 Suggested Plan for Vegetable Garden, 20 feet x 55 feet, Late Summer and Fall. This is how the garden in figure 37-3 appears after quickly maturing varieties have been harvested and succession plantings have been made.

and fall crops. Crops that were planted between tomatoes, such as broccoli, are harvested; however, no other crop is replanted in these places because the tomatoes need all the space.

GARDEN SOIL

The best garden soil is a healthy loamy one. It must have good drainage so that oxygen is available to roots, enabling the roots to penetrate the soil. A good supply of organic matter should be be available to hold moisture and provide plant food. The soil should also have a good supply of plant food. The pH range (soil acidity) should be from 6.3 to 7.0 for most vegetables.

Preparing the Soil

Clayey Soils. It is best to plow heavier garden soils (those containing large amounts of clay) in the fall of the year. Fall plowing allows the heavier soils to be planted earlier in the spring because plowed soil dries more quickly than soil left unplowed. Barnyard manure, crop residue, or leaves, should be plowed under in these soils to add organic matter. Where a steep slope could result in erosion, it is best to wait until spring to plow the garden.

Sandy Soils. Sandy soils are best planted with a fall cover crop such as rye to grow green manure and then plowed in early spring. This increases the organic matter in these soils. Sandy soils require a high level of organic matter to assist in holding moisture and plant food. These soils dry more quickly in spring and can be plowed earlier in the spring than heavy soils. Gardens in dry land areas should be plowed in the fall and left rough to absorb and retain moisture that falls during the winter.

Never plow a soil when it is wet. A good way to test the wetness of soil is to squeeze a handful of the soil. If it sticks together in a tight ball and does not easily crumble under slight pressure by the thumb and finger, it is too wet for plowing or other working. When checking for wetness, check at the lower plow depth, since soil may be dry on top and too wet on the bottom.

A soil test of the garden plot should be taken each year. Results of this test will tell if lime is needed to raise soil pH and how much of the three major plant food elements (nitrogen, phosphorus, and potash) are needed. Follow the recommendations given by the soil laboratory or the extension agent.

Liming (pH Adjustment)

If a soil test indicates a need for lime, it is best to *broadcast* (distribute) the lime on top of the plowed soil and harrow or otherwise mix it into the soil. When the pH is close to the level necessary for best growth, an application of 50 pounds of ground limestone per 100 square feet every two or three years should keep the pH near the proper level. This application will vary according to the soil type and the crop being grown. Plants that require high pH levels, such as beans, peas, and onions, need extra lime for good crop production. In a small garden where crops are rotated, the ideal pH cannot be obtained for each type of plant.

If soil pH is too high (alkaline), gypsum or sulfur may be used to lower the pH in the same way in which lime is used to raise it.

Fertilizing

Vegetable gardens need fertilizers added each year for best production. As stated before, a soil test will indicate the level of nitrogen, phosphorus, and potash available in the soil. From the soil test, the gardener can determine how much of each plant food element must be added to the garden soil.

To supply the major plant food elements, apply a good commercial fertilizer made for vegetable crops. The percentage of each of the three elements is given in the analysis on the bag. A *5-10-5* analysis indicates 5 percent nitrogen, 10 percent phosphorus, and 5 percent potash. The plant food elements are always listed in this order. Good vegetable fertilizers are those with an analysis of 5-10-5, 5-10-10, 5-10-15, 10-10-10, and 10-6-4.

In general, leafy vegetables and corn require larger amounts of nitrogen. A 10-10-10 or 10-6-4 analysis is good, depending upon how much phosphorus and potash the soil contains. The pod or fruit crops need more phosphorus — a fertilizer with a 5-10-5 or 1-2-1 ratio is good. Root crops need extra potash; a 5-10-15 or 5-10-10 is a good analysis for these crops. Unless

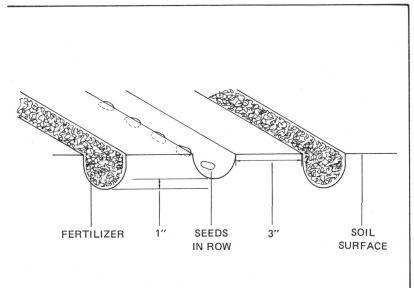

FERTILIZER   1″   SEEDS   3″   SOIL
                   IN ROW   SURFACE

Fig. 37-5 Banding fertilizer. The fertilizer is placed in bands 3 inches from each side of the seed row and 1 inch deeper than the depth of the seeds. Both the fertilizer and the seeds are covered by soil.

marked plant food deficiencies exist in the soil, it is not necessary to have special fertilizer for different crops in a small garden. In large commercial production, the analysis should match the crop needs as noted.

Applying Fertilizer

Broadcasting and Sidedressing. Vegetable crops may require fertilizer application as high as 5 pounds per 100 square feet. If this great an amount is added, it might be best to broadcast or spread evenly over the entire soil surface and mix half the amount into the soil before planting. Apply the rest later in the season as a *sidedress* on top of the soil on each side of the plants in the row, about 3 or 4 inches from the stem of the plant. (*Sidedressing* means placing fertilizer on the top of the soil around each plant.)

Banding. Another method for applying fertilizer is to place the fertilizer in rows dug 3 inches from each side of the row of seeds or plants and slightly deeper than the depth at which the seeds are planted. This is called *banding* fertilizer, figure 37-5.

Plowing Under. If a soil tests very low in phosphorus and/or potash, it is best to plow a fertilizer high in these elements under the soil. If applied on top of the soil, phosphorus and potash do not leach or wash down through the soil fast enough to be effective the same year. (If plowed under one year, these elements remain in the soil for the following year.)

When fertilizer recommendations are given in pounds per acre, it is often difficult to translate the amounts into those for small rows. Figure 37-6 gives conversion figures for determining how much fertilizer to use on smaller plots.

Measurement	Weight of fertilizer to apply when the weight to be applied per acre is:			
	100 pounds	400 pounds	800 pounds	1,200 pounds
Space between rows, and row length (in feet)	Pounds	Pounds	Pounds	Pounds
2 wide, 50 long......................	0.25	1.0	2.0	3.0
2 wide, 100 long.....................	.50	2.0	4.0	6.0
2½ wide, 50 long.....................	.30	1.2	2.4	3.6
2½ wide, 100 long....................	.60	2.4	4.8	7.2
3 wide, 50 long......................	.35	1.4	2.8	4.2
3 wide, 100 long.....................	.70	2.8	5.6	8.4
Area (in square feet)				
100..............................	.25	1.0	2.0	3.0
500..............................	1.25	5.0	10.0	15.0
1,000	2.50	10.0	20.0	30.0
1,500	3.75	15.0	30.0	45.0
2,000	5.00	20.0	40.0	60.0

Fig. 37-6 Approximate rates of fertilizer application per 50 or 100 feet of garden row, and per 100 to 2,000 square feet of garden area, corresponding to given rates per acre.

Summary

1. Decide what is to be planted.

2. Draw a garden plant showing how much of each crop is to be planted and the exact location in the garden of each crop.

3. Determine the amount of space between rows and whether or not interplanting between rows with early maturing crops is to be done.

4. Locate the garden close to water and in the sun on a loamy soil with good drainage.

5. Prepare the soil with organic matter by plowing under manure, compost, or cover crops.

6. Plow in fall or spring when the soil is dry enough to crumble.

7. Use lime, gypsum, or sulfur to correct pH. Dig into plowed or spaded soil.

8. Apply a good garden fertilizer. Apply the amount recommended in the soil test report.

STUDENT ACTIVITIES

1. Draw a garden plan for a family garden. Include succsssive crops of early and late plantings of short-season crops.

2. Study a vegetable variety recommendation fact sheet from a local university or extension agent. Write to an extension service in some other area of the country and request a fact sheet. Compare the two.

3. Study a garden seed catalog. Select and list the best vegetable varieties for your area.

4. Select a garden site to transform your plan into reality. Actually plant and care for the garden if possible. This may be a group or class project.

SELF-EVALUATION

A. Select the best answer from the choices offered to complete the statement or answer the question.

1. Perennial plants such as asparagus should be
   a. left out of the family garden.
   b. planted together on one side of the garden.
   c. rotated each year.
   d. none of the above

2. Two short-season crops that may be used in successive plantings are
   a. radishes and lettuce.
   b. horseradish and beets.
   c. tomatoes and snap beans.
   d. peppers and corn.

3. What is added to acidic soil (with a low pH) to raise the pH?
   a. organic matter
   b. fertilizer
   c. lime
   d. sand

4. Select the three long-season crops that should remain in the garden all season with no replanting.
   a. broccoli, snap beans, spinach
   b. spinach, kale, turnips
   c. beets, radishes, head lettuce
   d. tomato, eggplant, squash

5. Which of the following makes the best garden soil?

    a. sand   c. clay
    b. loam   d. silt

6. The best pH (soil acidity) range for most garden plants is

    a. 4.5 to 5.5.   c. 6.3 to 7.0.
    b. 5.5 to 6.5.   d. 6.5 to 8.5.

7. Heavier (clayey) soils are best plowed in the

    a. spring.   c. fall.
    b. summer.   d. dry season.

8. Sandy soils are best plowed in the

    a. spring.   c. fall.
    b. summer.   d. dry season.

9. Which vegetable crops use large amounts of nitrogen fertilizer?

    a. leaf   c. fruit
    b. root   d. all of these

10. Which vegetable crops use large amounts of potash fertilizer?

    a. pod   c. fruit
    b. leaf   d. root

11. Lime is applied

    a. as a side dress after vegetables are planted.
    b. on top of plowed soil and dug in.
    c. on top of soil and plowed under.
    d. in the planting hole.

12. Three vegetables that require extra lime and a relatively high soil pH are

    a. beans, peas, and onions.   c. eggplant, tomatoes, and corn.
    b. corn, celery, and beets.   d. corn, spinach, and peas.

13. Which vegetable crops use large amounts of phosphorous fertilizer?

    a. all   c. leaf
    b. root   d. pod or fruit

14. Which of the fertilizer analyses below is higher than the others in phosphorous?

    a. 5-10-5   c. 5-5-10
    b. 10-5-5   d. 2-3-4

15. Which of the fertilizer analyses below is highest in nitrogen?

    a. 5-10-5   c. 5-5-10
    b. 10-5-5   d. 16-8-4

B. Answer each of the following as instructed.

1. Explain in writing how to tell if a soil is too wet to plow.

2. Draw a sketch showing how to band fertilizer at planting time.

3. List the four most important items to consider when selecting a garden site.

OBJECTIVE

To demonstrate techniques for the planting of a small vegetable garden.

COMPETENCY TO BE DEVELOPED

After studying this unit, the student will be able to

- list the best varieties of ten vegetables for planting in the local area, including at least one disease-resistant variety for each.
- use the frost-free map and planting charts to determine planting dates for the ten vegetables.
- list five vegetables which are direct seeded in the garden and five which must be seeded indoors for transplanting.
- describe the steps in seeding vegetables for transplanting outdoors.

Materials List

listing of vegetable varieties recommended for the local area

seed catalogs

garden seeds

containers for starting seeds to grow plants for transplanting (flats, Jiffy 7, etc.)

garden rake, garden spade, garden hoe

string for lining out rows

garden hose

aluminum foil or cardboard for cutworm collars

SELECTING VARIETIES

One of the most important steps in the preparation of any garden is selecting the proper variety of vegetables to plant. For example, seed catalogs list dozens of varieties of sweet corn or tomatoes. Which is best for the area in which you live? An individual who sells seeds locally would probably be willing to offer advice concerning varieties. A local extension agent or horticulture teacher can also help determine which varieties are recommended for the area. Consider the use of hybrid varieties for more vigorous plants and a higher yield.

It pays to purchase seed from a reputable professional and not to depend on home supplies. Homegrown seed may carry diseases, or cross-pollination may result in seeds that do not come true to variety. The seed from hybrid vegetables does not produce plants that are true to seed. The cost of seed is very low compared to the total cost of producing vegetables; it pays to start with the best.

Disease-resistant varieties should be used whenever a disease is known to be a problem in the area. The selection of disease-resistant varieties can eliminate spraying and low yields at very little, if any, extra cost to the grower. Hybrid varieties are generally more vigorous and give higher yields for the same effort and cost. Hybrids should be used whenever possible.

Fresh seed should always be used. Read the date on the package to be sure that the seed was packed for the current year. Some seeds retain their vitality longer than others, and may grow after being held over for a year. Percent of germination will probably be lower than that listed on the package for first year planting, however.

Vegetable seeds may be divided into three classes: (1) *short-lived* (not effective after one or two years) — corn, leek, onion,

parsley, rhubarb, and salsify; (2) *moderately-long lived* (often good for three to five years) — asparagus, beans, brussels sprouts, cabbage, carrot, cauliflower, celery, kale, lettuce, okra, peas, pepper, radish, spinach, turnip, and watermelon; and (3) *long-lived* (may be good for more than five years) — beet, cucumber, eggplant, muskmelon, and tomato. Any seed over one year old should be given a germination test to see how many seeds actually sprout vigorously. This should be done before any of the seeds are planted in the garden.

> **Always buy seeds that are dated for planting that year.**

Seed stored in a dry, cool, dark place remains good longer than seed that is stored in an open container in a warm, moist place. How seeds are stored helps determine the length of active life and causes the figures given for the three classes of seed (short-lived, moderately long-lived, and long-lived) to vary considerably. These figures pertain to seed stored under good storage conditions.

DETERMINING PLANTING TIME

Planting seeds or transplanting plants at the proper time for the locality is very important. Temperatures can differ greatly between areas that are not many miles apart, causing the planting date for a particular vegetable to differ by as much as one or two weeks.

Vegetable crops may be grouped according to their hardiness (tolerance to cold). Figure 38-1 is a timetable listing the planting dates for some vegetable crops, based on the frost-free dates in spring and fall. The frost-free date in the spring of any locality is about the time that the oak trees in that area leaf.

It is important that the first planting of vegetables be as early in the spring as possible without danger of cold damage. Some vegetables can be planted before the frost-free date; others

Cold-hardy plants for early spring planting		Cold-tender or heat-hardy plants for later spring or early summer planting			
Very hardy (Plant 4 to 6 weeks before frost-free date.)	Hardy (Plant 2 to 4 weeks before frost-free date.)	Not cold-hardy (Plant on frost-free date.)	Requiring hot weather (Plant 1 week or more after frost-free date.)	Medium heat-tolerant (good for summer planting.)	Hardy plants for late summer or fall planting except in the northern region (Plant 6 to 8 weeks before first fall freeze.)
Broccoli	Beets	Beans, snap	Beans, lima	Beans, all	Beets
Cabbage	Carrots	Okra	Eggplant	Chard	Collard
Lettuce	Chard	New Zealand spinach	Peppers	Soybeans	Kale
Onions	Mustard	Soybeans	Sweet potatoes	New Zealand spinach	Lettuce
Peas	Parsnip	Squash	Cucumbers	Squash	Mustard
Potatoes	Radish	Sweet corn	Melons	Sweet corn	Spinach
Spinach		Tomatoes			Turnips
Turnips					

Fig. 38-1 Common vegetables grouped according to the approximate times they can be planted and their relative requirements for cool weather.

cannot. Other vegetables must be planted earlier than the frost-free date indicates because they grow best in cool weather, and cannot tolerate summer heat.

A gardener anywhere in the United States can determine the safe planting date for his or her particular area for different crops by using the information found in figures 38-2, 38-3, 38-4 and 38-5. The maps show the average dates of the last killing frost in spring (figure 38-2) and the first killing frost in fall (figure 38-3). Specific planting dates are determined by plugging these dates into figures 38-4 and 38-5.

HOW TO USE THE MAP AND TABLES

To determine the best time for spring planting of any vegetable in your locality:

1. Find your location on the map in figure 38-2 and then find the solid line on the map that comes closest to that location.

2. Find the date shown on the solid line. This is the average date of the last killing frost. The first number represents the month; the second number the day. (Example: 5-10 represents May 10.) Note this date and go to figure 38-5.

3. Find the column with this date at the top of it. This is the only column you will use. It gives dates for all vegetable crops listed for your area.

4. In this column, locate the vegetable that you wish to plant. The dates by the vegetable show the period of time during

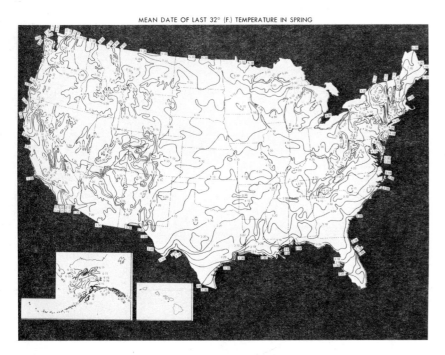

Fig. 38-2 Average dates of the last killing frost in spring. For full-page map, see p. 560 (from USDA *Home and Garden* Bulletin 202)

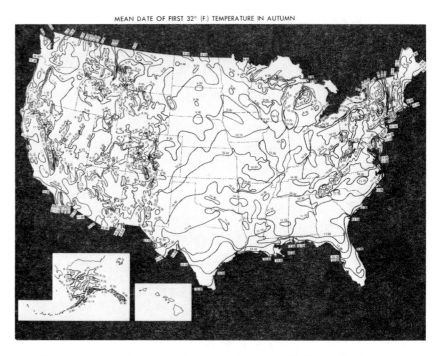

Fig. 38-3 Average dates of the first killing frost in fall. For full-page map, see page 561. (from USDA *Home and Garden* Bulletin 202)

which the crop can be safely planted. The best time is on or soon after the first date. The second date is the latest date for planting. Planting time becomes less desirable as you move toward the second date.

Figure 38-4 is used with the map in figure 38-3 in the same way to find planting dates for fall planting. A date about halfway between the two dates given in figure 38-4 is usually best.

Along the northern half of the Pacific Coast, warm weather crops should not be planted as late as the frost date and table indicate. Although frost comes late, very cool weather prevails for

some time before frost, retarding late growth of crops that prefer warm temperatures such as corn, beans, and tomatoes.

PLANTING SEEDS

Assume that the garden area is plowed and soil pH and fertilizer requirements necessary prior to planting have been taken care of. The next step is to harrow or rake the garden in preparation for planting. The soil should be worked to break all large clods of soil and to form a firm seedbed. For a seed to germinate, there must be good contact with the soil so that moisture can be pulled from the soil into the seed. Large, open air spaces result in poor germination.

	Planting dates for localities in which average date of first freeze is					
Crop	Aug. 30	Sept. 10	Sept. 20	Sept. 30	Oct. 10	Oct. 20
Asparagus[1]					Oct. 20-Nov. 15	Nov. 1-Dec. 15
Beans, lima				June 1-15	June 1-15	June 15-30
Beans, snap		May 15-June 15	June 1-July 1	June 1-July 10	June 15-July 20	July 1-Aug. 1
Beet	May 15-June 15	May 15-June 15	June 1-July 1	June 1-July 10	June 15-July 25	July 1-Aug. 5
Broccoli, sprouting	May 1-June 1	May 1-June 1	May 1-June 15	June 1-30	June 15-July 15	July 1-Aug. 1
Brussels sprouts	May 1-June 1	May 1-June 1	May 1-June 15	June 1-30	June 15-July 15	July 1-Aug. 1
Cabbage[1]	May 1-June 1	May 1-June 1	May 1-June 15	June 1-July 10	June 1-July 15	July 1-20
Cabbage, Chinese	May 15-June 15	May 15-June 15	June 1-July 1	June 1-July 15	June 15-Aug. 1	July 15-Aug. 15
Carrot	May 15-June 15	May 15-June 15	June 1-July 1	June 1-July 10	June 1-July 20	June 15-Aug. 1
Cauliflower[1]	May 1-June 1	May 1-July 1	May 1-July 1	May 10-July 15	June 1-July 25	July 1-Aug. 5
Celery[1] and celeriac	May 1-June 1	May 15-June 15	May 15-July 1	June 1-July 5	June 1-July 15	June 1-Aug. 1
Chard	May 15-June 15	May 15-July 1	June 1-July 1	June 1-July 5	June 1-July 20	June 1-Aug. 1
Chervil and chives	May 10-June 10	May 1-June 15	May 15-June 15	(2)	(2)	(2)
Chicory, witloof	May 15-June 15	May 15-June 15	May 15-June 15	June 1-July 1	June 1-July 1	June 15-July 15
Collards[1]	May 15-June 15	May 15-June 15	May 15-June 15	June 15-July 15	July 1-Aug. 1	July 15-Aug. 15
Cornsalad	May 15-June 15	May 15-July 1	June 15-Aug. 1	July 15-Sept. 1	Aug. 15-Sept. 15	Sept. 1-Oct. 15
Corn, sweet			June 1-July 1	June 1-July 1	June 1-July 10	June 1-July 20
Cress, upland	May 15-June 15	May 15-July 1	June 15-Aug. 1	July 15-Sept. 1	Aug. 15-Sept. 15	Sept. 1-Oct. 15
Cucumber			June 1-15	June 1-July 1	June 1-July 1	June 1-July 15
Eggplant[1]				May 20-June 10	May 15-June 15	June 1-July 1
Endive	June 1-July 1	June 1-July 1	June 15-July 15	June 15-Aug. 1	July 1-Aug. 15	July 15-Sept. 1
Fennel, Florence	May 15-June 15	May 15-July 15	June 1-July 1	June 1-July 1	June 15-July 15	June 15-Aug. 1
Garlic	(2)	(2)	(2)	(2)	(2)	(2)
Horseradish[1]	(2)	(2)	(2)	(2)	(2)	(2)
Kale	May 15-June 15	May 15-June 15	June 1-July 1	June 15-July 15	July 1-Aug. 1	July 15-Aug. 15
Kohlrabi	May 15-June 15	June 1-July 1	June 1-July 15	June 15-July 15	July 1-Aug. 1	July 15-Aug. 15
Leek	May 1-June 1	May 1-June 1	(2)	(2)	(2)	(2)
Lettuce, head[1]	May 15-July 1	May 15-July 1	June 1-July 15	June 15-Aug. 1	July 15-Aug. 15	Aug. 1-30
Lettuce, leaf	May 15-July 15	May 15-July 15	June 1-Aug. 1	June 1-Aug. 1	July 15-Sept. 1	July 15-Sept. 1
Muskmelon			May 1-June 15	May 15-June 1	June 1-June 15	June 15-July 20
Mustard	May 15-July 15	May 15-July 15	June 1-Aug. 1	June 15-Aug. 1	July 15-Aug. 15	Aug. 1-Sept. 1

Fig. 38-4 Latest dates, and range of dates, for safe fall planting of vegetables in the open (from USDA *Home and Garden* Bulletin 202)

Crop	Planting dates for localities in which average date of first freeze is					
	Aug. 30	Sept. 10	Sept. 20	Sept. 30	Oct. 10	Oct. 20
Okra			June 1-20	June 1-July 1	June 1-July 15	June 1-Aug. 1
Onion[1]	May 1-June 10	May 1-June 10	(2)	(2)	(2)	(2)
Onion, seed	May 1-June 1	May 1-June 10	(2)	(2)	(2)	(2)
Onion, sets	May 1-June 1	May 1-June 10	(2)	(2)	(2)	(2)
Parsley	May 15-June 15	May 1-June 15	June 1-July 1	June 1-July 15	June 15-Aug. 1	July 15-Aug. 15
Parsnip	May 15-June 1	May 1-June 15	May 15-June 15	June 1-July 1	June 1-July 10	(2)
Peas, garden	May 10-June 15	May 1-July 1	June 1-July 15	June 1-Aug. 1	(2)	(2)
Peas, black-eye					June 1-July 1	June 1-July 1
Pepper[1]			June 1-June 20	June 1-July 1	June 1-July 1	June 1-July 10
Potato	May 15-June 1	May 1-June 15	May 1-June 15	May 1-June 15	May 15-June 15	June 15-July 15
Radish	May 1-July 15	May 1-Aug. 1	June 1-Aug. 15	July 1-Sept. 1	July 15-Sept. 15	Aug. 1-Oct. 1
Rhubarb[1]	Sept. 1-Oct. 1	Sept. 15-Oct. 15	Sept. 15-Nov. 1	Oct. 1-Nov. 1	Oct. 15-Nov. 15	Oct. 15-Dec. 1
Rutabaga	May 15-June 15	May 1-June 15	June 1-July 1	June 1-July 1	June 15-July 15	July 10-20
Salsify	May 15-June 1	May 10-June 10	May 20-June 20	June 1-20	June 1-July 1	June 1-July 1
Shallot	(2)	(2)	(2)	(2)	(2)	(2)
Sorrel	May 15-June 15	May 1-June 15	June 1-July 1	June 1-July 15	July 1-Aug. 1	July 15-Aug. 15
Soybean				May 25-June 10	June 1-25	June 1-July 5
Spinach	May 15-July 1	June 1-July 15	June 1-Aug. 1	July 1-Aug. 15	Aug. 1-Sept. 1	Aug. 20-Sept. 10
Spinach, New Zealand				May 15-July 1	June 1-July 15	June 1-Aug. 1
Squash, summer	June 10-20	June 1-20	May 15-July 1	June 1-July 1	June 1-July 15	June 1-July 20
Squash, winter			May 20-June 10	June 1-15	June 1-July 1	June 1-July 1
Sweetpotato					May 20-June 10	June 1-15
Tomato	June 20-30	June 10-20	June 1-20	June 1-20	June 1-20	June 1-July 1
Turnip	May 15-June 15	June 1-July 1	June 1-July 15	June 1-Aug. 1	July 1-Aug. 1	July 15-Aug. 15
Watermelon			May 1-June 15	May 15-June 1	June 1-June 15	June 15-July 20

[1] Plants.
[2] Generally spring-planted

Fig. 38-4 Latest dates, and range of dates, for safe fall planting of vegetables in the open (from USDA *Home and Garden* Bulletin 202) (continued)

Crop	Planting dates for localities in which average date of first freeze is					
	Oct. 30	Nov. 10	Nov. 20	Nov. 30	Dec. 10	Dec. 20
Asparagus[1]	Nov. 15-Jan. 1	Dec. 1-Jan. 1				
Beans, lima	July 1-Aug. 1	July 1-Aug. 15	July 15-Sept. 1	Aug. 1-Sept. 15	Sept. 1-30	Sept. 1-Oct. 1
Beans, snap	July 1-Aug. 15	July 1-Sept. 1	July 1-Sept. 10	Aug. 15-Sept. 20	Sept. 1-30	Sept. 1-Nov. 1
Beet	Aug. 1-Sept. 1	Aug. 1-Oct. 1	Sept. 1-Dec. 1	Sept. 1-Dec. 15	Sept. 1-Dec. 31	Sept. 1-Dec. 31
Broccoli, sprouting	July 1-Aug. 15	Aug. 1-Sept. 1	Aug. 1-Sept. 15	Aug. 1-Oct. 1	Aug. 1-Nov. 1	Sept. 1-Dec. 31
Brussels sprouts	July 1-Aug. 15	Aug. 1-Sept. 1	Aug. 1-Sept. 15	Aug. 1-Oct. 1	Aug. 1-Nov. 1	Sept. 1-Dec. 31
Cabbage[1]	Aug. 1-Sept. 1	Sept. 1-15	Sept. 1-Dec. 1	Sept. 1-Dec. 31	Sept. 1-Dec. 31	Sept. 1-Dec. 31
Cabbage, Chinese	Aug. 1-Sept. 15	Aug. 15-Oct. 1	Sept. 1-Oct. 15	Sept. 1-Nov. 1	Sept. 1-Nov. 15	Sept. 1-Dec. 1
Carrot	July 1-Aug. 15	Aug. 1-Sept. 1	Sept. 1-Nov. 1	Sept. 15-Dec. 1	Sept. 15-Dec. 1	Sept. 15-Dec. 1
Cauliflower[1]	July 15-Aug. 15	Aug. 1-Sept. 1	Aug. 1-Sept. 15	Aug. 15-Oct. 10	Sept. 1-Oct. 20	Sept. 15-Nov. 1
Celery[1] and celeriac	June 15-Aug. 15	July 1-Aug. 15	July 15-Sept. 1	Aug. 1-Dec. 1	Sept. 1-Dec. 31	Oct. 1-Dec. 31
Chard	June 1-Sept. 10	June 1-Sept. 15	June 1-Oct. 1	June 1-Nov. 1	June 1-Dec. 1	June 1-Dec. 31
Chervil and chives	(2)	(2)	Nov. 1-Dec. 31	Nov. 1-Dec. 31	Nov. 1-Dec. 31	Nov. 1-Dec. 31
Chicory, witloof	July 1-Aug. 10	July 10-Aug. 20	July 20-Sept. 1	Aug. 15-Sept. 30	Aug. 15-Oct. 15	Aug. 15-Oct. 15
Collards[1]	Aug. 1-Sept. 15	Aug. 15-Oct. 1	Aug. 25-Nov. 1	Sept. 1-Dec. 1	Sept. 1-Dec. 31	Sept. 1-Dec. 31
Cornsalad	Sept. 15-Nov. 1	Oct. 1-Dec. 1	Oct. 1-Dec. 1	Oct. 1-Dec. 31	Oct. 1-Dec. 31	Oct. 1-Dec. 31
Corn, sweet	June 1-Aug. 1	June 1-Aug. 15	June 1-Sept. 1			
Cress, upland	Sept. 15-Nov. 1	Oct. 1-Dec. 1	Oct. 1-Dec. 1	Oct. 1-Dec. 31	Oct. 1-Dec. 31	Oct. 1-Dec. 31
Cucumber	June 1-Aug. 1	June 1-Aug. 15	June 1-Aug. 15	July 15-Sept. 15	Aug. 15-Oct. 1	Aug. 15-Oct. 1
Eggplant[1]	June 1-July 1	June 1-July 15	June 1-Aug. 1	July 1-Sept. 1	Aug. 1-Sept. 30	Aug. 1-Sept. 30
Endive	July 15-Aug. 15	Aug. 1-Sept. 1	Sept. 1-Oct. 1	Sept. 1-Nov. 15	Sept. 1-Dec. 31	Sept. 1-Dec. 31
Fennel, Florence	July 1-Aug. 1	July 15-Aug. 15	Aug. 15-Sept. 15	Sept. 1-Nov. 15	Sept. 1-Dec. 1	Sept. 1-Dec. 1
Garlic	(2)	Aug. 1-Oct. 1	Aug. 15-Oct. 1	Sept. 1-Nov. 15	Sept. 15-Nov. 15	Sept. 15-Nov. 15
Horseradish[1]	(2)	(2)	(2)	(2)	(2)	(2)
Kale	July 15-Sept. 1	Aug. 1-Sept. 15	Aug. 15-Oct. 15	Sept. 1-Dec. 1	Sept. 1-Dec. 31	Sept. 1-Dec. 31
Kohlrabi	Aug. 1-Sept. 1	Aug. 15-Sept. 15	Sept. 1-Oct. 15	Sept. 1-Dec. 1	Sept. 15-Dec. 31	Sept. 1-Dec. 31
Leek	(2)	(2)	Sept. 1-Nov. 1	Sept. 1-Nov. 1	Sept. 1-Nov. 1	Sept. 15-Nov. 1
Lettuce, head[1]	Aug. 1-Sept. 15	Aug. 15-Oct. 15	Sept. 1-Nov. 1	Sept. 1-Dec. 1	Sept. 15-Dec. 31	Sept. 15-Dec. 31
Lettuce, leaf	Aug. 15-Oct. 1	Aug. 25-Oct. 1	Sept. 1-Nov. 1	Sept. 1-Dec. 1	Sept. 15-Dec. 31	Sept. 15-Dec. 31

Fig. 38-4 Latest dates, and range of dates, for safe fall planting of vegetables in the open (from USDA *Home and Garden* Bulletin 202) (continued)

Crop	Planting dates for localities in which average date of first freeze is					
	Oct. 30	Nov. 10	Nov. 20	Nov. 30	Dec. 10	Dec. 20
Muskmelon	July 1-July 15	July 15-July 30				
Mustard	Aug. 15-Oct. 15	Aug. 15-Nov. 1	Sept. 1-Dec. 1	Sept. 1-Dec. 1	Sept. 1-Dec. 1	Sept. 15-Dec. 1
Okra	June 1-Aug. 10	June 1-Aug. 20	June 1-Sept. 10	June 1-Sept. 20	Aug. 1-Oct. 1	Aug. 1-Oct. 1
Onion[1]		Sept. 1-Oct. 15	Oct. 1-Dec. 31	Oct. 1-Dec. 31	Oct. 1-Dec. 31	Oct. 1-Dec. 31
Onion, seed			Sept. 1-Nov. 1	Sept. 1-Nov. 1	Sept. 1-Nov. 1	Sept. 15-Nov. 1
Onion, sets		Oct. 1-Dec. 1	Nov. 1-Dec. 31	Nov. 1-Dec. 31	Nov. 1-Dec. 31	Nov. 1-Dec. 31
Parsley	Aug. 1-Sept. 15	Sept. 1-Nov. 15	Sept. 1-Dec. 31	Sept. 1-Dec. 31	Sept. 1-Dec. 31	Sept. 1-Dec. 31
Parsnip	(2)	(2)	Aug. 1-Sept. 1	Sept. 1-Nov. 15	Sept. 1-Dec. 1	Sept. 1-Dec. 1
Peas, garden	Aug. 1-Sept. 15	Sept. 1-Nov. 1	Oct. 1-Dec. 1	Oct. 1-Dec. 31	Oct. 1-Dec. 31	Oct. 1-Dec. 31
Peas, black-eye	June 1-Aug. 1	June 15-Aug. 15	July 1-Sept. 1	July 1-Sept. 10	July 1-Sept. 20	July 1-Sept. 20
Pepper[1]	June 1-July 20	June 1-Aug. 1	June 1-Aug. 15	June 15-Sept. 1	Aug. 15-Oct. 1	Aug. 15-Oct. 1
Potato	July 20-Aug. 10	July 25-Aug. 20	Aug. 10-Sept. 15	Aug. 1-Sept. 15	Aug. 1-Sept. 15	Aug. 1-Sept. 15
Radish	Aug. 15-Oct. 15	Sept. 1-Nov. 15	Sept. 1-Dec. 1	Sept. 1-Dec. 31	Aug. 1-Sept. 15	Oct. 1-Dec. 31
Rhubarb[1]	Nov. 1-Dec. 1					
Rutabaga	July 15-Aug. 1	July 15-Aug. 15	Aug. 1-Sept. 1	Sept. 1-Nov. 15	Oct. 1-Nov. 15	Oct. 15-Nov. 15
Salsify	June 1-July 10	June 15-July 20	July 15-Aug. 15	Aug. 15-Sept. 30	Aug. 15-Oct. 15	Sept. 1-Oct. 31
Shallot	(2)	Aug. 1-Oct. 1	Aug. 15-Oct. 1	Aug. 15-Oct. 15	Sept. 15-Nov. 1	Sept. 15-Nov. 1
Sorrel	Aug. 1-Sept. 15	Aug. 15-Oct. 1	Aug. 15-Oct. 15	Sept. 1-Nov. 15	Sept. 1-Dec. 15	Sept. 1-Dec. 31
Soybean	June 1-July 15	June 1-July 25	June 1-July 30	June 1-July 30	June 1-July 30	June 1-July 30
Spinach	Sept. 1-Oct. 1	Sept. 15-Nov. 1	Oct. 1-Dec. 1	Oct. 1-Dec. 31	Oct. 1-Dec. 31	Oct. 1-Dec. 31
Spinach, New Zealand	June 1-Aug. 1	June 1-Aug. 15	June 1-Aug. 15			
Squash, summer	June 1-Aug. 1	June 1-Aug. 10	June 1-Aug. 20	June 1-Sept. 1	June 1-Sept. 15	June 1-Oct. 1
Squash, winter	June 10-July 10	June 20-July 20	July 1-Aug. 1	July 15-Aug. 15	Aug. 1-Sept. 1	Aug. 1-Sept. 1
Sweetpotato	June 1-15	June 1-July 1	June 1-July 1	June 1-July 1	June 1-July 1	June 1-July 1
Tomato	June 1-July 1	June 1-July 15	June 1-Aug. 1	Aug. 1-Sept. 1	Aug. 15-Oct. 1	Sept. 1-Nov. 1
Turnip	Aug. 1-Sept. 15	Sept. 1-Oct. 15	Sept. 1-Nov. 15	Sept. 1-Nov. 15	Oct. 1-Dec. 1	Oct. 1-Dec. 31
Watermelon	July 1-July 15	July 15-July 30				

[1] Plants.
[2] Generally spring-planted

Fig. 38-4 Latest dates, and range of dates, for safe fall planting of vegetables in the open (from USDA *Home and Garden* Bulletin 202) (continued)

Crop	Planting dates for localities in which average date of last freeze is						
	Jan. 30	**Feb. 8**	**Feb. 18**	**Feb. 28**	**Mar. 10**	**Mar. 20**	**Mar. 30**
Asparagus[1]					Jan. 1-Mar. 1	Feb. 1-Mar. 10	Feb. 15-Mar. 20
Beans, lima	Feb. 1-Apr. 15	Feb. 10-May 1	Mar. 1-May 1	Mar. 15-June 1	Mar. 20-June 1	Apr. 1-June 15	Apr. 15-June 20
Beans, snap	Feb. 1-Apr. 1	Feb. 1-May 1	Mar. 1-May 1	Mar. 10-May 15	Mar. 15-May 15	Mar. 15-May 25	Apr. 1-June 1
Beet	Jan. 1-Mar. 15	Jan. 10-Mar. 15	Jan. 20-Apr. 1	Feb. 1-Apr. 15	Feb. 15-June 1	Feb. 15-May 15	Mar. 1-June 1
Broccoli, sprouting[1]	Jan. 1-30	Jan. 1-30	Jan. 15-Feb. 15	Feb. 1-Mar. 1	Feb. 15-Mar. 15	Feb. 15-Mar. 15	Mar. 1-20
Brussels sprouts[1]	Jan. 1-30	Jan. 1-30	Jan. 15-Feb. 15	Feb. 1-Mar. 1	Feb. 15-Mar. 15	Feb. 15-Mar. 15	Mar. 1-20
Cabbage[1]	Jan. 1-15	Jan. 1-Feb. 10	Jan. 1-Feb. 25	Jan. 15-Feb. 25	Jan. 25-Mar. 1	Feb. 1-Mar. 1	Feb. 15-Mar. 10
Cabbage, Chinese	(2)	(2)	(2)	(2)	(2)	(2)	(2)
Carrot	Jan. 1-Mar. 1	Jan. 1-Mar. 1	Jan. 15-Mar. 1	Feb. 1-Mar. 1	Feb. 10-Mar. 15	Feb. 15-Mar. 20	Mar. 1-Apr. 10
Cauliflower[1]	Jan. 1-Feb. 1	Jan. 1-Feb. 1	Jan. 10-Feb. 10	Jan. 20-Feb. 20	Feb. 1-Mar. 1	Feb. 10-Mar. 10	Feb. 10-Mar. 20
Celery and celeriac	Jan. 1-Feb. 1	Jan. 10-Feb. 10	Jan. 20-Feb. 20	Feb. 1-Mar. 1	Feb. 10-Mar. 20	Mar. 1-Apr. 1	Mar. 15-Apr. 15
Chard	Jan. 1-Apr. 1	Jan. 10-Apr. 1	Jan. 20-Apr. 15	Feb. 1-May 1	Feb. 15-May 15	Feb. 20-May 15	Mar. 1-May 25
Chervil and chives	Jan. 1-Feb. 1	Jan. 1-Feb. 1	Jan. 1-Feb. 1	Jan. 15-Feb. 15	Feb. 1-Mar. 1	Feb. 10-Mar. 10	Feb. 15-Mar. 15
Chicory, witloof					June 1-July 1	June 1-July 1	June 1-July 1
Collards[1]	Jan. 1-Feb. 15	Jan. 1-Feb. 15	Jan. 1-Mar. 15	Jan. 15-Mar. 15	Feb. 1-Apr. 1	Feb. 15-May 1	Mar. 1-June 1
Cornsalad	Jan. 1-Feb. 15	Jan. 1-Feb. 15	Jan. 1-Mar. 15	Jan. 1-Mar. 1	Jan. 1-Mar. 15	Jan. 1-Mar. 15	Jan. 15-Mar. 15
Corn, sweet	Feb. 1-Mar. 15	Feb. 10-Apr. 1	Feb. 20-Apr. 15	Mar. 1-Apr. 15	Mar. 10-Apr. 15	Mar. 15-May 1	Mar. 25-May 15
Cress, upland	Jan. 1-Feb. 1	Jan. 1-Feb. 15	Jan. 15-Feb. 15	Feb. 1-Mar. 1	Feb. 10-Mar. 15	Feb. 20-Mar. 15	Mar. 1-Apr. 1
Cucumber	Feb. 15-Mar. 15	Feb. 15-Apr. 1	Feb. 15-Apr. 15	Mar. 1-Apr. 15	Mar. 15-Apr. 15	Apr. 1-May 1	Apr. 10-May 15
Eggplant[1]	Feb. 1-Mar. 1	Feb. 10-Mar. 15	Feb. 20-Apr. 1	Mar. 10-Apr. 15	Mar. 15-Apr. 15	Apr. 1-May 1	Apr. 15-May 15
Endive	Jan. 1-Mar. 1	Jan. 1-Mar. 1	Jan. 15-Mar. 1	Feb. 1-Mar. 1	Feb. 15-Mar. 15	Mar. 1-Apr. 1	Mar. 10-Apr. 10
Fennel, Florence	Jan. 1-Mar. 1	Jan. 1-Mar. 1	Jan. 15-Mar. 1	Feb. 1-Mar. 1	Feb. 15-Mar. 15	Mar. 1-Apr. 1	Mar. 10-Apr. 10
Garlic	(2)	(2)	(2)	(2)	(2)	Feb. 1-Mar. 1	Feb. 10-Mar. 10
Horseradish[1]							Mar. 1-Apr. 1
Kale	Jan. 1-Feb. 1	Jan. 10-Feb. 1	Jan. 20-Feb. 10	Feb. 1-20	Feb. 10-Mar. 1	Feb. 20-Mar. 10	Mar. 1-20
Kohlrabi	Jan. 1-Feb. 1	Jan. 10-Feb. 1	Jan. 20-Feb. 10	Feb. 1-20	Feb. 10-Mar. 1	Feb. 20-Mar. 10	Mar. 1-Apr. 1
Leek	Jan. 1-Feb. 1	Jan. 1-Feb. 1	Jan. 1-Feb. 15	Jan. 15-Feb. 15	Jan. 25-Mar. 1	Feb. 1-Mar. 1	Feb. 15-Mar. 15
Lettuce, head[1]	Jan. 1-Feb. 1	Jan. 1-Feb. 1	Jan. 1-Feb. 1	Jan. 15-Feb. 15	Feb. 1-20	Feb. 15-Mar. 10	Mar. 1-20
Lettuce, leaf	Jan. 1-Feb. 1	Jan. 1-Feb. 1	Jan. 1-Mar. 15	Jan. 1-Mar. 15	Jan. 15-Apr. 1	Feb. 1-Apr. 1	Feb. 15-Apr. 15
Muskmelon	Feb. 15-Mar. 15	Feb. 15-Apr. 1	Feb. 15-Apr. 15	Mar. 1-Apr. 15	Mar. 15-Apr. 15	Apr. 1-May 1	Apr. 10-May 15

Fig. 38-5 Earliest dates, and range of dates for safe spring planting of vegetables in the open (from USDA *Home and Garden* Bulletin 202)

| | Planting dates for localities in which average date of last freeze is | | | | | | |
Crop	Jan. 30	Feb. 8	Feb. 18	Feb. 28	Mar. 10	Mar. 20	Mar. 30
Mustard	Jan. 1-Mar. 1	Jan. 1-Mar. 1	Feb. 15-Apr. 15	Feb. 1-Mar. 1	Feb. 10-Mar. 15	Feb. 20-Apr. 1	Mar. 1-Apr. 15
Okra	Feb. 15-Apr. 1	Feb. 15-Apr. 15	Mar. 1-June 1	Mar. 10-June 1	Mar. 20-June 1	Apr. 1-June 15	Apr. 10-June 15
Onion[1]	Jan. 1-15	Jan. 1-15	Jan. 1-15	Jan. 1-Feb. 1	Jan. 15-Feb. 15	Feb. 10-Mar. 10	Feb. 15-Mar. 15
Onion, seed	Jan. 1-15	Jan. 1-15	Jan. 1-15	Jan. 1-Feb. 15	Feb. 1-Mar. 1	Feb. 10-Mar. 10	Feb. 20-Mar. 15
Onion, sets	Jan. 1-15	Jan. 1-15	Jan. 1-15	Jan. 1-Mar. 1	Jan. 15-Mar. 10	Feb. 1-Mar. 20	Feb. 15-Mar. 20
Parsley	Jan. 1-30	Jan. 1-30	Jan. 1-30	Jan. 15-Mar. 1	Feb. 1-Mar. 10	Feb. 15-Mar. 15	Mar. 1-Apr. 1
Parsnip			Jan. 1-Feb. 1	Jan. 15-Feb. 15	Jan. 15-Mar. 1	Feb. 15-Mar. 15	Mar. 1-Apr. 1
Peas, garden	Jan. 1-Feb. 15	Jan. 1-Feb. 15	Jan. 1-Mar. 1	Jan. 15-Mar. 1	Jan. 15-Mar. 15	Feb. 1-Mar. 15	Feb. 10-Mar. 20
Peas, black-eye	Feb. 15-May 1	Feb. 15-May 15	Mar. 1-June 15	Mar. 10-June 20	Mar. 15-July 1	Apr. 1-July 1	Apr. 15-July 1
Pepper[1]	Feb. 1-Apr. 1	Feb. 15-Apr. 15	Mar. 1-May 1	Mar. 15-May 1	Apr. 1-June 1	Apr. 10-June 1	Apr. 15-June 1
Potato	Jan. 1-Feb. 15	Jan. 1-Feb. 15	Jan. 15-Mar. 1	Jan. 15-Mar. 1	Feb. 1-Mar. 1	Feb. 10-Mar. 15	Feb. 20-Mar. 20
Radish	Jan. 1-Apr. 1	Jan. 1-Apr. 1	Jan. 1-Apr. 1	Jan. 1-Apr. 1	Jan. 1-Apr. 15	Jan. 20-May 1	Feb. 15-May 1
Rhubarb[1]							
Rutabaga				Jan. 1-Feb. 1	Jan. 15-Feb. 15	Jan. 15-Mar. 1	Feb. 1-Mar. 1
Salsify	Jan. 1-Feb. 1	Jan. 10-Feb. 10	Jan. 15-Feb. 20	Jan. 15-Mar. 1	Feb. 1-Mar. 1	Feb. 15-Mar. 1	Mar. 1-15
Shallot	Jan. 1-Feb. 1	Jan. 1-Feb. 10	Jan. 1-Feb. 20	Jan. 1-Mar. 1	Jan. 15-Mar. 1	Feb. 1-Mar. 10	Feb. 15-Mar. 15
Sorrel	Jan. 1-Mar. 1	Jan. 1-Mar. 1	Jan. 15-Mar. 1	Feb. 1-Mar. 10	Feb. 10-Mar. 15	Feb. 10-Mar. 20	Feb. 20-Apr. 1
Soybean	Mar. 1-June 30	Mar. 1-June 30	Mar. 10-June 30	Mar. 20-June 30	Apr. 10-June 30	Apr. 10-June 30	Apr. 20-June 30
Spinach	Jan. 1-Feb. 15	Jan. 1-Feb. 15	Jan. 1-Mar. 1	Jan. 1-Mar. 1	Jan. 15-Mar. 10	Jan. 15-Mar. 15	Feb. 1-Mar. 20
Spinach, New Zealand	Feb. 1-Apr. 15	Feb. 15-Apr. 15	Mar. 1-Apr. 15	Mar. 15-May 15	Mar. 20-May 15	Apr. 1-May 15	Apr. 10-June 1
Squash, summer	Feb. 1-Apr. 15	Feb. 15-Apr. 15	Mar. 1-Apr. 15	Mar. 15-May 15	Mar. 15-May 1	Apr. 1-May 15	Apr. 10-June 1
Sweetpotato	Feb. 15-May 15	Mar. 1-May 15	Mar. 20-June 1	Mar. 20-June 1	Apr. 1-June 1	Apr. 10-June 1	Apr. 20-June 1
Tomato	Feb. 1-Apr. 1	Feb. 20-Apr. 10	Mar. 1-Apr. 20	Mar. 10-May 1	Mar. 20-May 10	Apr. 1-May 20	Apr. 10-June 1
Turnip	Jan. 1-Mar. 1	Jan. 1-Mar. 1	Jan. 10-Mar. 1	Jan. 20-Mar. 1	Feb. 1-Mar. 1	Feb. 10-Mar. 10	Feb. 20-Mar. 20
Watermelon	Feb. 15-Mar. 15	Feb. 15-Apr. 1	Feb. 15-Apr. 15	Mar. 1-Apr. 15	Mar. 15-Apr. 15	Apr. 1-May 1	Apr. 10-May 15

[1] Plants.
[2] Generally fall-planted

Fig. 38-5 Earliest dates, and range of dates, for safe spring planting of vegetables in the open (from USDA *Home and Garden* Bulletin 202) (continued)

Crop	Planting dates for localities in which average date of last freeze is						
	Apr. 10	Apr. 20	Apr. 30	May 10	May 20	May 30	June 10
Asparagus[1]	Mar. 10-Apr. 10	Mar. 15-Apr. 15	Mar. 20-Apr. 15	Mar. 10-Apr. 30	Apr. 20-May 15	May 1-June 1	May 15-June 1
Beans, lima	Apr. 1-June 30	May 1-June 20	May 15-June 15	May 25-June 15			
Beans, snap	Apr. 10-June 30	Apr. 25-June 30	May 10-June 30	May 10-June 30	May 15-June 30	May 25-June 15	
Beet	Mar. 10-June 1	Mar. 20-June 1	Apr. 1-June 15	Apr. 15-June 15	Apr. 25-June 15	May 1-June 15	May 15-June 15
Broccoli, sprouting[1]	Mar. 15-Apr. 15	Mar. 25-Apr. 20	Apr. 1-May 1	Apr. 15-June 1	May 1-June 15	May 10-June 10	May 20-June 10
Brussels sprouts[1]	Mar. 15-Apr. 15	Mar. 25-Apr. 20	Apr. 1-May 1	Apr. 15-June 1	May 1-June 15	May 10-June 10	May 20-June 10
Cabbage[1]	Mar. 1-Apr. 1	Mar. 10-Apr. 1	Mar. 15-Apr. 10	Apr. 1-May 15	May 1-June 15	May 10-June 15	May 20-June 1
Cabbage, Chinese	(2)	(2)	(2)	Apr. 1-May 15	May 1-June 15	May 10-June 15	May 20-June 1
Carrot	Mar. 10-Apr. 20	Apr. 1-May 15	Apr. 10-June 1	Apr. 20-June 15	May 1-June 1	May 10-June 1	May 20-June 1
Cauliflower[1]	Mar. 1-Mar. 20	Mar. 15-Apr. 20	Apr. 10-May 10	Apr. 15-May 15	May 10-June 15	May 20-June 1	June 1-June 15
Celery and celeriac	Apr. 1-Apr. 20	Apr. 10-May 1	Apr. 15-May 1	Apr. 20-June 15	May 10-June 15	May 20-June 1	June 1-June 15
Chard	Mar. 15-June 15	Apr. 1-June 15	Apr. 15-June 15	Apr. 20-June 15	May 10-June 15	May 20-June 1	June 1-June 15
Chervil and chives	Mar. 1-Apr. 1	Mar. 10-Apr. 10	Mar. 20-Apr. 20	Apr. 1-May 1	Apr. 15-May 15	May 1-June 1	May 15-June 1
Chicory, witloof	June 10-July 1	June 15-July 1	June 15-July 1	June 1-20	June 1-15	June 1-15	June 1-15
Collards[1]	Mar. 1-June 1	Mar. 10-June 1	Apr. 1-June 1	Apr. 15-June 1	May 1-June 1	May 10-June 1	May 20-June 1
Cornsalad	Feb. 1-Apr. 1	Feb. 15-Apr. 15	Mar. 1-May 1	Apr. 1-June 1	Apr. 15-June 1	May 1-June 15	May 15-June 15
Corn, sweet	Apr. 10-June 1	Apr. 25-June 15	May 10-June 15	May 10-June 1	May 15-June 1	May 20-June 1	
Cress, upland	Mar. 10-Apr. 15	Mar. 20-May 1	Apr. 10-May 10	Apr. 20-May 20	May 1-June 1	May 15-June 1	May 15-June 15
Cucumber	Apr. 20-June 1	May 1-June 15	May 15-June 15	May 20-June 15	June 1-15		
Eggplant[1]	May 1-June 1	May 10-June 1	May 15-June 10	May 20-June 15	June 1-15		
Endive	Mar. 15-Apr. 15	Mar. 25-Apr. 15	Apr. 1-May 1	Apr. 15-May 15	May 1-30	May 1-30	May 15-June 1
Fennel, Florence	Mar. 15-Apr. 15	Mar. 25-Apr. 15	Apr. 1-May 1	Apr. 15-May 15	May 1-30	May 1-30	May 15-June 1
Garlic	Feb. 20-Mar. 20	Mar. 10-Apr. 1	Mar. 15-Apr. 15	Apr. 1-May 1	Apr. 15-May 15	May 1-30	May 15-June 1
Horseradish[1]	Mar. 10-Apr. 10	Mar. 20-Apr. 20	Apr. 1-30	Apr. 15-May 15	Apr. 20-May 20	May 1-30	May 15-June 1
Kale	Mar. 10-Apr. 1	Mar. 20-Apr. 10	Apr. 1-20	Apr. 10-May 1	Apr. 20-May 10	May 1-30	May 15-June 1
Kohlrabi	Mar. 10-Apr. 10	Mar. 20-May 1	Apr. 1-May 10	Apr. 10-May 15	Apr. 20-May 20	May 1-30	May 15-June 1
Leek	Mar. 1-Apr. 1	Mar. 15-Apr. 15	Apr. 1-May 1	Apr. 15-May 15	May 1-May 20	May 1-15	May 1-15
Lettuce, head[1]	Mar. 10-Apr. 1	Mar. 20-Apr. 15	Apr. 1-May 1	Apr. 15-May 15	May 1-June 30	May 10-June 30	May 20-June 30
Lettuce, leaf	Mar. 15-May 15	Mar. 20-May 15	Apr. 1-June 1	Apr. 15-June 15	May 1-June 30	May 10-June 30	May 20-June 30
Muskmelon	Apr. 20-June 1	May 1-June 15	May 15-June 15	June 1-June 15			

Fig. 38-5 Earliest dates, and range of dates, for safe spring planting of vegetables in the open (from USDA *Home and Garden* Bulletin 202) (continued)

this ↓

Crop	Planting dates for localities in which average date of last freeze is						
	Apr. 10	Apr. 20	Apr. 30	May 10	May 20	May 30	June 10
Mustard	Mar. 10-Apr. 20	Mar. 20-May 1	Apr. 1-May 10	Apr. 15-June 1	May 1-June 30	May 10-June 30	May 20-June 30
Okra	Apr. 20-June 15	May 1-June 1	May 10-June 1	May 20-June 10	June 1-20		
Onion[1]	Mar. 1-Apr. 1	Mar. 15-Apr. 10	Apr. 1-May 1	Apr. 10-May 1	Apr. 20-May 15	May 1-30	May 10-June 10
Onion, seed	Mar. 1-Apr. 1	Mar. 15-Apr. 1	Mar. 15-Apr. 15	Apr. 1-May 1	Apr. 20-May 15	May 1-30	May 10-June 10
Onion, sets	Mar. 1-Apr. 1	Mar. 10-Apr. 1	Mar. 10-Apr. 10	Apr. 10-May 1	Apr. 20-May 15	May 1-30	May 10-June 10
Parsley	Mar. 10-Apr. 10	Mar. 20-Apr. 20	Apr. 1-May 1	Apr. 15-May 15	May 1-20	May 10-June 1	May 20-June 10
Parsnip	Mar. 10-Apr. 10	Mar. 20-Apr. 20	Apr. 1-May 1	Apr. 15-May 15	May 1-20	May 10-June 1	May 20-June 10
Peas, garden	Feb. 20-Mar. 20	Mar. 10-Apr. 10	Mar. 20-May 1	Apr. 1-May 15	Apr. 15-June 1	May 1-June 15	May 10-June 15
Peas, black-eye	May 1-July 1	May 10-June 15	May 15-June 1				
Pepper[1]	May 1-June 1	May 10-June 1	May 15-June 10	May 20-June 10	May 25-June 15	June 1-15	
Potato	Mar. 10-Apr. 1	Mar. 15-Apr. 10	Mar. 20-May 10	Apr. 1-June 1	Apr. 15-June 15	May 1-June 15	May 15-June 1
Radish	Mar. 1-May 1	Mar. 10-May 10	Mar. 20-May 10	Apr. 1-June 1	Apr. 15-June 15	May 1-June 15	May 15-June 1
Rhubarb[1]	Mar. 1-Apr. 1	Mar. 10-Apr. 10	Mar. 20-Apr. 15	Apr. 1-May 1	Apr. 15-May 10	May 1-20	May 15-June 1
Rutabaga			May 1-June 1	May 1-June 1	May 1-20	May 10-20	May 20-June 1
Salsify	Mar. 10-Apr. 15	Mar. 20-May 1	Apr. 1-May 15	Apr. 15-June 1	May 1-June 1	May 10-June 1	May 20-June 1
Shallot	Mar. 1-Apr. 1	Mar. 15-Apr. 15	Apr. 1-May 1	Apr. 10-May 1	Apr. 20-May 10	May 1-June 1	May 10-June 1
Sorrel	Mar. 1-Apr. 15	Mar. 15-May 1	Apr. 1-May 15	Apr. 15-June 1	May 1-June 1	May 10-June 10	May 20-June 10
Soybean	May 1-June 30	May 10-June 20	May 15-June 15	May 25-June 10			
Spinach	Feb. 15-Apr. 1	Mar. 1-Apr. 15	Mar. 20-Apr. 20	Apr. 1-June 15	Apr. 10-June 15	Apr. 20-June 15	May 1-June 15
Spinach, New Zealand	Apr. 20-June 1	May 1-June 15	May 1-June 15	May 10-June 15	May 20-June 15	June 1-15	
Squash, summer	Apr. 20-June 1	May 1-June 15	May 1-30	May 10-June 10	May 20-June 15	June 1-20	June 10-20
Sweetpotato	May 1-June 1	May 10-June 10	May 20-June 10				
Tomato	Apr. 20-June 1	May 5-June 10	May 10-June 15	May 15-June 10	May 25-June 15	June 5-20	June 15-30
Turnip	Mar. 1-Apr. 1	Mar. 10-Apr. 1	Mar. 20-May 1	Apr. 1-June 1	Apr. 15-June 1	May 1-June 15	May 15-June 15
Watermelon	Apr. 20-June 1	May 1-June 15	May 15-June 15	June 1-June 15	June 15-July 1		

[1] Plants.
[2] Generally fall-planted

Fig. 38-5 Earliest dates, and range of dates, for safe spring planting of vegetables in the open (from USDA *Home and Garden* Bulletin 202) (continued)

> **The smaller the seed the more important it is to have a fine, firm seedbed.**

To be sure that the rows of seeds are planted straight, use a string stretched across the garden. To get the proper number of seed per foot of row, a wooden marker measured off in one foot lengths may be used, figure 38-6. If there is no problem estimating distance, use of the marker pole is not necessary. To establish the proper distance between rows, and the depth at which to plant seed, see figure 38-7, page 456.

The planting row should be made at least deep enough to cover the seed at the proper depth. Small seeds, such as carrot and lettuce, should be covered with only 1/4 to 1/2 inch of soil. The hoe handle can be used to form these shallow rows. For deeper planting, use the corner of the hoe blade or a wheel hoe. Read the directions on the seed package for the proper depth and distance apart to plant the seed, or use figure 38-7 as a guide. Fertilizer may be banded on each side of the row.

A hoe or garden rake, figure 38-8, page 459, is used to cover the seed. Try to cover the seed with fine soil that is free of lumps, stones, and clods. Cover all seeds in the row with the same depth of soil. Firm the soil around the seeds with the rake or hoe so that close contact is made between soil and seed. If the soil is a little wet when planting, do not pack or press soil down on seeds. This causes a hard crust to form and seeds will have difficulty breaking through the soil surface. Do not plant when the soil is wet.

PLANTING VEGETABLE PLANTS

Seeding

Some vegetables are not direct seeded in the garden. These seeds are planted in flats or other containers before the frost-free

Fig. 38-6 Using a line and marker pole to establish a straight row and the proper number of seeds per foot of row. Notice the even texture of the soil. (USDA photo)

date and later transplanted to the garden. Figure 38-9, page 459, shows a flat with rows marked off being seeded with vegetables for later transplanting. Soil in the flat should be a fine, carefully screened mixture of sand, soil, and peat moss, or one of the artificial soil mixes. Seeds may also be planted in individual peat pots or Jiffy 7 peat pellets. Figure 38-10, page 460, shows a Jiffy 7 peat pellet planted with a single seed.

Seeds must be planted before the frost-free date or the date for transplanting to the garden. The length of time needed to

	Requirement for 100 feet of row		Distance apart			
				Rows		
Crop	Seed	Plants	Depth for planting seed	Horse- or tractor-cultivated	Hand-cultivated	Plants in the row
			Inches	*Feet*		
Asparagus	1 ounce	75	1-1 1/2	4-5	1 1/2 to 2 feet	18 inches.
Beans:						
Lima, bush	1/2 pound		1-1 1/2	2 1/2-3	2 feet	3 to 4 inches.
Lima, pole	1/2 pound		1-1 1/2	3-4	3 feet	3 to 4 feet.
Snap, bush	1/2 pound		1-1 1/2	2 1/2-3	2 feet	3 to 4 inches.
Snap, pole	4 ounces		1-1 1/2	3-4	2 feet	3 feet.
Beet	2 ounces		1	2-2 1/2	14 to 16 inches	2 to 3 inches.
Broccoli:						
Heading	1 packet	50-75	1/2	2 1/2-3	2 to 2 1/2 feet	14 to 24 inches.
Sprouting	1 packet	50-75	1/2	2 1/2-3	2 to 2 1/2 feet	14 to 24 inches.
Brussels sprouts	1 packet	50-75	1/2	2 1/2-3	2 to 2 1/2 feet	14 to 24 inches.
Cabbage	1 packet	50-75	1/2	2 1/2-3	2 to 2 1/2 feet	14 to 24 inches.
Cabbage, Chinese	1 packet		1/2	2-2 1/2	18 to 24 inches	8 to 12 inches.
Carrot	1 packet		1/2	2-2 1/2	14 to 16 inches	2 to 3 inches.
Cauliflower	1 packet	50-75	1/2	2 1/2-3	1 to 2 1/2 feet	14 to 24 inches.
Celeriac	1 packet	200-250	1/8	2 1/2-3	18 to 24 inches	4 to 6 inches.
Celery	1 packet	200-250	1/8	2 1/2-3	18 to 24 inches	4 to 6 inches.
Chard	2 ounces		1	2-2 1/2	18 to 24 inches	6 inches.
Chervil	1 packet		1/2	2-2 1/2	14 to 16 inches	2 to 3 inches.
Chicory, witloof	1 packet		1/2	2-2 1/2	18 to 24 inches	6 to 8 inches.
Chives	1 packet		1/2	2 1/2-3	14 to 16 inches	In clusters.
Collards	1 packet		1/2	3-3 1/2	18 to 24 inches	18 to 24 inches.
Cornsalad	1 packet		1/2	2 1/2-3	14 to 16 inches	1 foot.
Corn, sweet	2 ounces		2	3-3 1/2	2 to 3 feet	Drills, 14 to 16 inches; hills, 2 1/2 to 3 feet.
Cress Upland	1 packet		1/8-1/4	2-2 1/2	14 to 16 inches	2 to 3 inches.

Fig. 38-7 Quantity of seed and number of plants required for 100 feet of row, depths of planting, and distances apart for rows and plants (from USDA *Home and Garden* **Bulletin 202)**

Crop	Requirement for 100 feet of row			Distance apart			
				Depth for planting seed	Rows		Plants in the row
	Seed	Plants			Horse- or tractor-cultivated	Hand-cultivated	
				Inches	*Feet*		
Cucumber	1 packet			1/2	6-7	6 to 7 feet	Drills, 3 feet; hills, 6 feet.
Dasheen	5 to 6 pounds	50		2-3	3 1/2-4	3 1/2 to 4 feet	2 feet.
Eggplant	1 packet	50		1/2	3	2 to 2 1/2 feet	3 feet.
Endive	1 packet			1/2	2 1/2-3	18 to 24 inches	12 inches.
Fennel, Florence	1 packet			1/2	2 1/2-3	18 to 24 inches	4 to 6 inches.
Garlic	1 pound			1-2	2 1/2-3	14 to 16 inches	2 to 3 inches.
Horseradish	Cuttings	50-75		2	3-4	2 to 2 1/2 feet	18 to 24 inches.
Kale	1 packet			1/2	2 1/2-3	18 to 24 inches	12 to 15 inches.
Kohlrabi	1 packet			1/2	2 1/2-3	14 to 16 inches	5 to 6 inches.
Leek	1 packet			1/2-1	2 1/2-3	14 to 16 inches	2 to 3 inches.
Lettuce, head	1 packet	100		1/2	2 1/2-3	14 to 16 inches	12 to 15 inches.
Lettuce, leaf	1 packet			1/2	2 1/2-3	14 to 16 inches	6 inches.
Muskmelon	1 packet			1	6-7	6 to 7 feet	Hills, 6 feet.
Mustard	1 packet			1/2	2 1/2-3	14 to 16 inches	12 inches.
Okra	2 ounces			1-1 1/2	3-3 1/2	3 to 3 1/2 feet	2 feet.
Onion:							
Plants		400		1-2	2-2 1/2	14 to 16 inches	2 to 3 inches.
Seed	1 packet			1/2-1	2-2 1/2	14 to 16 inches	2 to 3 inches.
Sets	1 pound			1-2	2-2 1/2	14 to 16 inches	2 to 3 inches.
Parsley	1 packet			1/8	2-2 1/2	14 to 16 inches	4 to 6 inches.
Parsley, turnip-rooted	1 packet			1/8-1/4	2-2 1/2	14 to 16 inches	2 to 3 inches.
Parsnip	1 packet			1/2	2-2 1/2	18 to 24 inches	2 to 3 inches.
Peas	1/2 pound			2-3	2-4	1 1/2 to 3 feet	1 inch.
Pepper	1 packet	50-70		1/2	3-4	2 to 3 feet	18 to 24 inches.
Physalis	1 packet			1/2	2-2 1/2	1 1/2 to 2 feet	12 to 18 inches.
Potato	5 to 6 pounds, tubers			4	2 1/2-3	2 to 2 1/2 feet	10 to 18 inches.

Fig. 38-7 Quantity of seed and number of plants required for 100 feet of row, depths of planting, and distances apart for rows and plants (from USDA *Home and Garden* Bulletin 202) (continued)

	Requirement for 100 feet of row			Distance apart			
					Rows		
Crop	Seed	Plants	Depth for planting seed	Horse- or tractor-cultivated	Hand-cultivated	Plants in the row
			Inches	*Feet*		
Pumpkin	1 ounce		1-2	5-8	5 to 8 feet	3 to 4 feet.
Radish	1 ounce		1/2	2-2 1/2	14 to 16 inches	1 inch.
Rhubarb		25-35		3-4	3 to 4 feet	3 to 4 feet.
Salsify	1 ounce		1/2	2-2 1/2	18 to 26 inches	2 to 3 inches.
Shallots	1 pound (cloves)		1-2	2-2 1/2	12 to 18 inches	2 to 3 inches.
Sorrel	1 packet		1/2	2-2 1/2	18 to 24 inches	5 to 8 inches.
Soybean	1/2 to 1 pound		1-1 1/2	2 1/2-3	24 to 30 inches	3 inches.
Spinach	1 ounce		1/2	2-2 1/2	14 to 16 inches	3 to 4 inches.
Spinach, New Zealand	1 ounce		1-1 1/2	3-3 1/2	3 feet	18 inches.
Squash:						
Bush	1/2 ounce		1-2	4-5	4 to 5 feet	Drills, 15 to 18 inches; hills, 4 feet.
Vine	1 ounce		1-2	8-12	8 to 12 feet	Drills, 2 to 3 feet; hills, 4 feet.
Sweetpotato	5 pounds, bedroots	75	2-3	3-3 1/2	3 to 3 1/2 feet	12 to 14 inches.
Tomato	1 packet	35-50	1/2	3-4	2 to 3 feet	1 1/2 to 3 feet.
Turnip greens	1 packet		1/4-1/2	2-2 1/2	14 to 16 inches	2 to 3 inches.
Turnips and rutabagas	1/2 ounce		1/4-1/2	2-2 1/2	14 to 16 inches	2 to 3 inches.
Watermelon	1 ounce		1-2	8-10	8 to 10 feet	Drills, 2 to 3 feet; hills, 8 feet.

Fig. 38-7 Quantity of seed and number of plants required for 100 feet of row, depths of planting, and distances apart for rows and plants (from USDA *Home and Garden* Bulletin 202) (continued)

reach transplanting size varies from vegetable to vegetable. A good general rule is to plant the seed six to eight weeks before plants are to be transplanted to the garden.

After the seeds are planted, the flat or pot should be watered to settle the media around the seeds. Close contact between the soil and seed is important for germination. Cover the flat or pot with newspaper, glass, or plastic to hold in moisture. The seeded container should never be allowed to dry out, since newly germinating seeds are easily killed by lack of moisture. Uncover the seedlings as soon as shoots appear through the soil. Water as often

Fig. 38-8 A garden rake is used to cover seeds evenly. Lima beans are being planted. (USDA photo)

as necessary to keep the growing media moist but not wet. Fertilize the seedlings as recommended on the fertilizer container label. Do not use too much fertilizer — new plants are easily burned. (Some soilless mixes have a slow-release fertilizer in the mix, making additional fertilizer unnecessary.)

To determine which vegetable crops are usually planted in the garden as seeds and which ones may be transplanted to the garden as plants, see figure 38-7. The column titled *Requirement for 100 feet of row* lists both the amount of seed and the number of plants for those crops that are transplanted to the garden as plants.

First Transplanting

If seeds are sown as directed on the package, two transplantings will be necessary. The first transplanting must be made soon after the seedlings develop the first pair of true leaves. At this first transplanting, each plant should be spaced about 2 inches apart in a flat or planted in individual peat pots or Market Packs. The soil mix used is the same mix the seeds were planted in —

Fig. 38-9 Seeds being planted in a flat to be grown indoors for later transplanting to the garden. Notice the labels placed for seed identification. (It is possible that the seeds shown in this photo are placed too close to one another, and may require thinning.) (USDA photo)

Fig. 38-10 Jiffy 7 pellet used to start individual plants from seed. (Left) pellet before soaking in water; (center) plant breaking soil; (right) plant ready for transplanting to the garden. (USDA photo)

one-third soil, one-third sand or perlite, and one-third peat moss, or an artificial soil mix. A good soluble fertilizer should be used within a few days after transplanting and repeated according to directions. Some of the soilless planting mixes contain fertilizer, making the addition of fertilizer unnecessary.

Hardening Off

As the date nears for the plants to be transplanted to the garden, the atmosphere must be gradually changed to resemble as closely as possible the conditions of the outside garden. If plants are in a greenhouse in which temperature and humidity are high, the following should be done.

1. Place the plants where the temperature can gradually be lowered over a period of ten days to two weeks. This slows the growth rate and toughens the plant.

2. Lower the humidity and allow the soil to become dryer. Do not allow the plants to wilt, however. This also slows growth and toughens the plant.

3. If planted in flats, cut a square around the plant with a sharp knife. Use all the area of the flat, and make the blocks as big as possible. In the period of ten days to two weeks, the plant will grow new roots in a tighter root ball and be able to withstand transplanting shock better.

These three steps slow the growth rate to prepare the plant for chilling weather, drying winds, a shortage of water, or high temperatures. Tender, succulent growth must be hardened for best growth of plants.

The three steps of hardening off can be done in a greenhouse if all the plants it contains are being hardened off, or in a special cold frame. Cold frames are usually used for hardening off. Permanent cold frames are usually constructed of concrete block or wood with a glass sash on top. A temporary cold frame is shown in figure 38-11. Cold frames are never heated — the frame is opened enough during the day to keep the sun from raising the temperature too high and closed at night to hold heat in and prevent freezing of the plants. The cold frame is gradually opened more and more each day until the plants are hardened off and living in the outdoor environment. Most plants are then ready for transplanting. (Some plants, such as tomatoes, must not be transplanted to the garden until all danger of frost is past. On the other hand, other crops, such as cabbage, are more tolerant of cold weather and can be transplanted to the garden earlier.)

Second Transplanting

The vegetable plants are now properly hardened off and ready to transplant to the garden. The following are steps in transplanting:

1. Water the plants to be transplanted the evening before, or at least several hours before transplanting.

2. If plants are grown in peat pots, peel off the top half of the pot and break off the bottom. If roots are already growing through the pot, this is not necessary. This practice helps speed the spread of roots into the garden soil, especially if dry weather follows transplanting.

3. If plants are grown in flats, cut the soil between the plants and lift them out with a putty knife or other small tool. Leave as much soil around the roots as possible. Handle carefully.

4. Check the distance apart at which the plants are to be placed and set them immediately in the garden soil. Place the plants a little deeper in the garden soil than they grew in the original container. Tall, spindly plants, especially tomatoes, may be planted deeper.

> **Do not allow plant roots to be exposed to the sun or drying wind.**

5. Pull soil toward the plant and firm slightly. Fill the hole only two-thirds full. Stop and water enough to settle the soil around the roots. Pull in enough dry soil to level the hole and cover the wet area. This prevents moisture loss and baking of the soil.

Fig. 38-11 Temporary cold frame used for hardening off flower and vegetable plants. (USDA photo)

Fig. 38-12 Transplanting a tomato plant to the garden. Notice the soil that has been left on the plant's roots. A string is being used to keep rows straight. (USDA photo)

Figure 38-12 shows tomato plants being transplanted from pots to the garden.

If possible, transplant on a cloudy day or in late afternoon or evening. Plants will not wilt as much if they are not exposed to bright sunlight immediately after transplanting, resulting in a higher percentage of healthy, growing plants.

A strip of cardboard, aluminum foil, or newspaper wrapped around the stem and extending about 1 inch underground and about 2 inches aboveground will prevent cutworms from cutting off and killing the new plant, figure 38-13.

Double Cropping

Since some of the vegetables reach maturity earlier than others, consideration should be given to planting something in their place as they are harvested. Replanting the same crop or planting other crops are both possibilities. Vegetables may be grouped according to how long they grow in the garden and their use as replanted or successive crops, figure 38-14.

Fig. 38-13 Placement of a cardboard collar around a tomato plant to prevent cutworms from cutting off the plant just above soil level. (USDA photo)

Successive Plantings

Crops Occupying the Ground All Season

asparagus	okra
beans, pole lima	onions (from seeds)
beans, pole snap	parsnips
beets, late	peppers
carrots, late	pumpkins
corn, late	rhubarb
cucumbers	rutabagas
eggplant	salsify
melons	squash
	tomatoes

Successive Crops Which May Be Replanted or Followed by Other Crops

beans, dwarf	peas
lettuce	radishes
mustard	spinach
parsley	turnips

Early Crops Which May Be Followed by Others

beets, early	mustard
cabbage, early	onion sets
carrots, early	spinach
corn, early	turnips, early

Late Crops Which May Follow Others

beets, late	kale
brussels sprouts	mustard
cabbage, late	peas, late
cauliflower	spinach
endive	turnips

Fig. 38-14 By grouping vegetables in the garden according to their use as replanted or successive crops, harvesting is simplified.

GARDEN TOOLS

Very few tools are necessary for cultivation of a small garden, figure 38-15. In most cases, the only tools necessary include:

- a spade or spading fork for digging up soil

- a steel bow garden rake for leveling and preparing the seed bed

- a garden hoe for laying off rows and controlling weeds

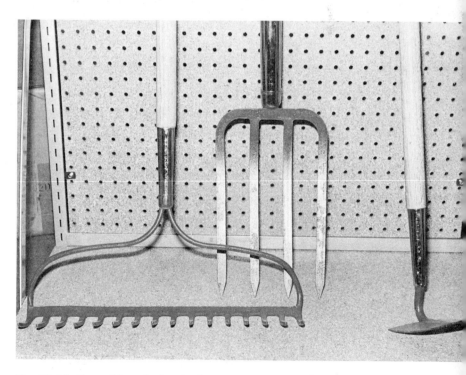

Fig. 38-15 Essential tools for the home garden. From left to right: garden rake, fork spade, and garden hoe. (Richard Kreh, photographer)

- a strong cord for laying off rows

- a wheelbarrow for hauling fertilizer and mulch

- a garden hose long enough for watering the garden

For larger gardens measuring about 2,000 to 4,000 square feet, a wheel hoe is helpful in saving labor.

For gardens measuring over 4,000 square feet, a rotary tiller rather than a rake is useful for preparing the soil for planting and cultivating to control weeds.

STUDENT ACTIVITIES

1. Plant seeds of vegetable crops in flats for sale or for use in a home or school garden.

2. Give a demonstration on the proper planting techniques of a home garden. Be sure to mention that the class has plants that are for sale.

3. Construct a temporary cold frame structure.

SELF-EVALUATION

1. Specify whether the following vegetables are direct seeded in the garden or transplanted to the garden as small plants.

beans	corn	peas	spinach
beets	cucumbers	peppers	squash
cabbage	eggplant	potatoes	sweet potatoes
carrots	onions	radishes	tomatoes

2. Specify whether the following vegetables are cold-tender or cold-hardy.

broccoli	cucumber	peas	sweet corn
beans	eggplant	radishes	tomatoes
cabbage	onions		

3. List the five steps in transplanting to the garden.

4. List five crops which
   (a) occupy the soil all season.
   (b) are successive crops.
   (c) are late crops which may follow others.

unit 39
Caring for the Vegetable Garden

OBJECTIVE

To care for a vegetable garden, producing a marketable-quality crop.

COMPETENCY TO BE DEVELOPED

After studying this unit, the student will be able to

- determine watering needs of vegetable plants and list three ways in which water is applied.
- determine the type and amount of fertilizer for a specific crop by using a soil test recommendation.
- establish weed control programs using mulches, cultivation, and herbicides.
- list five vegetables grown in the area, a pest which commonly attacks each vegetable, and one method of control for each.

Materials List

a rain gage

garden hose and sprinkler

garden fertilizer

garden tools (listed in Unit 38)

weed killers if used

mulching material

insecticides (malathion, Sevin)

fungicides (captan, zineb)

sprayer and/or duster

rubber gloves and respirator

Planting the garden is only a small part of the work involved in growing vegetables. In planning the size of the garden, the time available for its care should be considered. It is better to have a small, well-kept garden than a large, weedy one. The best preparation, planning, and planting is of little value if the garden is not cared for properly.

WATERING

All plants must be watered when they are transplanted to the garden. There may also be other times during the growing season when lack of rainfall makes watering desirable for best yield, or even necessary for best plant growth.

When plants wilt for any length of time, production is reduced; watering helps prevent serious damage to the plants at these times. If the plants do not recover soon after the sun is down, the need for water is even more serious.

Application of Water

If the garden is watered, it must be done properly — most plants are actually harmed by shallow watering. Soak the soil to a depth of at least 6 inches. This requires a great deal of water. To be sure the proper depth has been reached, dig down into the soil to see how far the water has penetrated. Another way to tell if enough water has been added is to place a can with straight sides or a rain gage in the garden and measure how much water has been applied as a result of the sprinkling. The plants should receive no less than 1 inch of water per week. This amount includes any rainfall which may occur. Frequent, light waterings encourage shallow root systems and may actually reduce crop yield.

Water may be applied with sprinklers, a hose, by trench, or by ditch irrigation. Ditch or trench irrigation can be used only if the garden is fairly level.

The use of natural mulches 3 to 4 inches thick can reduce the need for water in the garden. A good mulch shades the soil and keeps it cool, reducing the rate of evaporation of water from the soil. Mulches also slow down runoff of rainwater. Black plastic mulch does a good job in moisture retention and also warms the soil during the early season, encouraging growth.

FERTILIZERS (ADDITIONAL)

At this point, probably some fertilizer has been applied to the garden by broadcasting and plowing or digging in. Additional fertilizer application after the plants are growing is made by spreading the fertilizer around the plants (band application) or along the row (side dress application), figure 39-1. If earlier application of fertilizer was not made to the garden, the amount of fertilizer applied after the plants have established growth will be increased slightly. In these cases, refer again to the soil test recommendations for the amount to apply. Never apply more than 3 pounds of fertilizer per 100 square feet in a single application.

Plants requiring high amounts of nitrogen, such as corn and the leafy vegetables, respond best to applications of fertilizer to the soil surface, such as by side dressing or band application. This is because nitrogen leaches down to the roots much faster than phosphorus or potash.

WEED CONTROL

Weed control is the single most important cultural practice in most gardens. All other effort is useless if weeds are allowed to overtake the garden. Weeds rob garden vegetables of water, plant food, and light. Some weeds help spread diseases, insects, and nematodes. Weeds are most easily controlled before they break through the ground or while they are quite small.

Fig. 39-1 Fertilizer being placed in a wide ring on the soil surface around a tomato plant several weeks after planting. (USDA photo)

Cultivation

When weeds are small (less than 1 inch tall) shallow cultivation (turning the soil over 1 or 2 inches deep) will control the weeds and prevent damage to crop roots. Deep cultivation damages many crop roots and brings moist soil to the surface, which dries out and causes moisture loss from the soil. If weeds are allowed to grow taller than 1 inch, deep cultivation is necessary to kill the weeds, and in the process, many vegetable plant roots are damaged.

The secret to good weed control is to kill the weeds when they are very small by using a hoe or other shallow cultivator to kill them. The scuffle hoe is an excellent tool for shallow cultivation. By sliding it back and forth just under the soil surface, small weeds are cut off and killed. This shallow cultivation is quicker, requires less muscle, and does not damage vegetable crop roots.

As soon as the soil is dry enough to work after a rain, it should be hoed or cultivated to kill any weed seeds that have sprouted. This also leaves the soil surface loose to absorb later rainfall. Another advantage to the loose surface soil is that the weed seeds do not germinate and grow as quickly. Hoeing and cultivation should be done often enough to eliminate weeds while they are less than 1 inch tall. Hoeing or cultivation is unnecessary if weeds are not present, unless the soil has a hard crust on it that would prevent rainwater from soaking in.

Mulching

Mulching is another excellent method of weed control in the garden, figure 39-2, page 468. Organic material such as straw, leaves, sawdust, grass clippings, ground corn cobs, and wood chips, make excellent mulch. Apply several inches of mulch around plants to cover the entire soil surface. The application of materials that pack tightly, such as grass clippings, should be more shallow (two inches deep) than coarse materials such as straw.

If weeds grow up through the mulch, the mulch is either not thick enough or not applied uniformly. It is important that the mulch keep all light from the soil surface. Organic mulches are best applied only after the soil has warmed in late spring. Too early an application shades the soil, thereby preventing it from warming up in the spring and early summer.

Fig. 39-2 Grass mulch being applied around a garden plant. Notice that the mulch is thick enough to completely shade the soil. (USDA photo)

> Be sure the material used for mulch
> does not contain weed seeds.

When plant material is used for mulch, it is necessary to apply extra nitrogen fertilizer. As the soil organisms rot the mulch, the soil is robbed of nitrogen, which is used as food by microorganisms. Sawdust uses a great deal of nitrogen in this rotting process. To compensate, add extra nitrogen around the plants as a sidedress when organic mulches are used.

Fig. 39-3 Black polyethylene plastic used as a mulch around tomato plants. (USDA photo)

Black polyethylene plastic makes an excellent mulch for vegetable gardens. The black color warms the soil and may be used early in the spring when plants are first planted in the garden. Polyethylene is available in rolls which may be rolled down the garden row on each side of the plants. Soil is placed over the edge of the plastic to hold it in place, figure 39-3. Plants may be planted by punching a hole through the plastic and placing the plant in the soil below.

Mulches also conserve moisture in the soil, reducing the need for watering. If properly mulched, the garden need not be cultivated, saving a great deal of work in weed control.

Chemical Weed Control

Modern herbicides provide safe, dependable weed control where hand labor was once required. Herbicides require careful application since too much may injure the crop and too little may not control the weeds.

Herbicide Chart

The following herbicides are suggested for safe and effective use in the home garden. The chemicals listed will provide control of many broadleaf weeds and grasses. However, since a few weeds are resistant to these herbicides, some hand hoeing may be necessary. Read the labels carefully to determine which weeds are controlled by which chemicals and for instructions for application. Use the herbicides only for controlling weeds in those crops listed on the label.

HERBICIDE	RATE
Dacthal* (75% Wettable Powder)	10-15 level tablespoons per 1,000 square feet

- Apply at time of seeding or transplanting of dry beans, mung beans, snap beans, broccoli, brussels sprouts, cabbage, cauliflower, collards, eggplant, garlic, kale, mustard greens, onions, potatoes, sweet potatoes, southern peas, strawberries, turnips.
- Apply 4-6 weeks after planting of potatoes. (Dacthal may be used on cucumber, watermelon, cantaloupe, and squash for weed control only between rows and hills.) Do not apply within 10 to 12 inches of the stems or injury to the stems may occur. Dacthal application should be followed by rainfall or irrigation (1/2-1 inch) within one to two days of application for best results. Soil should be loose and weed free when Dacthal is applied. Use low rate for sandy soils and high rate for clayey soils.

HERBICIDE	RATE
Telvar* (80% Wettable Powder)	3 level tablespoons per 1,000 square feet

- Apply to asparagus (established plantings) in spring to freshly worked soil prior to weed and spear emergence. Repeat at end of harvest season after weed growth has been removed.

HERBICIDE	RATE
Treflan*	2-3 level teaspoons per 1,000 square feet

- Apply prior to planting of snap beans, lima beans, okra, southern peas, carrots, peas.
- Apply prior to transplanting of broccoli, brussels sprouts, cabbage, cauliflower, peppers, and tomatoes. Treflan must be thoroughly mixed 2 to 3 inches into the soil immediately after application. A rotary tiller gives the best results. Plant any time within two weeks.

HERBICIDE	RATE
Bladex*	4-6 level tablespoons per 1,000 square feet

- Apply immediately after seeding corn. Corn or potatoes may be planted in the same crop year following corn, and all vegetables may be planted the following crop year without a risk of injury.

*Trademarks of Diamond-Shamrock Corporation, E.I. DuPont DeNemours and Company, Eli Lilly and Company, and Shell Chemical Company, respectively.

Dacthal and Treflan can also be used for controlling weeds around flowers, trees, and shrubs. See your county extension agent for more information.

Application

A sprinkling can may be used to apply these herbicides. Determine the rate of application by applying only water first; then add the chemical and apply to the measured area. Generally, it is more desirable to make two trips over the area and walk in a different direction each time while applying one-half of the total amount of chemical. This technique provides a more uniform application. A similar application technique should be used when applying herbicides with hand sprayers.

> **Caution: Weed control chemicals are considered poisonous even though many of them are relatively nontoxic to humans. Read the label on the container and follow all directions and safety precautions. Avoid prolonged or repeated contact with the skin and be sure to wash thoroughly after use. Store herbicides out of reach of children and pets and away from foodstuffs.**

THINNING

Plants that are allowed to grow too close together compete with one another for such things as food and water. For this reason, it is sometimes necessary to remove a number of plants. Check directions on the seed package for thinning instructions. Never plant transplanted plants more closely together than recommended.

INSECT AND DISEASE CONTROL

Insects and diseases can cause a serious decrease in production of the garden if not controlled.

Prevention

Some of the best control measures are preventives rather than cures.

Dispose of Crop Residue. Dispose of all crop residue that may carry diseases or insects over to the next year's planting. Compost this material so that it rots before the next season.

Rotate Crops. Rotation of crops is another good preventative. Avoid planting the same crop in the same area each year.

Plant Treated Seeds. Plant seeds which are treated to control disease as a good step in prevention of certain diseases.

Purchase Resistant Varieties. Buy resistant varieties whenever possible. Many new varieties are disease resistant; some are insect or nematode resistant. A good example is the tomato Better Boy. It is labeled *VFN*, which means that it resists verticillium wilt, fusarium wilt, and nematodes.

Purchase Healthy Plants. Plants that are vigorously growing because of good fertilizer and other cultural practices are more able to survive disease and insect attacks and produce good crops.

If disease resistant hybrids are planted and good cultural practices are followed, many garden vegetables will need no further insect or disease control. Corn, tomatoes, peas, beets, carrots, lettuce, onions, and turnips rarely need much additional help. Other vegetables, especially melons, cucumbers, beans, and cabbage, generally need some additional insect control. This varies with location and local problems.

Chemical Control

Most garden insects can be effectively controlled with two relatively safe insecticides, malathion and Sevin. Most fungus diseases can be controlled with zineb or maneb and captan fungicides. These insecticides and fungicides can be mixed together and applied at the same time as a spray. Captan should not be mixed with an emulsifiable concentrate of malathion, but is compatible with the wettable powder of malathion. Read the label for mixing, use, and safety precautions. Check to see how close to harvest date the sprays or dusts can be applied. The chemical must have time to break down and become harmless before the vegetables are eaten.

> **Always apply pesticides only to those crops for which they are recommended. Apply only the amount that is specified on the label. Wash all treated vegetables before eating.**

Application

Most garden pesticides are applied as dusts using a hand duster. Dusts are available in diluted form and ready to apply. Sprays are often applied with a knapsack sprayer, figure 39-4. Material that is sprayed is always diluted with water. Application should be made on a quiet day when the wind is not blowing so that the chemical does not drift to other areas.

A spray schedule for the vegetable garden can be obtained at your local extension agent's office or from a State Agriculture College Extension Service office. Another guide, *Insects and Diseases of Vegetables in the Home Garden*, (Bulletin 46) is available from the USDA.

Fig. 39-4 Typical hand duster for application of dusts to garden crops (left) and knapsack sprayer for application of sprays (right). (Richard Kreh, photographer)

VEGETABLE SPRAY SCHEDULE

Figure 39-5 lists vegetables that most often require spraying or dusting, the pests that commonly attack the plant, and recommended chemicals for control. Remember that many vegetables never require chemical sprays or dusts. Many others that do require the use of chemicals respond to a mixture of malathion and Sevin for insect control and captan or zineb for control of fungus diseases.

> If a chemical must be used, read the recommendation on the pesticide label before purchase.

Plant		Pest and Description	When to Spray	Chemical (Per Gallon of Water)
Asparagus	Asparagus Beetle	Asparagus Beetle.* Metallic blue to black beetle with orange markings. Larvae olive green, 1/3 inch long. Feeds on leaves and shoots. (*Only the most damaging insects are listed for each crop in the chart.)	When larva or adult first begin to feed Check label for days to wait before harvesting crop.	carbaryl (Sevin) (50% WP) 2 tablespoons or malathion (50% EC) 2 level teaspoons or rotenone (5% WP) 4 level tablespoons
Bean	Mexican Bean Beetle	Mexican Bean Beetle. Feeds on underside of leaves. Adult is orange to yellow with 16 black spots on back. Larvae are fuzzy, 1/3 inch long.	When larva or adult begins to feed; repeat in 7 to 10 days	malathion (50% EC) 2 level tablespoons or carbaryl (Sevin) 2 level tablespoons or rotenone (5% WP) 4 level tablespoons
Beet Swiss Chard	Beet Webworm	Leaf Miner. Adult is slender, gray, black-haired fly 1/4 inch long. Larvae pale green or whitish. Leaves have blotches. Beet Webworm. Larvae are yellow to green, with black stripe and numerous black spots on back. Up to 1 1/4 inches long.	When blisters first appear on leaves Repeat in 7 days When leaves start to roll or fold	malathion (50% WP) 2 level teaspoons pyrethrum spray mixed with water according to directions

Fig. 39-5 Spray Schedule for Vegetables Most Often Requiring Spray

Plant	Pest and Description	When to Spray	Chemical (Per Gallon of Water)
Broccoli Brussels Sprouts Cabbage Cauliflower Kale Kohlrabi	**Cabbage Looper.** Pale green worm, white stripes on back which loops as it crawls. Up to 1 1/2 inches long. Chews holes in leaf. **Imported Cabbage Worm.** Velvet green, size up to 1 1/4 inches long. Chews holes in leaves. **Aphid.** Small, green to powdery blue, soft. Causes curling of leaves. **Club Root Disease.** (swelling on roots causing wilting of plants)	When worms first begin to feed; repeat in 7 days. Continue up to 1 week before harvest if necessary. When aphids first appear Do not apply within 7 days of harvest. at time of planting	carbaryl (Sevin) (50% WP) 2 level tablespoons malathion (25% WP) 4 level tablespoons Terrachlor, 6 level tablespoons per gallon of water. Apply 1 cupful of diluted chemical in each planting hole.

Cabbageworm

Celery Celery Leaf Tier	**Leaf Tier.** Larvae are greenish caterpillars 3/4 inch long. Eats holes in leaves and stalks and ties leaves together with webs.	When webs are first seen. Repeat in 1/2 hr. First dust drives them out of webs; second application kills them.	pyrethrum dust (0.2%) comes ready to use (no water)
Corn (Sweet) Corn Earworm	**Corn Earworm.** Larvae are green, brown, or pink worms; light stripes along sides and on back. Up to 1 3/4 inches long. Feeds on early corn shoot in season. Later feeds on kernels near top of ear.	For shoot damage, spray early in season. For ear damage, spray silks the day after silks appear. Spray until silk is wet. Repeat 4 times at 2-day intervals.	carbaryl (Sevin) (50% WP) 3 level tablespoons

Fig. 39-5 Spray Schedule for Vegetables Most Often Requiring Spray (continued)

Plant	Pest and Description	When to Spray	Chemical (Per Gallon of Water)
Cucumber **Pumpkin** **Squash**	**Cucumber Beetle, Flea Beetle, Leaf Hopper, Pickle Worm**	When insect damage is first seen. Repeat in 7 days. Do not spray open blossoms.	malathion (50% WP) 2 level teaspoons or methoxychlor (50% WP) 2 level tablespoons
	Anthracnose. Reddish brown spots on leaves, tan cankers on stems, round sunken spots on fruit.	When damage is first seen. Repeat every 7 to 10 days.	zineb or captan according to directions on label Plant rotation of not less than 3 years is helpful.
Cucumber Beetle			
Eggplant	**Colorado Potato Beetle.** Adult is yellow, black striped, 3/8 inch long. Larva is brick red, hunchbacked, 3/5 inch long. Eats leaves.	When insect is seen and leaf damage is first noticed	carbaryl (Sevin) (50% WP) 2 or 3 level teaspoons
Colorado Potato Beetle	**Eggplant Lacebug.** Adult is grayish to light brown; flat lacelike wings, 1/16 inch long. Nymph: yellowish, louselike; up to 1/10 inch long. Feeds on underside of leaves; leaves turn yellow and brown.	Apply as soon as insect or damage is noticed. Repeat in 7-10 days. Do not apply within 7 days of harvest.	malathion (25% WP) 4 level tablespoons
Eggplant Lacebug			
Lettuce	**Aphid.** Adult is tiny, green, soft bodied. Feeds on underside of leaves.	When insect is first seen. Repeat in 7 days. Use up to 14 days before harvest if needed.	malathion (57% EC) 2 level teaspoons
Aphid	**Cabbage Looper.** Larva is pale green worm which loops as it crawls; white stripes down back. Size up to 1 1/2 inches long. Feeds on underside of leaves, leaving ragged holes.	When insect or damage is first seen. Repeat in 7 days. Do not apply within 7 days of harvest.	malathion (25% WP) 2 level teaspoons

Fig. 39-5 Spray Schedule for Vegetables Most Often Requiring Spray (continued)

Plant	Pest and Description	When to Spray	Chemical (Per Gallon of Water)
Muskmelon	**Aphid.** Same as for lettuce.	Same as for lettuce.	Same as for lettuce.
	Downy Mildew and Leaf Spot. Brownish spots on leaves; leaves die. Fruits are not affected.	When injury is first noticed. Repeat in 7 to 10 days.	zineb or captan as recommended on the label
Aphid	**Mosaic.** (viruses) Mottled green and yellow curled leaves.	Control aphids and striped cucumber beetles to prevent spread. See control already given.	See control for aphids and striped cucumber beetles. Destroy weeds and diseased plant. Do not plant seed from infected plants.
Okra	**Corn Earworm.** Larva is green or pink worm; light stripes along sides and on back; up to 1 3/4 inches long. Eats holes in pods.	As pods form; repeat at least twice at 7-day intervals	carbaryl (Sevin) (50% WP) 2 or 3 level tablespoons
	Stink Bug. Black, green, or brown shield-shaped bug; 5/8 inch long.	As pods form; repeat at least twice at 7-day intervals	malathion (25% WP) 2 level teaspoons
Southern Green Stink Bug	**Wilt.** Causes yellow and wilted leaves.		Rotate crops at least 3 years apart.
Onion	**Maggot.** White legless worm 3/8 inch long. Burrows into young bulbs.	At planting time	diazinon (25% EC) 2 level teaspoons. Apply in row before planting onion sets.
Thrip	**Thrip.** Adult is tiny, active yellow or brownish, winged insect. Larva is white, same appearance as adult, only smaller. Sucks juices and causes blotches on leaves. Tips may turn brown and die.	When injury is first noticed Repeat 2 or 3 times at 7- to 10-day intervals.	malathion (25% WP) 4 level tablespoons

Fig. 39-5 Spray Schedule for Vegetables Most Often Requiring Spray (continued)

Plant		Pest and Description	When to Spray	Chemical (Per Gallon of Water)
Pea	\n\nAphid	**Aphid.** Small, green to powdery blue, soft. Feeds on underside of leaf, causing leaves to curl.	When insects first appear	malathion (25% WP)\n4 level tablespoons\nor\nmalathion (57% EC)\n2 teaspoons
Potato	\n\nColorado Potato Beetle	**Colorado Potato Beetle.** Same as eggplant.	Same as eggplant	Same as eggplant
		Flea Beetle. Black, brown, or striped jumping beetles; about 1/16 inch long. Leaves appear to have been shot full of small holes.	As soon as beetle is seen	carbaryl (Sevin) (50% WP)\n2 level tablespoons
		Early Blight. Leaves show small, irregular, dark brown spots. Fungus is carried in soil.	As soon as damage is seen	zineb or captan according to directions\nPlant only disease-free tubers.
	Potato Flea Beetle	**Late Blight.** Dark irregular dead areas on leaves and stems. Infected tubers rot in storage; plants may die early in season.	Same as for early blight	Same as for early blight\nUse blight resistant varieties.
Tomato	\n\nFlea Beetle	**Flea Beetle.** Same as potato.	Same as for potato	Same as for potato
		Tomato Fruit Worm. Larva is green, brown, or pink; light green stripes along sides and on back; up to 1 3/4 inches long. Same as corn earworm. Eats holes in fruit and buds.	When fruits first begin to set\nRepeat 3 times at 2-week intervals	carbaryl (Sevin) (50% WP)\n2 level tablespoons
		Fusarium and Verticillium Wilts.		Plant resistant varieties.

Fig. 39-5 Spray Schedule for Vegetables Most Often Requiring Spray (continued)

Summary

1. Choose fertile, well-drained soil and apply proper fertilizer to produce vigorous plants.

2. Plant crops adapted to the area.

3. Keep weeds under control.

4. Use only disease-free seed, certified if possible.

5. Treat seed with chemicals to protect against decay and damping-off.

6. Purchase disease-free plants from a reputable dealer.

7. Plant disease resistant varieties whenever possible.

8. Destroy all plant refuse as soon as harvest is over to prevent carry-over of diseases and insects.

9. Use insecticides as a last resort, and then only according to directions.

10. Select insecticides that do the least harm to helpful insects and the environment.

HARVESTING

Harvest vegetable crops at the peak of quality. Some crops, such as sweet corn, peas, and asparagus, lose quality rapidly. With these vegetables, just a few days makes a great difference in quality. Other crops, such as beets, cabbage, and pepper, retain their quality over a long period of time.

STUDENT ACTIVITY

1. Actually plant a garden. Demonstrate the techniques of planting and caring for the garden.

SELF-EVALUATION

A. Select the best answer from the choices offered to complete the statement or answer the question.

1. How much rain does the average garden plant require per week?
   a. 1 foot
   b. 1 inch
   c. 1/2 inch
   d. 3/4 inch

2. A signal which indicates that plant production is being reduced and water should be added is
   a. the plants turn brown.
   b. the plants fall over.
   c. the plants wilt.
   d. all of these

3. Organic mulches prevent water loss from the soil by reducing

   a. evaporation of water.
   b. water runoff.

   c. soil temperature.
   d. all of these

4. Plants requiring more nitrogen than most plants respond best to extra application of fertilizer to the soil surface because

   a. they are heavy feeders.
   b. nitrogen leaches down to the roots quickly.
   c. they are grasses.
   d. all of the above

5. Fertilizer should not be applied heavier than

   a. 10 pounds per 100 square feet.
   b. 7 pounds per 100 square feet.

   c. 5 pounds per 100 square feet.
   d. 3 pounds per 100 square feet.

6. The single most important cultural practice in most gardens is

   a. weed control.
   b. watering.

   c. insect control.
   d. disease control.

7. A garden should be cultivated or hoed for weed control

   a. every 10 days.
   b. after every rain.
   c. while weeds are less than 1 inch tall.
   d. just before harvest time.

8. Cultivation of the garden should be shallow (1 or 2 inches deep) because

   a. shallow cultivation controls small weeds.
   b. deep cultivation damages vegetable crop roots.
   c. soil moisture is brought to the surface and lost in deep cultivation.
   d. all of the above

9. Organic mulches used for weed control, such as straw, should be applied to the soil in a heavy enough layer that

a. mud does not come through.

c. it is soft to the touch.

b. no light can reach the soil.

d. none of these

10. When sawdust mulch is used

a. it tends to pack too tightly.

b. good weed control is assured.

c. extra nitrogen fertilizer must be applied to the soil around the plants.

d. the soil becomes acidic.

11. Chemical pesticides should be applied

a. to all garden vegetable crops.

b. with sprayers.

c. early in the morning.

d. only to those crops recommended on the label.

12. Straw mulch is best applied late in the spring because

a. it shades the soil and keeps it from warming up.

b. that is when the weeds are worst.

c. there is plenty of moisture in the early spring.

d. it is easier to obtain straw then.

13. Two disease control methods which do not involve the use of chemicals are

a. crop rotation and use of resistant varieties.

b. seed treatment and the use of resistant varieties.

c. burning crop residue and seed treatment.

d. none of the above

B. 1. List the ten key points for insect and disease control.

2. Select two vegetables and list the pest which most commonly attacks them, a description of the pest, and the chemicals used to control the pest.

unit 40
Favorite
Garden Vegetables

OBJECTIVE

To cultivate and harvest five vegetables commonly grown in the local area.

COMPETENCY TO BE DEVELOPED

After studying this unit, the student will be able to

- list the requirements for planting, fertilizing, harvesting, and storing five garden vegetables that grow successfully in the local area.

- select five locally grown vegetables and plant and grow them to marketable quality.

Materials List

    seed catalogs

    seed

    planting site

    fertilizer

    minimal garden tools (spade, rake, hoe)

    duster or sprayer

    chemical sprays or dusts if necessary

Notes on the Unit

In making fertilizer recommendations, the two preplant methods of application mentioned for each vegetable are broadcasting and digging in, and band application. The standard fertilizer application consists of about 3 pounds of a 10-10-10 fertilizer per 100 square feet or 75 feet of row area. Additional needs are given for each vegetable.

ASPARAGUS

Asparagus is among the earliest of spring vegetables to appear. It grows best in areas where the soil freezes to a depth of at least a few inches in winter. Asparagus, a perennial, may live for many years.

The asparagus crop may be grown on almost any well-drained fertile soil. First dig a trench 16 inches wide, laying aside the topsoil. Plow or spade lots of manure, leaf mold, rotted leaves, or peat into the subsoil to a depth of 14 to 16 inches (below the normal soil surface.) Mix in 5 to 10 pounds of a complete fertilizer per 75 feet of row (length). Fill the trench with topsoil to within 6 inches of the natural soil level. Set the asparagus crowns and cover with 1 to 2 inches of soil. As the shoots grow during the first season, gradually replace the topsoil in the trench until it reaches the natural soil level. Set the plants at least 1 1/2 feet apart in the row. Rows should be 4 to 5 feet apart.

Fertilizer

The first year after planting, side-dress with 1 pound of 5-10-10 fertilizer per 100 square feet of bed or 75 feet of row at the first cultivation. In established beds, apply 3 or 4 pounds of a 5-10-5 fertilizer per 100 square feet of bed or 75 feet of row just before harvesting begins in the spring.

Harvest

Remove no shoots the first two years after planting. Each following year, remove all the shoots until about July 1, when cutting must be stopped to allow the plant to replenish root reserves of plant food. Harvest when the spears are 6 to 10 inches long. To harvest, cut the spear at the base with a sharp knife. Harvest frequently to prevent tough spears.

Use

Most asparagus is cooked fresh. It may be frozen or canned.

Storage

Store by freezing or canning. It may be kept fresh for short periods of time at 32°F (0°C) and 85 to 90 percent relative humidity.

> In the fall, remove the dead plant tops and work them into the soil.

LIMA BEANS

Lima beans, figure 40-1, page 482, are a warm-season crop and should not be planted until the soil and the air have warmed sufficiently. Limas are able to grow in most soils. In heavy soils that tend to turn crusty on top, the seed should be covered with sand rather than soil so that the seeds can push through the soil more easily.

Lima beans require a growing season of about 4 months at a relatively high temperature. They need a richer soil than snap or green beans.

Limas grow both on poles and in bush form. The pole limas need some means of support such as poles or trellises. When poles are used, plant three or four plants per hill and space the hills 2 to 3 feet apart. When planted to grow on trellises, space the plants 6 to 12 inches apart. Bush varieties are planted 3 to 4 inches apart in rows spaced 3 feet apart.

Fig. 40-1 Burpee's Fordhook Lima beans (Courtesy W. Atlee Burpee Company)

Fertilizer

The ordinary application of fertilizer at planting time is usually enough. Lima beans should not be treated with a high nitrogen fertilizer and should not be side-dressed in late season.

Harvest

Harvest lima beans when the seeds are close to full size but before the color changes from green to white.

Use

Fresh, frozen, canned, dried.

Storage

Limas remain fresh at 32°F (0°C) and 85 to 90 percent relative humidity. They can also be frozen, canned, or dried.

SNAP BEANS

Snap beans, figure 40-2, should not be planted until the ground is warm. Successive plantings can be made every 2 to 3 weeks until 8 weeks before the fall frost date. Snap beans can be grown in a wide variety of soils, providing that the soil is well drained and not so heavy that it crusts, preventing seeds from emerging from the soil.

Fertilizer

The general garden fertilizer program before planting is usually sufficient. On light sandy soils, a sidedressing of a 5-10-5 fertilizer may be beneficial.

Fig. 40-2 Snap beans (Courtesy W. Atlee Burpee Company)

Fig. 40-3 Burpee's Redhart beet (Courtesy W. Atlee Burpee Company)

Harvest

Pick when the seeds in the pods are one-third their mature size and when the pods are still young and tender.

Use

Fresh cooked, frozen, canned.

Storage

Ideal fresh storage should be at 45° to 50°F (7° to 10°C) and at 85 to 90 percent relative humidity. They can also be frozen or canned.

BEETS

Beets, figure 40-3, are very hardy vegetables and grow under a wide range of climatic conditions. They will not however, tolerate severe freezing. Beets require a loose, loamy soil for best results.

Beets are planted in rows as close as 16 inches apart for hand hoeing. Plants are spaced 3 inches apart in the row. Successive plantings may be made at three-week intervals during the season. Seeds planted in hot, dry weather may not germinate well. Mulching or watering to keep the soil moist will help resolve this problem. Beets do not like acid soils. Check the soil pH and add lime if needed.

Fertilizer

Beets should be encouraged to maintain a steady growth pattern. A 5-10-10 fertilizer should be applied at the time of planting and another application made at one of the early cultivations.

Harvest

For tender beets, harvest when they are 1 1/2 to 2 inches in diameter. A fall crop which is to be stored may be allowed to grow a little larger.

Use

Fresh cooked, canned, pickled.

Storage

Store fresh at 32°F (0°C) and 90 to 95 percent relative humidity for best results.

Fig. 40-4 Broccoli (USDA photo)

BROCCOLI

If a spring crop of broccoli is desired, plant seeds indoors and set plants in the garden in March or April. Set the plants 12 to 18 inches apart in rows 3 feet apart. Plants begin to yield sprouts about 10 weeks after planting, figure 40-4. A fall crop may be direct seeded by sowing seed in June.

Fertilizer

Being a leafy vegetable, broccoli responds well to additional fertilizer. In addition to the fertilizer applied at the time of planting, broccoli responds to a sidedressing soon after planting. The sidedressing should be done with a high nitrogen fertilizer such as a 10-10-10 at a rate of 1 or 2 pounds per 100 feet of row.

Harvest

The broccoli head, which is a cluster of flower buds, should be harvested before the buds begin to open. Small side clusters develop after the main head is cut, and may be harvested over a long period of time.

Use

Fresh (as a greens crop), frozen.

Storage

For best results, store fresh broccoli at 32°F (0°C) and at 90 to 95 percent relative humidity.

> Do not plant in the same area in which cabbage, cauliflower, or brussels sprouts have been planted, since the same pests attack all these plants.

BRUSSELS SPROUTS

Being hardier than cabbage, brussels sprouts can survive outdoors over winter in all the mild sections of the country. They may be grown as a winter crop in the southern regions of the United States and as early and late cabbage is grown in northern sections. As the heads begin to form, break the lower leaves from the stem to give them more room. Do not remove top leaves.

Fertilizer

See Broccoli.

Harvest

Remove the sprouts when they are 1 to 1 1/4 inches in diameter by cutting them close to the stem.

Use

Fresh, frozen.

Storage

See Broccoli.

CABBAGE

Cabbage, figure 40-5, is one of the most important of the home garden crops. It can be grown throughout almost the entire United States. For a spring crop, set the plants in the garden as early as the soil can be prepared. For fall crops, plant again in late summer.

Cabbage may be planted after early potatoes, peas, beets, spinach, or other early crops, or it may be set between rows of crops before these crops are harvested. Set plants about 1 foot apart in rows 3 feet apart. Plantings must be very shallow.

Fig. 40-5 Cabbage (variety Earliana) (Courtesy W. Atlee Burpee Company)

Fertilizer

Fast growth of plants is necessary for good quality cabbage. Being a leaf crop, cabbage requires more nitrogen than many other plants. An additional sidedressing of a high nitrogen fertilizer, such as a 10-10-10, when the heads are half grown is very helpful.

Harvest

Harvest when the heads are firm, but before they split.

Use

Fresh (in salads and coleslaw), cooked as sauerkraut.

Storage

Fall cabbage may be stored in an outside pit or trench or in a cool vegetable cellar. Storage at 32°F (0°C) and at 90 to 95 percent relative humidity is best.

> Be sure to employ methods for control of the cabbageworm.

CARROT

Carrots, figure 40-6, are cold-hardy. Seed may be planted as early in spring as the ground can be worked. Carrot seed does not

Fig. 40-6 Carrots (Courtesy W. Atlee Burpee Company)

germinate well in hot, dry weather. The soil requirements for carrots are very similar to those of beets. They prefer a deep, loose, sandy loam which is free of stones.

Space plants no closer than 1 inch apart in the row. Cultivate by pulling loose soil in toward the plant to keep the crown covered. This prevents the top of the carrot from turning green.

Fertilizer

Carrots respond to nitrogen fertilizer. Apply 2 pounds of 10-10-10 fertilizer by band application beside the seed row before planting. Side-dress with 1 pound of a 10-10-10 or 10-6-4 fertilizer after the second or third cultivation.

Harvest

Harvest carrots when the plants are 1 to 1 1/2 inches in diameter, depending on the variety.

Use

Raw (alone or in salads), cooked fresh, canned.

Storage

Carrots must be dug and stored before frost kills the roots. Fresh carrots are stored best at 32°F (0°C) and at 90 to 95 percent relative humidity. They may also be stored outdoors covered with straw or other material to prevent freezing.

CAULIFLOWER

Cauliflower, figure 40-7, requires cultural conditions similar to that of cabbage. Cauliflower will not tolerate as much cold weather as cabbage, however. High temperatures cause poor quality heads. Cauliflower grows on any type of well-drained, fertile soil. Set the plants 1 1/2 to 2 feet apart in 3-foot rows. When the heads are 1 to 2 inches in diameter, tie the outer leaves around the heads to *blanch* them (remove color).

Fertilizer

Side-dress with a high nitrogen fertilizer. Add Borax to the fertilizer to supply the minor element boron, which cauliflower uses in large amounts.

Harvest

Harvest when the heads are firm and compact. Do not allow the heads to become "ricey" appearing.

Use

Raw, cooked, fresh.

Storage

Fresh storage at 32°F (0°C) and 85 to 90 percent relative humidity is best. May also be frozen and canned.

CELERY

Celery is a cool-season crop grown as a spring or fall crop in the upper southern and in northern sections of the country. It is a good winter crop in lower southern regions.

Soil which is rich, moist, well drained, deeply prepared, and loose is a must for celery. Plenty of organic material is also necessary. About 10 weeks is required to grow healthy plants from seed to transplanting size. Set plants 18 to 24 inches apart for hand cultivation. For tractor cultivation, space plants 6 inches apart in rows 3 feet apart.

Fig. 40-7 Cauliflower (Courtesy W. Atlee Burpee Company)

Fertilizer

At planting time, dig in 5 pounds of a 10-10-10 fertilizer per 100 feet of row. Mix thoroughly with the soil.

Harvest

Harvest when large enough to eat.

Use

Fresh, alone, in salads and other dishes.

Storage

For early winter use, bank soil around the plants and cover the tips with straw or leaves to prevent freezing. May also be stored in a cold frame or cellar with high humidity. Storage at 34°F (2°C) with 90 to 95 percent relative humidity is best.

CELERIAC

The cultivation of celeriac known as a turnip-rooted celery, is the same as that of celery. The root, rather than the top, is eaten.

Fertilizer

See Celery.

Harvest

Use the roots any time after they are big enough.

Storage

Store in the ground with straw or in cold storage.

CHARD

See Beets. The only difference is that the plants grow larger and require spacing of at least 6 inches apart in the row. Chard needs a rich, loose soil. It is sensitive to soil acidity.

CHERVIL

Salad chervil is grown as parsley is. The seed must be bedded in damp sand for a few weeks before planting, or germination is very slow.

Turnip-rooted chervil is planted in the fall in southern regions and may not appear until spring. In northern regions, it is either planted in the fall or seeds are started in flats to grow transplants for spring planting. For cultivation, see beets.

CHICORY, WITLOOF

Witloof chicory is grown for both its roots and tops. It is a hardy plant which tolerates both hot and cold climates. A rich, loamy soil without much organic matter is best.

Sow seeds in rows in late spring. Thin so that they are kept 6 to 8 inches apart in the rows. If sown too early, the plants go to seed, preventing later forcing.

Fertilizer

Work a good garden fertilizer into the soil at planting time. No additional fertilizer is needed.

Harvest

The tops are sometimes harvested when young.

Use

Top is used as cooking greens.

Storage

The roots are lifted in the fall and placed in moist soil in a warm cellar for forcing. Roots are covered with a few inches of sand. Under this covering, the leaves form a solid head known as a *witloof*.

THE ONION GROUP (CHIVES, GARLIC, LEEK, SHALLOT, ONION)

Onions thrive under a wide variety of climatic and soil conditions. A moist, warm climate is best.

The soil should be fertile, moist, and loose. It should also contain plenty of compost, have a fine texture, and be free of clods and foreign matter. Soil pH should be high for best results.

Onions are generally started as sets or as small dry onions grown the previous year from seed. Seedling onions have an advantage over sets in that they rarely go to seed and thus produce better bulbs.

Plant onion sets 3 inches apart in rows 1 foot apart. Plant 2 inches deep.

Fertilizer

A fertilizer high in phosphorus and potash, such as a 5-10-10, should be applied at the rate of 5 pounds per 100 square feet. Side-dress if the tops do not appear to be healthy, green, and actively growing.

Harvest

Onion stalks may be harvested young for salad greens or left in the ground to mature. The mature onions are pulled for storage after the tops have died down.

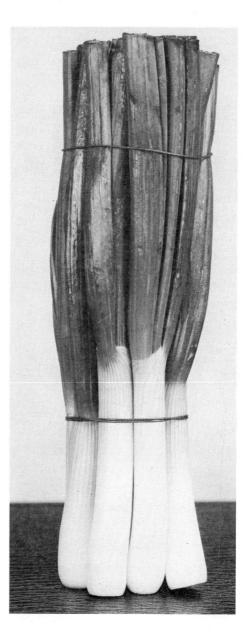

Fig. 40-8 Leeks harvested young for fresh consumption. (Courtesy W. Atlee Burpee Company)

Use

Fresh (in salads or alone), cooking, seasoning.

Storage

Store mature onions at 32°F (0°C) and in a relatively dry, well-aerated location at 70 percent relative humidity for best results.

The following is an explanation of members of the onion group only as they differ from onions.

Chives. Chives are perennials and should be planted where they can be left for more than one year. Divide and reset if plants become too thick. They are used primarily for seasoning.

Garlic. Large bulbs must be separated before planting, as each contains about ten small bulbs. Garlic is used primarily for seasoning.

Leek. Rather than forming a bulb, leek produces a thick, fleshy, stemlike onion, figure 40-8. Leek is started from seed and harvested any time it is large enough to eat. Store as celery is.

Shallot. The shallot is a small onion. Its bulbs have a milder flavor than that of most onions. They seldom form seeds and are propagated by small divisions into which the plant splits during growth. The shallot should be lifted and separated, and the smaller ones replanted, each year.

COLLARDS

Collards are grown and used similarly to cabbage. They withstand heat better than other members of the cabbage group.

Collards do not form a true head, but rather a rosette of leaves which may be tied together for blanching.

CORN, SWEET

Sweet corn, figure 40-9, requires a great deal of garden space and is best used only in larger gardens. It is planted in 3-foot rows, 12 to 14 inches apart in the row, or in hills of three plants per hill spaced 3 feet apart.

Corn is a tender crop and should not be planted until the danger of frost has past. It should not be planted in single rows. Plantings four rows wide are desirable, even if rows must be short. Sweet corn grows in almost any well-drained garden soil.

Fertilizer

Corn is a grass and has a high nitrogen requirement. About half the fertilizer used (3 pounds per 100 square feet of 10-10-10)

Fig. 40-9 Sweet corn (Courtesy W. Atlee Burpee Company)

Fig. 40-10 Cucumbers used in pickling. (Courtesy W. Atlee Burpee Company)

should be dug into the soil before planting or banded beside the row. Apply the rest of the 10-10-10 fertilizer as a sidedress every 3 weeks until silking of ears. This greatly increases yields.

Harvest

Sweet corn is ready to eat when milk squirts from the kernel as the thumbnail is pushed into the kernel. The kernels should be full size. Pull and cook immediately. Sugar content of corn decreases rapidly after it is pulled from the plant.

Use

Fresh as roasting ears, frozen, canned.

Storage

Storage of sweet corn is by freezing and canning. Fresh ears are best held for a short time at 32°F (0°C) and at 85 to 90 percent relative humidity.

CUCUMBER AND MUSKMELON (CANTALOUPE)

Cucumbers, figure 40-10, and muskmelons are warm-season crops. Hot weather may be too severe and greatly lower yields. In southern sections, they may be grown in the fall. A very fertile, loose soil high in rotted organic matter is necessary for best yields.

Seeds are planted in hills 6 feet apart, followed by thinning to four plants per hill, or in rows 6 feet apart with plants spaced 2 to 3 feet apart in the row. Cover seed with 1/2 inch of soil. If soil is heavy and becomes crusty, cover seed with sand. Plants may be started indoors in peat pots and transplanted to the garden for earlier crops.

> Control of the cucumber beetle is necessary for good crop yields. The beetle spreads bacterial wilt, which kills the plants.

Fertilizer

Heavy application of fertilizer is necessary. Apply 5 pounds of a 10-10-10 fertilizer for every 50 feet of row or for every ten hills. Mix the fertilizer into the top 8 to 10 inches of soil before planting.

Harvest

Cucumbers are harvested for fresh slicing while they are crisp and fresh and before any yellowing begins. Cucumbers used for pickling may be harvested at any size desired but before any yellowing begins. Pull all mature fruit off the plant, even if it is not wanted. Yield is reduced if fruit is allowed to remain on the vine.

Cantaloupe or muskmelon is harvested when the fruit separates from the stem with a slight tug or push with the thumb.

Use

Cucumber is used in salads and pickling. Cantaloupe is eaten as fresh fruit.

Storage

Fresh storage should be at 45°F (7°C) and 85 to 90 percent relative humidity for best results.

Fig. 40-11 Eggplant (Courtesy W. Atlee Burpee Company)

EGGPLANT

The eggplant, figure 40-11, demands warm weather and a warm soil. A growing season of from 100 to 140 days is needed for crop maturity.

Culture and fertilization are the same as that for tomatoes, except that eggplant may be planted closer together — 18 inches apart in the row. Eggplant is a highly productive plant; six plants will furnish enough for the average family.

Harvest

Harvest any time after proper size is reached and while the fruit is still glossy.

ENDIVE

For cultural requirements, see head lettuce. Endive is less sensitive to heat than head lettuce. In the south, it is a winter crop and in the north, a spring, summer, and fall crop. Tie leaves over the heart to shade and blanch the plant.

Use

See Lettuce.

Storage

Dig plants with a root ball attached and place in a cold frame or cellar where they will not freeze. Use as needed.

KALE

Kale is a member of the cabbage family. It is a hardy plant and lives over winter as far north as southern Pennsylvania. It is also resistant to heat and may be grown in summer. It is best grown as a cool-weather crop.

Plant in rows 12 inches apart and keep plants thinned to 6 inches apart in the row. Sow seed in early spring or in early August for a fall crop. Seed may also be broadcast and raked in. Cover seed with only 1/2 inch of soil.

Fertilizer

Dig about 3 pounds of a 10-10-10 fertilizer into the soil when seeding. A sidedress of fertilizer high in nitrogen halfway through the growing season helps to produce a tender and larger crop.

Harvest

Kale may be harvested by pulling the entire plant from the soil or by cutting off the outside leaves as the proper size is reached. Old leaves are tough and stringy.

Use

Cooked as greens, fresh, frozen.

KOHLRABI

Kohlrabi is grown for its swollen stem. For cultural requirements and uses, see cabbage.

Harvest

Harvest kohlrabi when it is young and tender.

LETTUCE

Lettuce, a cool-season crop, is very sensitive to heat. In the southern U.S., the growing of lettuce is continued into late fall, winter, and spring. In most other areas, it is a spring and fall crop.

Lettuce grows in any good garden soil. The soil should not be highly acidic, however; check soil pH before planting. Head lettuce plants are grown indoors and transplanted to the garden. Leaf lettuce may be direct seeded. It may be planted very early since temperatures as low as 28°F (–2°C) do not harm plants that have been properly hardened off. Set head lettuce plants 12 inches apart in rows 3 feet apart. Leaf lettuce plants may be planted much closer to each other.

Fertilizer

A fertilizer high in nitrogen and phosphorus such as a 10-10-10 should be dug into the soil at planting time or side-dressed along the rows at a rate of 3 pounds per 100 feet of row.

Harvest

Harvest head lettuce when the heads are firm or slightly earlier. Leaf lettuce is picked before the leaves become tough. The older leaves should be used first.

Use

Fresh, in salads and other dishes.

Storage

Lettuce is stored at 32°F (0°C) and 90 to 95 percent relative humidity for best results.

MUSTARD

Mustard grows in almost any type of soil. It is a cool-season crop and matures rapidly. Prepare as a fresh cooked greens dish, such as kale.

OKRA

See Tomato.

PARSLEY

Parsley is hardy in cold weather, but sensitive to heat. It thrives under the same conditions as kale and lettuce. It is important that parsley be planted early in the season. It grows in any good, loose garden soil. Since young parsley plants are slow to start, it is best to start them indoors in flats and transplant to the garden. Soaking seed overnight in water helps speed up germination.

Plant parsley in rows 14 inches apart with plants 6 inches apart in the row. No special fertilizer is needed.

Turnip-rooted parsley is grown for the root and has much the same flavor as celeriac. The cultural conditions are the same as for parsley. It is stored as other root crops are.

PEAS

Peas, figure 40-12, are a cool-season crop and can be seeded in the spring as soon as the ground can be prepared. A soil high in organic matter is best.

Plant in double rows 6 to 8 inches apart with just enough space between double rows to allow cultivation. Sow seeds 1 inch apart in the rows.

Fig. 40-12 Peas (Courtesy W. Atlee Burpee Company)

Fertilizer

Peas respond to high applications of fertilizer. Apply 3 pounds of a 10-10-10 fertilizer on each side of the row before planting and 2 pounds of the same fertilizer at the early blossom stage.

Harvest

Harvest when the pods are well filled but just before the seeds reach full size. Cook or process immediately after picking for best results.

Use

Fresh cooked.

Storage

Peas are especially good fresh from the garden. If they must be stored, they remain freshest at 32°F (0°C) and 85 to 90 percent relative humidity. Freeze or can for permanent storage.

PEPPERS

The cultural practices for peppers, figure 40-13, are the same as those for tomatoes, except that warmer soil is required before planting. Set plants 1 1/2 to 2 feet apart in rows 3 feet apart.

Fertilizer

Apply 2 pounds of a 10-10-10 fertilizer and dig it into the soil, or band it beside rows at planting. Apply a sidedress of fertilizer at first blossom with a 10-10-10 fertilizer. Use enough to keep plants actively growing.

Fig. 40-13 Sweet peppers (USDA photo)

Harvest

Harvest peppers when they reach the desired size.

Use

Fresh (in salads and other dishes) and pickled.

Storage

Store at 45° to 50°F (7° to 10°C) and at 85 to 90 percent relative humidity for best results.

POTATO

The potato is .one of the highest producing vegetables in terms of food per area of land. A cool-season crop, potatoes are planted as early in spring as possible or in mid to late summer for a fall crop. Any loose, well-drained, fertile soil is suitable. The soil should be moderately acidic.

Plant small (1 1/2 to 2 ounce), chunky rather than thin, cut pieces of potato with one or more eyes. Place the seed piece 3 inches deep in the soil and space them 10 to 12 inches apart in rows spaced 3 feet apart.

Fertilizer

Apply 10 to 15 pounds of 10-10-10 fertilizer per 100 feet of row. The fertilizer can be broadcast and worked into the soil or applied as a sidedressing beside the row at planting time, but should never touch the seed pieces. Sidedress applications of fertilizer may be made later if the tops do not appear dark green.

Harvest

Potatoes may be dug for immediate use as soon as they are large enough to eat. Do not dig potatoes that are to be stored until the tops are dead. Handle them carefully and avoid skinning. Do not leave potatoes exposed to light or they will turn green and be unfit for use.

Use

Cooked fresh.

Storage

Store potatoes at 40° to 45°F (4° to 7°C) in a well-ventilated area. Keep the storage area dark to prevent greening. If possible, curing at 50° to 60°F (10° to 16°C) in moist air before storage is an advantage.

RADISH

Radishes, figure 40-14, are hardy in cold weather but cannot tolerate heat. They may be grown in spring or fall. The crop matures from 3 to 6 weeks after planting, so a number of plantings can be made. A rich, loose, well-drained soil is best.

There are two types of radishes — the mild, small, quickly maturing type which reaches edible size in twenty to forty days, and the large winter radishes which require seventy-five days for growth.

Fig. 40-14 Radishes (Courtesy W. Atlee Burpee Company)

Fertilizer

Fertilize by digging 2 pounds of a 10-10-10 fertilizer per 100 feet of row into the soil at planting time. Side-dress if needed to maintain steady growth.

Harvest

The early, fast growing radishes are harvested as soon as they have reached a desirable size for storage.

Use

Fresh (in salads).

Storage

Store as other root crops are stored.

RUTABAGA AND TURNIP

Rutabagas and turnips, cool-season crops, are similar in terms of soil and fertilizer requirements and ways in which they are used as food items. They are both grown in southern regions chiefly in the fall, winter, and spring; in northern parts of the country, they are grown mostly in the spring and autumn. Rutabagas grow best in the northern regions; turnips are better for gardens south of Indianapolis, Indiana or northern Virginia. A half ounce of seed of these vegetables allows for the broadcasting of 300 square feet.

Turnips require sixty to eighty days to mature; rutabagas require ninety to 110 days.

Fertilizer

If planted after early potatoes or other early crops which have been fertilized, no additional fertilizer is needed.

Harvest

Harvest as soon as the roots are large enough to eat.

Use

Fresh cooked.

Storage

Storage is the same as for other root crops — 32°F (0°C) at 90 to 95 percent relative humidity.

SPINACH

Spinach, figure 40-15, is a hardy, cool-weather crop. It is able to withstand winters in the southern regions and is an early spring and late fall crop in the northern regions. Seed should be planted as early in the spring as possible. Sow the seed in rows 12 to 30 inches apart (depending upon the method of cultivation). Plants should be thinned so that they stand 2 to 4 inches apart in the row.

Spinach will grow in almost any well-drained soil. Check the requirement for spinach and test the soil.

Fertilizer

Apply 3 or 4 pounds of a 10-10-10 fertilizer per 100 square feet at planting time and work into the soil. Side-dress with nitrogen to keep the plants actively growing.

Harvest

Harvest as leaves become mature enough to eat.

Use

Cook fresh (as greens).

Storage

Spinach keeps freshest at 32°F (0°C) and 90 to 95 percent relative humidity.

NEW ZEALAND SPINACH

New Zealand spinach is not related to common spinach. It is generally grown as a substitute for common spinach in hot weather. The plants are larger than common spinach and must be spaced 18 inches apart in rows that are 3 feet apart. Plant seed as soon as frost danger is past.

Fig. 40-15 Spinach (Courtesy W. Atlee Burpee Company)

Fig. 40-16 Golden zucchini squash (Courtesy W. Atlee Burpee Company)

SQUASH

The squash, figure 40-16, is a hardier crop than melons and cucumbers. It is, however, a warm-season crop.

Summer squash, which includes all bush types, should be eaten while young and tender. This group consists of scallops, yellow straightneck, yellow crookneck, and the vegetable marrows such as cocozelle and zucchini. Summer squash plants are heavy producers.

Winter squash consists principally of vine types. They have hard rinds, making them easy to store. Winter squash includes such varieties as Boston marrow, table queen, Hubbard, delicious, and butternut.

Plant bush squash 2 to 3 feet apart in rows 3 to 4 feet apart. Plant vine squash 4 to 6 feet apart in rows 6 to 8 feet apart.

Fertilizer

Squash responds well to applications of manure and fertilizer. Dig manure and liberal amounts of fertilizer high in potash, such as 5-10-15, into the soil before planting. Side-dress at the first blossom with a 10-10-10 fertilizer.

Harvest

Harvest summer squash while the skin is still tender. Winter squash must be allowed a longer time to mature and is harvested just before the first killing frost.

Use

Fresh cooked.

Storage

Store summer squash at 32° to 40°F (0° to 4°C) and at 85 to 90 percent relative humidity. Winter squash is stored at 50° to 55°F (10° to 13°C) and at 70 to 75 percent relative humidity.

SWEET POTATO

Sweet potatoes, figure 40-17, grow best in warm weather. They are grown in gardens as far north as southern New York and southern Michigan. A frost-free period of 150 days with relatively high temperatures is necessary for their cultivation. Sprouts or slips should not be set out in the garden until all danger of freezing is past. The home gardener should purchase the sprouts for planting, and not attempt to grow them.

Plant in rows 3 to 3 1/2 feet apart and space plants 12 to 15 inches apart in the row. Sweet potato plants are generally set on top of ridges. A well-drained, sandy loam soil which is moderately deep is best.

Fertilizer

Moderate use of a fertilizer high in potash, such as a 5-10-15, is best. Dig and mix half the fertilizer into the soil at planting time (1 pound per 50 feet of row) and apply the rest as a sidedress during July.

Harvest

Harvest sweet potatoes just before the first frost. Allow the roots to dry before storage. Be careful not to bruise or injure the potatoes in any way.

Use

Fresh cooked.

Fig. 40-17 Sweet potatoes (USDA photo)

Storage

Sweet potatoes should be allowed to cure for one to two weeks at a relatively high temperature (85°F). Storage is best at 50° to 55°F (10° to 13°C) and 85 to 90 percent relative humidity.

Fig. 40-18 Tomatoes (USDA photo)

TOMATO

The tomato, figure 40-18, is probably the most popular garden vegetable. It is a warm weather crop and a highly productive plant. Tomatoes are planted in the garden only after the danger of frost is past. Space the plants 3 feet apart in rows which are 3 feet apart. Set the plants deeper in the garden than they were growing before transplanting. Plants are sometimes staked or caged for protection. A mulch is of great value in conserving moisture. Installation of a cutworm collar around each plant is necessary. Try the new wire caging method along with a heavy mulch for almost carefree tomato production.

Fertilizer

Before planting, dig compost or other organic matter and 5 pounds of a 5-10-10 fertilizer per 100 square feet of area into the soil. Side-dress when early fruit sets with a 10-10-10 fertilizer. Limestone placed in the hole at planting time may help prevent blossom-end rot (caused by a calcium deficiency.)

Harvest

Harvest tomatoes as they turn red, but before the fruit becomes soft. For canning, the fruit should be fully red and fully ripe.

Use

Fresh (in salads), cooked fresh, canned, and for making catsup and tomato paste.

Storage

Ripe fruit is stored ideally at 50°F (10°C) and at 85 to 90 percent relative humidity. Green mature fruit stores best at 55° to 70°F (13° to 21°C)

WATERMELON

Watermelons, figure 40-19, require a great deal of garden space. They are more sensitive to cold than cucumbers or cantaloupes. Soil conditions are the same as for cucumbers.

Plant watermelons in hills 8 feet apart. Plant 4 to 6 seeds per hill and thin to two or three plants per hill. Mix about a bushel of organic material in the soil of each hill.

Fertilizer

Mix a handful of a 10-10-10 fertilizer with the organic matter applied to each hill.

Harvest

Watermelons are ready to harvest when the tendrils nearest the fruit die and the side of the fruit nearest the ground changes from white to creamy yellow.

Use

Fresh (as a dessert or in salads); rinds may be pickled.

Fig. 40-19 Watermelons (Courtesy W. Atlee Burpee Company)

STUDENT ACTIVITY

1. Plant a vegetable garden as a class project. Examine a seed catalog and select the varieties best suited for the area. (Local extension and experiment station bulletins give recommendations of varieties to plant.) Meet as a class the following fall to discuss reasons for the success or failure of the garden.

SELF-EVALUATION

Select five vegetables that are grown in the local area from the vegetables discussed in this unit. For each vegetable, list cultural requirements, including soil preparation and special needs of the plant, and how the vegetable is harvested.

Section 12
The Small
Fruit Garden

OBJECTIVE

To grow strawberries of marketable quality in the home garden and to become familiar with commercial strawberry-producing areas in the United States.

COMPETENCY TO BE DEVELOPED

After studying this unit, the student will be able to

- select a site for a strawberry planting and properly prepare the soil for planting.
- diagram four planting systems for strawberries.
- list the seven points to consider when selecting a strawberry variety to plant.
- list the seven steps in planting strawberries.
- explain the cultural practices of weed control, mulching, and fertilizing as they relate to care of strawberry plantings.

unit 41
Strawberries

Materials List	
extension bulletins on strawberry culture in the local area	
strawberry plants for identification of plant parts	
If a planting site is available:	
plants	plant labels
fertilizer	mulch or cultivator or hoe
string	tape measure
shovel	measuring sticks cut to proper plant spacing length

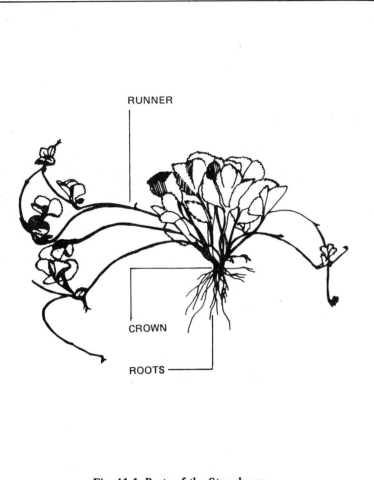

RUNNER

CROWN

ROOTS

Fig. 41-1 Parts of the Strawberry

The strawberry, figure 41-1, is a perennial plant. The top dies back to the crown each fall in areas where winter temperatures drop below freezing. The roots live from year to year and new leaves and blossoms grow from the crown each spring. The new fruit is produced from these blossoms.

Runners are the means by which strawberries naturally propagate. Plants normally send out runners from early summer until fall. Each runner roots and forms new plants. Everbearing plants that produce berries all summer have few if any runners.

STRAWBERRY PRODUCTION

The strawberry is one of the first fresh fruits to be marketed in spring and is usually in high demand at this time. Strawberry plants are high producers on a per acre basis and are considered to be a high cash crop.

Strawberries are grown nationwide, but there are two large commercial areas: (1) The Pacific Coast states and (2) the eastern United States. Approximately 65 percent of all strawberries grown in the U.S. are grown on the Pacific Coast. There are some major differences in cultural requirements between the two growing areas and within the Pacific Coast area.

GROWING AREAS ON THE PACIFIC COAST

California

Strawberries are grown in California by means of a hill system. The plants are kept in cold storage until summer, when they are set in the field. They are generally set in double rows. Production the first year is heavy.

Summer plantings are irrigated to keep salt accumulation in the soil minimal. All plantings are mulched with clear polyethylene.

Central Coast

This area includes Santa Cruz, Monterey, and Santa Clara counties. Plants are planted in early August and do not produce runners after the first year because of warm winters. Harvest season is from April until November.

Central Valleys (near Fresno)

This area experiences much cold weather in winter and hotter than average weather in summer. The harvest season is April, May, and part of June. Plants are set in the fields in mid-July.

Santa Maria

This area includes Santa Barbara and San Luis Obispo Counties. Harvest begins in April and continues throughout the summer and fall. Plants are set in the field in both winter and summer.

Southern California

This area includes the Oxnard coastal plain of Ventura County and the coastal regions of the counties of Los Angeles, Orange, and San Diego. Plants are harvested for only one year. Plants are set in the field both in summer and winter.

Oregon

The Northern Willamette Valley has the largest acreage of strawberries of any single area in the United States. Harvest is from May until early July. Many berries are also grown in the Hood River Valley.

Washington

The main strawberry producing area in Washington is the Puget Sound region. Harvest is from early June until mid-July. Southwestern Washington also produces strawberries commercially.

Strawberry varieties grown in the Pacific Coast states are the spring- or June-bearing varieties. These varieties produce flower buds in late summer and bear flowers and fruit the following spring.

Fig. 41-2 Strawberries ready for the market (USDA photo)

Since cultural requirements differ from area to area, specific recommendations should be obtained from the local extension service or a horticulturist. For a specific description of strawberry varieties, consult Bulletin 1043 *Strawberry Varieties in the United States*, available from local extension offices.

GROWING AREAS IN THE EASTERN UNITED STATES

This growing area includes all regions east of the Great Plains and north of the Southern Coastal Plains. Cultural practices are generally the same for this entire area.

In the eastern United States, dormant plants are usually set in early spring. Runners grow from the original plants starting in

June and a succession of new runners grow and take root around the original plant. Flowers are removed from the plants the first year, causing more vigorous growth. Fruit buds are formed in the fall and, by the end of October, can be seen in the crown. Fruit buds also develop in leaf axils of vigorous plants.

The number of leaves on a plant in the fall is a good indication of what the following year's production will be like. The more leaves there are, the more berries will result. The crop is generally harvested in late May and June. Most of these berries are sold fresh.

The following information applies to both the Pacific Coast areas and the eastern United States.

SELECTING THE PLANTING SITE

Soil

Strawberries grow best on a loam or sandy loam soil. Clayey loam or silty loam soils are satisfactory if well drained. If planted on a very sandy soil, plantings should be fruited (allowed to bear fruit) only for one season.

It is very important to supply the soil with an abundance of organic matter. This can be added by plowing under a crop such as rye or by applying barnyard manure. Care should be taken not to plow under weed seeds. (Chemical weed control has helped to decrease this problem.)

Drainage

Drainage is very important for good growth of strawberry plants. The fungus disease red stele is almost always present in wet soils. It attacks the plant roots and kills the plant.

Frost

Strawberries cannot be grown in low areas where frost is a problem during blossoming time. Cold air collects in low areas in the same way that water does; frost damage in these pockets can be severe.

Diseases

Avoid soils infested with red stele root rot or areas below infected fields where water may wash the disease into the planting site. Surecrop, Midway, and Sunrise are strawberry varieties that are resistant to the common types of red stele. These varieties may be grown where the disease is a problem. Check for other, more recently developed varieties for local areas.

If tomatoes, potatoes, or peppers have been grown in a soil recently, verticillium wilt is probably present. If this is the case, plant only a disease resistant variety such as the variety Surecrop.

SOIL PREPARATION

Soil preparation for planting strawberries should begin one or two years before the actual planting date. Soil tests should be made to determine soil pH or acidity. Strawberries grow best at a pH of 6.0 to 6.5. If it is necessary, lime or gypsum should be added to correct the soil pH at least six months or, better yet, one year before planting. This allows time for the lime or gypsum to dissolve and correct the soil pH before plants are set in the field. If magnesium is low, apply dolomitic limestone.

Fertile soils high in organic matter are necessary for top crop yield. Large quantities of animal or green manure should be plowed under to raise the organic matter level one to two years before planting. However, heavy green manure crops should not be plowed under just before planting strawberries — they should have time to partially decay so that large loose air pockets are not left in the soil.

In preparation for planting, the soil is plowed to a depth of 7 to 9 inches. If the soil test shows phosphorus and potassium to be low, both should be added to correct the deficiency before planting. Any phosphorus or potash should be either plowed under or mixed into the soil after plowing. The soil should be worked until it is firm and free of large clods and air pockets. The soil must pack firmly around the new plants.

Land that was in sod for years should not be directly planted with strawberries. Such soils may contain root-eating grubs which destroy new plants. It is best to plant a sod area with a cultivated crop, such as corn, for a year before planting strawberries.

Areas which contain nutgrass, quackgrass, Bermuda grass, Johnson grass, or other perennial weeds should be avoided. Chemical weed control can help solve this problem.

CHOOSING VARIETIES TO PLANT

The selection of varieties to plant depends on climate, soil, and the purpose for which the crop is grown. The following are other factors to be considered.

- A variety that produces firm berries is needed for shipping to market.
- Large, softer textured varieties for freezing have a bright red color, firm flesh, and a tart flavor.
- The fruit must ripen when the best market prices are available.
- Plants that bloom late in the season miss spring frosts that could destroy the crop.
- If planting for the wholesale market, varieties that are well known are planted, yielding enough volume to provide truckload lots. This ensures a better price for the fruit.

- For the home garden and local roadside market, three or more varieties that ripen at different times are planted. This supplies fresh fruit over a longer picking season.

PURCHASING PLANTS

Purchase plants from a reputable nursery and be certain they are disease and nematode free. Nursery plants should be certified true to variety and free of insects and signs of disease. Be sure that the plants are fresh upon arrival. The roots should appear bright and plump. If roots appear dry, soak the roots in water for an hour before planting or heeling in.

If plants cannot be planted immediately, either heel them in the soil outside or store them in cold storage with the roots packed in moist sphagnum moss or other packing.

Heeling in is done by making a V-shaped trench deep enough for the roots to spread out with the crowns at ground level. The plants are then placed along one edge of the trench so that the roots are separated and the trench is filled in and packed around the roots. The plants can be kept in the trench until planted to the field, figure 41-3, page 508.

If cold storage is to be used, store plants that are not dormant at 40°F (4°C) and dormant plants at 32° to 36°F (0° to 2°C). When stored at the proper temperature, dormant plants can be kept for a longer period of time. When using cold storage, first wrap the plants in polyethylene plastic to keep them from drying out.

PLANTING TIME

In the eastern United States, the best time to plant is early spring. Moisture is usually available from natural rainfall and the

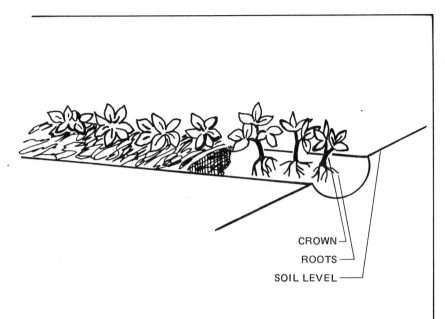

CROWN
ROOTS
SOIL LEVEL

Fig. 41-3 The Technique of Heeling in. The plants are placed in the trench with the crown at soil level. The trench is then filled with soil and packed down.

temperatures are cool at this time. Pacific Coast regions are planted at differing times, as discussed earlier.

PLANTING STRAWBERRIES

Planting Systems

There are three training systems commonly used in the eastern United States.

Hill System (No runners are allowed to grow.) In the hill system, plantings are made in double or triple rows. Rows are spaced 12 inches apart and plants in the row are 10 to 12 inches apart. A 24-inch space is left between the double rows, figure 41-4. Plant-

ing with triple rows is done in the same way except that there are three rows between the 24-inch alleys. The double-row system allows for 29,000 plants per acre; the triple row system, 32,670 plants.

The unwanted runners are generally cut off by machines that then throw them into the alley between rows where they are cut by sharp discs.

Spaced Matted-Row System (Some runners are allowed to grow to fill the rows in.) In this system, plants are set 18 to 24 inches apart in the row, and a 42-inch space is left between rows. An acre allows for 6,225 plants if they are spaced 24 inches apart in the row and 8,300 plants if the 18-inch spacing is used, figure 41-5.

Matted-Row System (Most runners are allowed to grow.) Plant spacing for the matted-row system is the same as for the spaced matted-row system. The difference is that all runners are allowed to grow freely and fill in the rows. Plants become crowded in this system, and yields and individual fruit size may be smaller than under the spaced matted-row system.

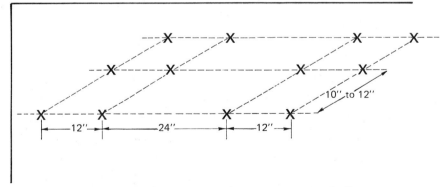

Fig. 41-4 Double-Row Hill System. In this system, a 24-inch alley separates two rows that are 12 inches apart. Plants in the rows are 10 to 12 inches apart.

Double Rows on Raised Beds. This is the most common method of planting in California. The system consists of raised rows spaced about 38 to 44 inches from center to center. The two rows in each bed are spaced 8 to 12 inches apart and plants are set in the rows 8 to 12 inches apart.

In the Pacific Northwest, plants are set in single rows on raised beds 39 to 42 inches apart with plants spaced 12 to 14 inches apart in the rows.

Steps in Planting

Regardless of the system of planting the general steps in the procedure are the same.

1. Cultivate the soil immediately before planting to kill germinating weed seeds. Be sure all clods are broken up and that the soil is firm.

2. Lay out rows according to spacing requirements. Rows must be opened deep enough to plant the roots straight down.

3. Keep the plant roots moist before planting. Never expose roots to the drying sun or wind.

4. Set the plant in the row at the proper depth with the crown just at ground level, figure 41-6. This is very important. Plants set too deep smother and die; plants set too shallow dry out.

5. Fan the roots out and straighten them. If roots are too long, they may be trimmed to a length as short as 4 inches.

6. Pack the soil around the roots by stepping on the soil and pressing. Do not cover the crown and do not leave any roots exposed.

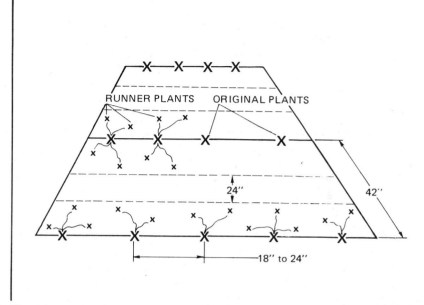

Fig. 41-5 Spaced Matted-Row System. Plants are spaced 18 to 24 inches apart in rows 42 inches apart. Runners are allowed to grow so that plants are eventually spaced 6 inches apart. A 24-inch alley is left between runner plants.

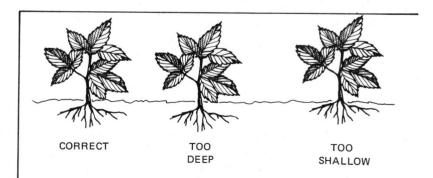

Fig. 41-6 Placing Strawberry Plants at the Proper Depth in the Ground.

7. If possible, water immediately after planting.

Many plants are set by machine in large operations. For small operations, a two-person team with a spade and buckets to carry plants is sufficient. One person inserts the spade in the ground and pushes it forward. The second person places a plant in the hole, after which the first person withdraws the spade and firms the soil around the plant with the foot.

Further watering may not be needed in the East, but frequent irrigation during the summer is generally needed in all areas of California.

CULTIVATION

Plantings should be cultivated immediately after they are placed in the soil. This loosens the soil and kills germinating weed seeds. Cultivation should be shallow (no more than 1 or 2 inches) and far enough away from the plant that the crowns are not exposed or covered too deeply by soil thrown by the cultivator. After this first cultivation, all commercial plantings should be treated with weed killers. Additional cultivation may be necessary when the chemical weed control begins to fail. If Dacthal is used for weed control, this occurs in about two months.

Any plants that die may be replaced by placing runners in the location and allowing them to take root.

WEED CONTROL

To obtain good yield of quality fruit, some method of weed control must usually be employed. Hand weeding is too expensive a method to use in strawberry production. The combination of mechanical cultivation and use of weed killers is the most practical method. Black plastic mulch is effective in controlling weeds, and also conserves moisture and helps control fruit rot.

In California, soil fumigation with methyl bromide before planting is used to control weeds and nematodes. Because of the many different climatic and soil conditions that affect weed control with chemicals, no specific control measures are given here. Contact a local extension service or horticulturist for recommendations.

MULCHING

Mulching is extensively used in the eastern growing regions. Mulching helps control freezing and thawing of the soil, which pushes plants out of the soil and exposes roots to drying.

Mulching with natural material also helps in weed control if it is applied so that it is thick enough, figure 41-7. It helps

Fig. 41-7 A strawberry patch planted with single rows of plants, well mulched for weed control.

conserve moisture and keeps the fruit clean. Small-grain straw makes the best mulch, but almost anything that does not contain weed seeds and stays in place is suitable. Black plastic also makes an excellent mulch.

Clear polyethylene bed mulch is essential to successful strawberry growing in California. Plastic mulching takes much of the danger out of winter planting by increasing soil temperature as much as 10 degrees. This results in earlier ripening of fruit, larger harvests, and longer harvest seasons.

FERTILIZING AFTER PLANTING

After planting, fertilizers should be used only in the late summer or early fall to increase the number of fruit buds in spring-bearing plants. Fertilizer containing nitrogen, applied in the spring, generally results in soft berries lacking in flavor, and should not be used.

FROST CONTROL

Planting on high ground prevents most frost damage to spring blossoms. If frost should occur during blooming time, an irrigation sprinkler system turned on during the time freezing temperatures exist will prevent the blossoms from freezing. Plants mulched with straw can be protected from frost by pulling the mulch back over the plants on nights when frost is expected.

WINTER KILL

In the eastern regions of the U.S., freezing and thawing of the soil heaves plants out of the ground, resulting in drying of roots and death of the plants. A heavy mulch applied after the plants have been hardened off to winter conditions (usually after a temperature of 20°F has been reached) will greatly reduce this loss. The plants should be completely covered to a depth of 3 or 4 inches.

HARVESTING

All strawberries are picked by hand since no machine is currently available that can do the job properly. Poor harvesting techniques can ruin even the best berries. If handled roughly, berries are bruised. If stems are too short, they puncture other fruit in the container. In eastern growing areas, strawberries are almost always picked with a 1/2-inch stem attached. The berry should not be pulled from the vine; the stem should be pinched off instead. Pulling breaks the skin on the berry and quality drops rapidly. In California, many berries are picked without stems for the freezing market. The fruit should be ripe for best flavor, but not too soft.

CARE OF THE PLANTING AFTER HARVEST

Weed Control

Weeds must be controlled after harvest either by mechanical cultivation and hand pulling, by mulching, or by chemical weed control. Failure to control weeds results in reduced crop production the following year.

Renewal (Revitalizing the Planting)

The root system of strawberries is often weakened while the fruit crop is maturing. To correct this situation, many growers mow the tops off the plants immediately after harvest. As soon as the tops dry, the area between the rows is narrowed down so that it is no wider than 18 inches. Plants are thinned out within the rows with a hoe or spike-toothed harrow. Proper weed control must follow this operation.

STUDENT ACTIVITY

Plan and care for a patch of strawberries at home or on the school grounds.

SELF-EVALUATION

Select the best answer from the choices offered to complete the statement.

1. Most of the strawberries grown in the eastern regions of the United States are sold

   a. on the fresh market.

   b. for processing.

   c. in summer and fall.

   d. in carload lots.

2. Strawberries grow best on a loamy soil that is high in

   a. calcium.

   b. phosphorus.

   c. organic matter.

   d. sand content.

3. If strawberries must be planted in a soil infested with red stele disease, the grower should

   a. drain the soil to remove excess water.

   b. add as much organic matter as possible.

   c. sterilize the soil.

   d. plant only disease-resistant varieties.

4. Strawberries should not be planted in soil recently planted with tomatoes, potatoes, peppers, or eggplant because

   a. these plants spread verticillium wilt which would attack the strawberry plants.

   b. these plants spread red stele disease which would attack the strawberry plants.

   c. these plants leave a toxic chemical in the soil.

   d. none of the above

5. Strawberries grow best at a soil pH of

   a. 5.5 to 6.0.

   b. 6.5 to 7.0.

   c. 6.0 to 6.5.

   d. 5.0 to 5.5.

6. If lime is necessary to raise the pH of soil in which strawberries are grown, it should be added

   a. two years before planting strawberries.
   b. from six months to a year before planting strawberries.
   c. at the time of planting.
   d. as a sidedress after plants are set.

7. If land that has been sodded for years is plowed and planted directly with strawberries,

   a. too much organic matter would be present.
   b. there is a danger of root damage from grubs.
   c. weed control would be almost impossible.
   d. the plant food supply will require a great increase.

8. In the hill system of planting strawberries, runners are

   a. hand placed to widen the hills.
   b. allowed to grow in limited numbers.
   c. allowed to grow freely.
   d. not allowed to root and grow.

9. In the matted-row system of planting strawberries in the East, the planting

   a. tends to become overcrowded with plants.
   b. requires a great deal of hand placement and removal of runners.
   c. usually produces the largest berries of any system.
   d. remains productive for more years than any other system.

10. The most important reason for cultivating or harrowing the soil just before planting strawberries is

    a. to level the soil.   c. to firm the soil.
    b. to mix in fertilizer and lime.   d. to kill germinating weed seeds.

11. After the first cultivation of newly set strawberry plants, all commercial plantings should be treated with

    a. a soil fumigant.   c. an insecticide.
    b. a weed killer.   d. a fungicide.

12. Mulching of strawberry plantings with straw or other material is used in eastern regions of the U.S.

    a. to prevent winter freezing and thawing and to control weeds.
    b. to keep fruit clean and control fungus diseases.
    c. to control insects and field mice.
    d. to force early blooming and protect against frost.

13. Nitrogen or a complete fertilizer is applied to spring-bearing strawberry plantings in early fall to

    a. produce larger berries.
    b. encourage runner formation.
    c. encourage setting of a greater number of fruit buds.
    d. none of the above

14. Strawberries for the fresh fruit market are generally picked with

    a. the cap and a 1/2-inch piece of stem connected.
    b. machines.
    c. the cap and stem removed.
    d. the cap on the stem removed.

15. In the renewal process, the tops are mowed off all plants in the strawberry planting and the plants are

    a. fertilized.   c. thinned out.
    b. sprayed for disease control.   d. none of these

OBJECTIVE

To plant and grow marketable blueberries.

COMPETENCY TO BE DEVELOPED

After studying this unit, the student will be able to

- list the six areas of the country in which blueberries grow naturally and the type of berry grown in each area.
- identify the two types of commercially grown blueberries and select the one best suited for growing in the local area.
- determine the soil conditions necessary for best highbush blueberry production by listing the three main soil requirements.
- draw to scale a diagram of a blueberry planting.
- plant, mulch, fertilize, spray, and prune one variety of blueberry.

unit 42 Blueberries

Materials List

rooting frame to propagate blueberry plants

blueberry cutting wood

propagating knife; pruner; sprayer

rooting hormone

rooting media (sphagnum moss and sand or perlite)

soil test kit or soil sample boxes

graph paper

Blueberries are one of the easiest and most rewarding small fruits to grow. If the soil pH can be adjusted so that it is low

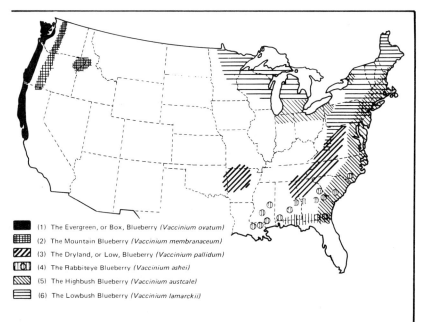

(1) The Evergreen, or Box, Blueberry *(Vaccinium ovatum)*
(2) The Mountain Blueberry *(Vaccinium membranaceum)*
(3) The Dryland, or Low, Blueberry *(Vaccinium pallidum)*
(4) The Rabbiteye Blueberry *(Vaccinium ashei)*
(5) The Highbush Blueberry *(Vaccinium austcale)*
(6) The Lowbush Blueberry *(Vaccinium lamarckii)*

Fig. 42-1 Map of the United States, showing areas in which blueberries are extensively harvested. (USDA map)

enough to support the acid soil requirement of blueberries, cultural practices should not be difficult. In addition to being a fruit plant, blueberries make an attractive landscape plant either as single plants or planted in a hedge. The spring blossoms are very fragrant and the fall foliage is a beautiful red. The berries attract birds who come to feed on them.

It is much easier to cultivate blueberries in areas of the country where they grow naturally. Any commercial production is usually restricted to these areas. Figure 42-1 is a map showing the areas in the United States where blueberries grow in the wild. Many blueberries are harvested for sale from wild plants growing in the regions shown in the map. There are two major cultivated species of blueberries, the highbush and the rabbiteye.

HIGHBUSH BLUEBERRIES

Location

The highbush, figure 42-2, grows naturally in an area from southeastern North Carolina to southern Maine and westward to southern Michigan. The rabbiteye blueberry has a natural range in southern Georgia, southern Alabama, and northern Florida.

Highbush blueberries are planted where the soil is moist and acid. The berries have better flavor where the days are long and the nights are cool during the ripening season. Since blueberries bloom early, plantings should not be made in low areas where late spring frosts may kill the blossoms. The plants are not winter-hardy to temperatures below −20°F (−7°C).

Highbush blueberries do require some winter cold to break the winter rest period and to blossom normally. They are not planted farther south than northern Georgia and northern Louisiana for this reason.

Soil

The highbush blueberry grows best in moist, acidic soil. A pH range of 4.3 to 4.8 is best. If soil pH is lower (more acidic) than this, add ground limestone to raise the pH. If the pH is higher than 4.8, add finely ground sulfur or fertilize with ammonium sulfate to lower the pH (make the soil more acid).

The soil must have good drainage. If water stands on the soil surface for several hours after a hard rain or for extended periods of time in winter or early spring, the area is not right for blueberries.

The best indication that blueberries may succeed in a soil is the presence of other native plants that grow best in acidic soil. If wild blueberries, huckleberries, azaleas, or laurel grow in an area, chances are good that cultivated blueberries will grow well there.

The soil should be high in organic matter. The plowing under of large amounts of humus or green manure crops is very helpful. This material helps in moisture retention and in the development of good soil structure. Natural sand-peat mixtures make excellent blueberry soils. They drain well and the peat content holds moisture. The soil should be plowed and harrowed and rows established with a layoff plow for easier planting. (A layoff plow opens a trenchlike row in the soil.)

Planting

Plants are spaced 4 or 5 feet apart in rows that are spaced 9 or 10 feet apart. Spacing plants as such means that there are about 1,100 plants per acre. Two-year-old plants are generally used and are planted at the same depth at which they grew in the nursery row.

The planting is usually done as early in the spring as the soil can be worked. In New Jersey, some plants are set in the fall, and in eastern North Carolina, planting is done in late fall and winter.

Plant roots must not be exposed to drying wind or sun during the planting process. To prevent this, keep plants covered with moist burlap or some other covering. When planted, the roots of the plant are placed in holes big enough for them to fit in without bending or twisting around in the hole. Firm the soil around the plant roots and water if possible. Plants are cut back one-third to one-half at planting time.

Fertilizer

Soil pH or acidity should be adjusted, if necessary, six months before planting by mixing either lime or sulfur into the soil, depending on which way the pH is to be adjusted. If the soil tests low in phosphorus or potash, these elements should also be

Fig. 42-2 The highbush blueberry variety Earliblue. Plants of this variety do not drop fruit as it ripens. (USDA photo)

plowed down or mixed into the soil prior to planting. Use a soil test to determine how much fertilizer is needed.

The following fertilizer schedule is suggested after planting.

1. After buds start to swell: 400-500 pounds of a 10-10-10 fertilizer per acre (not neutralized so that it is acidic). Acidic fertilizer helps to keep the soil acidic.

2. Six weeks later: If soil pH is 4.8 or higher, 100 pounds of ammonium sulfate adds nitrogen and helps keep the soil acidic. If plants are growing vigorously, this second application and additional fertilizer use may not be necessary. On very sandy soils where nitrogen leaches rapidly, a second application of ammonium sulfate six weeks after the application described in (2) is recommended.

The fertilizer is broadcast around the plant to within 6 inches of the plant and out as far as the roots extend. This is usually a foot or more beyond the spread of the outer branches. Older plant roots extend farther out than younger plants.

Irrigation

The highbush blueberry is not drought resistant. During periods of dry weather, 1 to 2 inches of water per acre should be added every ten days to two weeks during the picking season. Do not water any less than this amount at a single application. Shallow watering encourages shallow-rooted plants which fail to survive during periods of stress and winter months.

Mulches

Mulching with straw, sawdust, wood chips, or other material helps conserve soil moisture and control weeds and erosion. These materials should be added deeply enough to shade the ground and to prevent the surfacing of weeds. A fine material such as sawdust should be applied only 2 or 3 inches thick, but coarse, open material such as straw should be up to 4 or 5 inches thick.

When organic mulches are used, especially sawdust, additional nitrogen fertilizer must be applied to maintain good growth. This is because the rotting process of organic mulches robs the soil of nitrogen.

Pollination

Blueberries are not self-fruitful or self-pollinating. Two or more varieties must be planted within two rows of each other to ensure good pollination. Honeybees should be placed within 200 yards of the plants to carry out the pollination. Wild insect pollinators are generally not present in sufficient numbers to assure proper set of the fruit.

Pruning

Fruit is produced on wood of the previous season's growth. The largest fruit is produced on the most vigorous wood. This means that good vigorous growth must be maintained each year to produce a crop of large berries year after year.

Most varieties tend to bear too much fruit. Unless many of the fruit buds are removed, the berries are small and there is not enough vigorous growth for the next year's crop. To prune varieties that are growing erect, thin out the plants in the center. To prune spreading varieties, remove lower branches.

Heavy pruning should not be done in any one year. The heavier the pruning is, the smaller the crop will be. Too heavy pruning causes fruit to ripen earlier and over a shorter period of time. It is sometimes done to ripen berries earlier to obtain the highest price market, figure 42-3.

(A) (B)

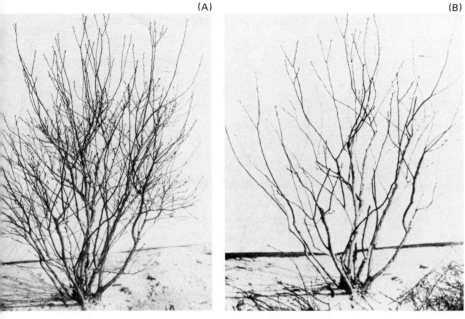

Fig. 42-3 (A) A four-year-old highbush blueberry bush before pruning. (B) The same blueberry bush after pruning. This plant was pruned heavily to produce an earlier crop of larger berries. Total production is reduced by such extensive pruning. In a more fertile soil, many more buds could have been left, resulting in a larger crop. (USDA photo)

Light pruning consists of thinning out the smaller, weaker branches and cutting a few stems to ground level. It is usually sufficient to ensure vigorous growth for the next year's crop and good-sized fruit during the current year.

> A vigorous plant requires less pruning than a weak, slow growing plant. Vigorous plants tend to produce large berries without much pruning.

Very little pruning is needed until the third year after planting. After that, pruning should be done each year.

Summary

1. The center of the bush is kept open by removing old and weak wood.

2. Low hanging branches are removed.

3. Most of the small slender branches are cut off at ground level, leaving only strong vigorous shoots and branches.

Propagation

Propagation of highbush blueberries is generally by hardwood cuttings. Cuttings 4 to 5 inches long are made from dormant shoots of the previous season's growth in winter. Only wood with leaf buds is used. If there are blossom buds on the tips of shoots, cut the ends off and discard them. Cuttings are usually grown in the nursery for a year after rooting and sold as two-year-old plants for planting to the field.

Varieties

For use in western North Carolina, Maryland, and New Jersey

Bluetta	Bluecrop	Lateblue	Jersey
Collins	Berkeley	Blueray	

For use in eastern North Carolina

Berkeley	Croatan	Jersey
Morrow	Murphy	Wolcott

For use in Michigan and New England

Bluecrop	Blueray	Collins	Coville
Earliblue	Jersey	Lateblue	

Fig. 42-4 (A) Shows the cutting wood and where it is cut to remove blossom buds. **(B)** shows the cutting as it begins to callus and establish top growth. **(C)** shows a well-rooted cutting. **(USDA photo)**

For use in western Oregon and Washington

Berkeley	Bluecrop	Blueray	Collins
Coville	Darrow	Dixi	Herbert

Most of these varieties are hybrids which tend to produce large fruit on vigorous plants.

RABBITEYE BLUEBERRIES

Rabbiteye blueberries are important mainly because they are able to grow in dry upland areas where the highbush will not, and they can be grown farther south because they require a very short rest period in winter. They are not as sensitive to soil conditions as the highbush, and resist heat and drought better. The rabbiteye grows in areas from eastern North Carolina to central Florida and west to Arkansas and east Texas.

Culture

Rabbiteye varieties are planted in midwinter in rows 12 feet apart with plants spaced 6 to 12 feet apart in the row. Small plants are sensitive to fertilizer; none should be applied the first year.

Pruning is not generally done but some pruning on older bushes is desirable. The older stems and smaller young shoots may require removal to prevent the bushes from becoming too thick. The rabbiteye blueberry is vigorous enough that pruning is not needed to produce strong new growth. The fruit ripens later and over a longer period of time than the highbush blueberry.

Plants are propagated by offshoots, or suckers, which grow up from the roots as far as 8 feet from the plant. They are dug up and cut free of the main root and grow in a nursery for one year before planting to the field. Propagation by softwood cuttings is also very successful.

Varieties

- Woodward is the earliest, largest and has good flavor.
- Tifblue has excellent flavor.
- Homebell exhibits large fruit, and very vigorous growth.
- Tifblue has the lightest blue color.

Many wild rabbiteye blueberries are harvested in the United States.

HARVESTING (ALL VARIETIES)

For the fresh market, blueberries are generally picked by hand. In eight hours, a picker can harvest 60 to 80 pints. Machine harvesting is used in some areas, employing machines that shake the berries into a net or hopper. Fruit harvested by machine must be sorted and cleaned. Most machine-harvested fruit is used for fruit that is processed before being sold.

Yields of blueberries range from 50 pints on a two-year-old planting to 6,000 pints per acre on a six-year-old planting.

PEST CONTROL

Weeds

Weeds may be controlled by mulching, shallow cultivation, or use of weed killers. For weed killer recommendations for your area, see a local horticulturist or extension agent.

Diseases

North Carolina suffers the most severe losses of blueberries due to disease. Stem canker causes the most damage. Foliage diseases, stunt virus, stem blight, and root rot are also serious enough to consider a control program.

In New Jersey, the most serious diseases are stunt virus and mummy berry. In Michigan, fusicoccum canker, phomopsis canker and mummy berry are serious problems. In the Pacific Northwest, mummy berry is difficult to control and is very serious.

> A well-drained soil and good sanitation help control many diseases.

Insects (See Spray Schedule)

A spray schedule for the home blueberry grower follows in figure 42-5, page 522. More detailed information for commercial growers is available from the local extension service in USDA Bulletin No. 2254.

Disease or Insect and When To Control	Description	Control	Remarks
Scale insects (delayed dormant)	small round spots on stem	superior oil applied in delayed dormant stage	Plants must be completely covered with spray.
Mummy berry (when buds first break)	stems and blossoms blighted. As berries approach maturity, they become mummified and drop to the ground.	Clean cultivation in the spring after each rain to disturb the mummies. Spray with ferbam 1 lb. and 2 oz. per 100 gal. of water or benomyl 4 oz. per 100 gal. of water. Spray at bud break and in 2 weeks.	Fungus lives over winter in the mummified berries and spreads spores in early spring.
Other fungus diseases (continue use of ferbam every two weeks)	leaf, fruit, and stem fungus diseases	ferbam at 2-week intervals Prune and burn diseased plant parts.	Sanitation is very important.
Plum curculio (2 weeks after bloom)	dark brown beetles- 1/4 inch long, larvae grayish white, legless, curved body, small brown head	malathion 25% WP 6 lbs. per 100 gal. of water 2 weeks after bloom	Weed in areas where the adult lives during winter months.
Fruit worms (2 weeks after bloom)	Small gray moths lay eggs after bloom has fallen. Larvae are brownish red on top. Some are green on sides and underneath.	same as plum curculio or use Sevin as directed	More than one species of fruit worms may be present.
Leaf hoppers (late May to June 5; repeat in 10 days)	Small, tent-shaped insects	malathion 57% EC 1 qt. per 100 gal. or Guthion 50% WP 1 lb. per 100 gal.	Spread virus Spray dates given are for Mid-Atlantic states.
Blueberry maggot (June 22; repeat every 10 days until harvest)	Adults look like small houseflies and are 3/16 inch long with a brown face and shiny black body, white on the side and rear of thorax. Maggots hollow berry by eating the flesh.	malathion 57% EC 1 qt. to 100 gal.	Check label for days before harvest and cease application as necessary.

Fig. 42-5 Blueberry Spray Schedule

STUDENT ACTIVITIES

1. Propagate blueberry plants from ten hardwood cuttings. See Unit 11 for information on propagation of hardwood cuttings.

2. Take a soil test of a possible site for a blueberry planting at home or on the school grounds. Determine if the pH is suitable for growing blueberries. Report to the class on the findings. If suitable soil can be located, prepare the site for a blueberry planting, using the plants propagated in activity 1.

3. If blueberries are cultivated locally, visit a local grower for tips on growing the crop.

SELF-EVALUATION

Select the best answer from the choices offered to complete the statement or answer the question.

1. Blueberries prefer a moderately acid soil. The best pH range for cultivation of blueberries is

   a. 6.5 to 7.0.
   b. 6.0 to 6.5.

   c. 3.5 to 4.0.
   d. 4.3 to 4.8.

2. If the soil is too acidic for blueberries, what material should be added to raise the pH?

   a. lime
   b. sulfur

   c. phosphorus
   d. potash

3. Raising the soil pH from 4.3 to 4.8 makes the soil

   a. less acidic.
   b. more acidic.

   c. neutral.
   d. alkaline.

4. The highbush blueberry prefers a soil that is

   a. moist and acidic.
   b. dry and acidic.

   c. moist, acidic, and well drained.
   d. moist, alkaline, and well drained.

5. Fertilizer for blueberries should not be neutralized so that it will

   a. help keep the soil acidic.

   b. be more readily available.

   c. be more economical to use.

   d. be easier to spread.

6. If highbush blueberries need irrigation during the harvest season, what is the least amount of water which should be applied at one time?

   a. 1 acre foot

   b. 1/2 inch

   c. 3 inches

   d. 1 inch

7. If a soil test indicates a need for phosphorus and potash, the chemicals are applied

   a. by plowing and mixing them into the soil before planting.

   b. in bands beside the plants at planting time.

   c. as a sidedress after planting.

   d. two weeks before picking the first harvest.

8. Ammonium sulfate is a good source of nitrogen for blueberries. Knowing that nitrogen leaches through the soil rather rapidly, the best application method is to

   a. plow the soil and make the application.

   b. place it in bands beside the plants at planting.

   c. spread it on the soil surface around the plants.

   d. place it in the bottom of the hole at planting time.

9. When organic or natural mulches are used around blueberry plants, the rotting of these materials robs the soil of the plant food element

   a. nitrogen.

   b. calcium.

   c. phosphorus.

   d. potash.

10. Blueberry varieties must be planted near each other to

   a. lengthen the picking season.

   b. ensure pollination and fruit set.

   c. help control the spread of disease.

   d. all of these

11. Blueberries are pruned to

   a. remove some of the fruit buds and promote vigorous new growth.
   b. increase production per acre.
   c. let the sun shine inside the plant.
   d. help control fungus diseases.

12. Blueberry plants that grow erect are pruned by

   a. pruning off lower branches.
   b. cutting back the tips of branches.
   c. thinning out the center of the plant.
   d. removing only tall growing tops.

13. Heavy pruning of blueberry plants results in

   a. large crops of large berries.
   b. fruit ripening earlier and over a shorter period of time.
   c. too much new growth.
   d. a reduction in plant vigor.

14. Highbush blueberries are propagated by

   a. root suckers.
   b. layering.
   c. budding.
   d. hardwood cuttings.

15. Rabbiteye blueberries are propagated by

   a. root suckers and softwood cuttings.
   b. hardwood cuttings and layering.
   c. grafting.
   d. budding.

16. Root rot of blueberries, a fungus disease, can largely be prevented by

   a. planting on a well-drained soil.
   b. planting only rot-resistant varieties.
   c. spraying with a fungicide.
   d. shallow cultivation.

17. Since blueberries bloom early, a frost-free site is important. Which site would give the best protection from spring frost?

   a. a high, level area
   b. a high, slightly sloping area
   c. a low, level area
   d. a low area surrounded by hills

OBJECTIVE

To outline a program for the propagation, cultivation, and maintenance of a bramble fruit planting.

COMPETENCY TO BE DEVELOPED

After studying this unit, the student will be able to

- list the five key points to consider when selecting a site for a planting of bramble fruits.
- outline a soil preparation plan for a bramble fruit planting.
- demonstrate the planting procedure for bramble fruits, including proper spacing of plants.
- outline a fertilizer and pruning program for bramble fruits.
- recognize insect and disease pests which infest brambles in the local area.
- list the seven measures for disease prevention in bramble fruit plantings.

Materials List

one of each of the following bramble plants for identification purposes:

red raspberry	erect blackberry
black raspberry	trailing blackberry

---

unit 43
The Bramble Fruits

The bramble fruits are some of the most delicious and useful of the small fruits, figure 43-1. Although this unit concerns the blackberry, red raspberry, and black raspberry, most of the information also applies to the boysenberry, dewberry, and loganberry, which are other brambles. Any differences in pruning or other practices concerning brambles will be noted.

LOCATION OF THE PLANTING

Good air circulation is important for bramble plantings. Because of this, a location that is higher than the surrounding area or with a slight slope must be chosen. A site that is above the level of the surrounding ground is not as subject to spring frosts that could kill the blossoms or cause winter injury of plants.

Do not plant in an area close to wild raspberry and blackberry plants. These wild plants may be diseased and will spread disease to the cultivated crop. Wild plants may be destroyed or removed from the area prior to planting.

Red raspberries should be planted at least 1,000 feet from black raspberries. Reds carry a virus disease which can be passed on to the black raspberries, causing severe damage to the blacks.

Brambles should not be planted in a soil that has been planted with potatoes, tomatoes, peppers, eggplant, or tobacco in the last three or four years. These plants may leave a fungus wilt disease in the soil that affects the planting of bramble fruits. The wilt organism can live only three or four years in the soil away from its host plant (a plant on or in which the disease lives).

Soil for Planting Site

The soil should have good drainage capacity and be high in organic matter. Subsoil drainage is more important than soil type. Organic matter increases the moisture-holding capacity of the soil. This promotes good growth, but does not interfere with good drainage. The subsoil must be well drained with no hard layer that will stop roots from penetrating the soil or slow water movement through the soil. Plants that cannot establish a deep root

Fig. 43-1 A blackberry branch with a typical fruit cluster ready for picking. (USDA photo)

system cannot produce a crop during dry weather. The brambles grow in a wide variety of soils, from sandy loams to clay loams.

Soil Preparation

If possible, begin preparing the soil for blackberries or raspberries one or two years before planting. Plow under green manure crops or animal manure to build up the organic matter level of the soil. The year the plants are to be set, plow the soil to a depth of 7 to 9 inches. Condition the soil by harrowing to break up all large clods and to level the soil surface.

A soil test should be taken prior to plowing. If lime or phosphorus and potash fertilizer are needed, they are applied and

Fig. 43-2 Black raspberry tip plant being set in the ground. A spade is being used to make the hole in which the plant is placed. Note that the plant is held by the cane from the parent plant. (USDA photo)

either plowed under or harrowed into the surface. Do not apply lime and fertilizer in the same application, however. It is probably best to plow under the fertilizer and harrow the lime into the surface after plowing. Lime may not be needed since bramble fruit grows well in a pH as low as 6.0 to 6.5.

THE PLANTING PROCESS

Always start with plants that are certified to be disease free and true to variety. Do not allow the plant roots to dry out during the planting process. Start with wet roots and keep the roots covered with wet burlap or plastic until planted. Do not place plants in the row so far ahead of the planters that they dry.

Establishing rows with a layoff plow at the proper distance apart makes the planting process easier. Place the plant in the row and cover with soil, firming around the roots. The plants are set at the same depth in the ground at which they grew before. Red raspberries are planted 1 to 2 inches deeper than they grew in the nursery. Do not cover new sprouts that may have started to grow. Instead, pull soil up around them later, after the sprouts have had a chance to grow more. A spade may be used to plant the brambles, figure 43-2. To plant, push the spade in the ground, pull it forward, place the plant in the opened slit spreading and keeping the roots straight, pull the shovel out, step on the soil to force the slit closed, and firm the soil around the roots.

The 6-inch piece of cane on the new plant that is used as a handle is cut level with the ground and removed from the field. This is a disease control measure for black raspberries and blackberries; it is not necessary with red raspberries.

Planting System

Most of the bramble fruits are best grown in rows, either free standing, on wire trellises, or staked to hold them up. Figure

43-3 gives recommended planting distances between rows of brambles.

Plants that grow erect need only one wire about 30 inches from the ground fastened to posts spaced 15 to 30 feet apart in the row. Trailing plants such as the boysenberry should have a two-wire trellis with the first wire 30 inches from the ground and the second wire 5 feet from the ground. The canes are tied to the wire with string. Some growers prune erect blackberries and raspberries back to an initial heading of 24 inches. Most canes of this height will stand without support. This practice is used commercially because of the cost of trellis construction. Some canes fall and fruit is lost. If a trellis is not used, plants must be pruned shorter so that the canes are able to stand.

If trellis rows are 400 to 500 feet long, it is best to break the trellis system to provide a cross-lane through the center. Figure 43-4 illustrates a trellis system for erect blackberries or black raspberries. Figure 43-5 shows a two-wire trellis system for trailing blackberries or boysenberries.

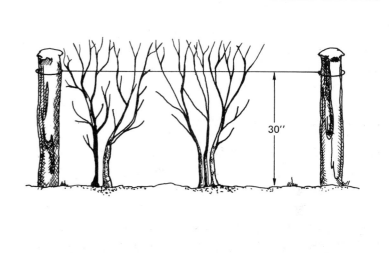

Fig. 43-4 Single-wire trellis for black raspberry and erect growing blackberry. The posts are set 15 to 20 feet apart.

Plant	Distance Between Rows*	Distances Between Plants in the Row
Black raspberry	7 to 10 feet	4 to 5 feet
Blackberry	7 to 10 feet	5 feet
Boysenberry (trailing types)	7 to 10 feet	8 feet
Red raspberry	7 to 10 feet	2 to 3 feet
Purple raspberry	7 to 10 feet	4 to 5 feet
Yellow raspberry	7 to 10 feet	2 to 3 feet

*Distance between rows varies with the method of cultivation that will be used. A small garden tractor works well at 7 feet apart; a large tractor requires 10 feet.

Fig. 43-3 Planting Distances for Bramble Fruits

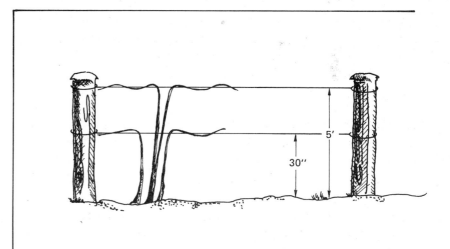

Fig. 43-5 Two-wire trellis system for trailing blackberries and boysenberries.

PRUNING AND TRAINING

The top growth of brambles lives for two years. Fruit is formed on wood of the previous season; these canes die soon after the fruit is harvested. The roots of plants are perennial and continue to live for many years.

Pruning and Training Free-standing Plants

Pruning Blackberries during Growing Season. The first season plants are pruned to a height of about 15 inches. After the first year, the height is increased to 24 inches since the plants are more vigorous and are able to produce larger, stronger canes that stand better. Not all canes reach the correct size at the same time, so the planting should be checked once a week and all canes 24 inches or longer headed back. Heading back causes the plants to form many side branches or laterals, which increases fruit production.

Pruning Blackberries in Early Spring. Before buds swell in the spring, remove all canes that are 1/2 inch or less in diameter by cutting off at ground level. At least two canes must be left, but most plants should have four or five canes. Shorten all of the laterals to 6 inches in length. Fruit from pruned laterals is larger and of better quality.

Pruning Blackberries after Harvest. Immediately after the crop is harvested, prune all old canes that fruited by cutting off at ground level. This promotes new cane growth from buds at the base of the plant and helps control diseases. These old canes should be removed from the field and destroyed to kill disease organisms. At this time, continue the growing cycle by cutting back all new canes to the 24-inch height as soon as this height is reached. Purple raspberries are pruned the same way except that the new canes are headed back to 30 to 36 inches instead of 24 inches.

If plants are trained to grow on a trellis, the heading back of new canes should be kept about 12 inches higher than recommended for nonsupported plants. Other pruning practices are the same.

Pruning Red Raspberries during Growing Season. Red raspberries may be tied to a trellis or left free standing. If free standing, head back all canes to a height of 30 inches; if on a trellis, cut back to 36 to 48 inches. All canes that are diseased or do not reach about 1/2 inch in diameter are cut off at ground level. The canes are thinned to stand no closer than 4 to 6 inches apart, leaving about seven or eight strong canes per hill.

Pruning Red Raspberries after Harvest. All old canes are removed as with black raspberries. New canes are thinned and all root suckers growing into the rows or between rows are removed.

Pruning Upright Growing Blackberries during Growing Season. The first season after planting, all canes are cut back when they reach a height of 24 inches. Each year after the first, canes are pruned to 36 inches. This cutting back forces branching of the canes and formation of a greater number of fruit buds. It also causes canes to grow thicker and stronger, making them more able to stand erect. All suckers that sprout between the rows should be pulled as soon as possible to prevent the planting from becoming too thick.

Pruning Upright Growing Blackberries in Early Spring. Before growth starts in the spring, prune all laterals to a length of 12 inches.

Pruning Upright Growing Blackberries after Harvest. All old canes that fruited are cut off at ground level and removed immediately after harvest. These canes should be burned or otherwise destroyed to help control diseases. Thin any new canes that surface, leaving four of the most vigorous canes per plant.

Blackberry canes with a great deal of fruit often fall due to lack of support. A one-wire trellis support may be necessary. Tie canes to the wire with soft string.

Pruning Semitrailing Blackberries and Boysenberries during Growing Season. Trailing blackberries are neither headed back nor summer pruned in any manner except to remove the old canes after fruiting. The new canes are allowed to grow along the ground.

Pruning Semitrailing Blackberries and Boysenberries in Early Spring. Laterals are shortened on all canes to 10 inches and the long trailing vines are tied to a two-wire trellis with soft string.

Pruning Semitrailing Blackberries and Boysenberries after Harvest. Remove the old canes that have fruited immediately after harvest and destroy them.

CULTIVATION

Cultivate the bramble fruits often enough to control weeds. Cultivation should be shallow — no more than 2 inches deep. Stop cultivation during harvest time to prevent fruit loss. Brambles should not be cultivated in the fall since this stimulates late growth and plants do not harden off properly for winter.

Weeds may also be controlled with mulches and weed killers. Contact the local extension service for chemical weed control recommendations.

FERTILIZING

It is necessary to apply lime, phosphorus, and potash before planting, as mentioned earlier. After planting, side-dress a nitrogen fertilizer 6 inches from the plant at a rate of 40 pounds of actual nitrogen per acre.

Each year after, apply a 5-10-5 fertilizer as a sidedress just before blossoming time at a rate of about 600 pounds per acre, or about one-half cupful around each plant. Keep fertilizer at least 6 inches away from the crown of the plant. After harvesting, apply nitrate of soda (a nitrogen fertilizer) at a rate of 200 to 300 pounds per acre or ammonium nitrate (a nitrogen fertilizer) at a rate of 80 to 100 pounds per acre.

HARVESTING

Berries are picked when they are fully ripe but still firm. To keep ahead of the fruit so that none gets too ripe, raspberries should be picked every two or three days. Hot weather ripens fruit more quickly and could cause picking dates to change. Blackberries should be picked every three or four days.

Only the firm berries are picked; all injured or soft berries are thrown away. Do not hold handfuls of berries during picking. This practice causes bruising of fruit which results in early spoilage and a product that cannot be marketed.

Place the fruit in the shade or some other cool spot as soon as it is picked. Heating in the container causes rapid spoilage. Berries picked in the morning when they are cool remain fresh longer than berries picked after the sun has heated the fruit.

Many growers are operating pick-your-own operations. This is one of the most economical systems for harvesting fruit if organized properly. Hand picking for the fresh market is very expensive. Processed berries are often picked by machines. This process calls for special pruning and good disease control.

Fig. 43-6 A quart box of black raspberries, Starking Black Giant variety (Courtesy Stark Brothers Nurseries, Incorporated)

Average yields of fruit are about 6,000 pounds of blackberries per acre. Raspberries yield about 2,000 quarts per acre.

PROPAGATION

Black and purple raspberries are propagated by tip layering. Red raspberries are propagated from root suckers and root cuttings. The rooted suckers are dug up and separated from the parent plant and planted in early spring.

Root cuttings from red raspberries are taken in early spring. Pieces of root 2 or 3 inches in length are cut from the parent plant and covered with 2 inches of soil and placed in the nursery. The roots send up shoots which are set in the field the following spring. When old plants are dug up, roots left in the ground send up many shoots which form new plants. This is an easy way to propagate red raspberries.

Upright blackberries are propagated from root suckers or root cuttings, as red raspberries are. Trailing blackberries are propagated by tip layering, just as black raspberries are.

VARIETIES

There are many varieties of the bramble fruits and new varieties are constantly being developed. Some varieties grow best in colder climates, others survive better in southern regions of the U.S. Recommended varieties for the local area may be obtained from the local agricultural extension service.

INSECTS AND DISEASES

Insects are not usually as destructive to the bramble fruits as diseases are. If control measures are used, attack by disease organisms can be kept to a minimum without the use of an extensive spray schedule. Some spraying is necessary, however, depending on local conditions and the needs of that particular year.

The following are recommended disease prevention measures.

1. Start with the proper site.
   a. Do not plant black or purple raspberries in fields that were planted with tomatoes, potatoes, peppers, or eggplant.
   b. Keep plantings of black and red raspberries separated by at least 1,000 feet.
   c. Destroy all wild raspberry plants in or near the planting site one year before planting.

2. Choose disease-resistant varieties.

3. Buy and plant only disease-free plants.

4. Remove all diseased plants and burn them.

5. Remove old canes after harvest.

6. Keep the planting free of weeds and fallen leaves.

7. Use pesticides when necessary. Obtain a spray schedule from the local extension service or the USDA Farmers' Bulletin 2208, *Controlling Diseases of Raspberries and Blackberries.*

STUDENT ACTIVITIES

1. Visit a small fruit grower and discuss all aspects of the operation with him or her.

2. Propagate and plant at least one bramble fruit. Include proper pruning techniques in the program.

SELF-EVALUATION

Select the best answer from the choices offered to complete each statement.

1. Wild raspberry and blackberry plants are removed from areas planted with commercial plantings of brambles because
   a. the plants cross-pollinate, resulting in small fruits.
   b. disease is spread from the wild plants to the new planting.
   c. new plants selected from the planting may become mixed with the wild planting.
   d. all of the above

2. A fungus wilt disease is generally present in soils that have been planted with potatoes, tomatoes, peppers, eggplant, or tobacco. This disease also attacks bramble fruits. It is necessary to wait _____ years after any of these disease-carrying plants are grown on a soil before planting brambles.
   a. three or four
   b. four or five
   c. five to ten
   d. two or three

3. Subsoil type is more important than soil type for bramble fruits because
   a. brambles grow in a wide variety of soils.
   b. an open subsoil allows deep rooting.
   c. soil drainage through the subsoil is more important.
   d. all of the above

4. If it is found that potash or phosphorus fertilizer is needed as a result of a soil test,

   a. it is applied as a band application on each side of the plants.
   b. it is plowed under or harrowed into the soil before planting.
   c. it is spread in a circle around the plants.
   d. it is applied after picking the crop.

5. When planting black raspberry plants, they are set

   a. at least 2 inches deeper than the depth at which they grew in the nursery.
   b. at least 1 inch above the top of the soil level.
   c. at the same depth at which they grew in the nursery.
   d. at about 3 inches deeper than the depth at which they grew in the nursery.

6. The planting depth of red raspberry plants is

   a. 1 or 2 inches deeper than the depth at which they grew in the nursery.
   b. at the same depth at which they grew in the nursery.
   c. 3 inches deeper than the depth at which they grew in the nursery.
   d. at least 2 inches deeper than the depth at which they grew in the nursery.

7. The 6-inch piece of plant used as a handle to hold the new plant during planting is cut off black raspberry and blackberry plants after planting and removed from the field to

   a. help control disease.
   b. make the planting more attractive.
   c. prevent the cultivator from hooking the new plant.
   d. all of the above

8. Organic matter is important in the soil in which brambles are planted because

   a. it stimulates earthworm activity.
   b. it is cheaper than fertilizer in promoting plant growth.
   c. it helps in weed control.
   d. it helps the soil to hold moisture.

9. If a trellis is not used to support black raspberry plants, the plants must be cut back or topped

   a. more often.
   b. to a shorter total length.
   c. at a higher total height.
   d. twice a year instead of once a year.

10. When purchasing bramble fruit plants, be sure that the plants are certified

    a. to be of proper planting size.
    b. free of insects.
    c. to grow.
    d. free of disease and true to variety.

11. Pruning during the growing season to cut back the canes on raspberry plants results in

    a. the formation of more side branches.
    b. the production of more fruit.
    c. canes that can support themselves better.
    d. all of the above

12. Pruning of black raspberries and blackberries in early spring consists of

    a. shortening all laterals and removing small, weak canes.
    b. topping or cutting back canes.
    c. removing the old canes that fruited the previous summer.
    d. thinning the number of canes.

13. Immediately after harvest, pruning of all bramble fruits consists primarily of

    a. thinning out the number of canes.
    b. heading back or shortening canes.
    c. removing the old canes that just finished fruiting.
    d. shortening laterals to promote the following year's fruiting.

14. During the growing season, red raspberry canes are thinned to leave about

    a. four strong canes per hill.
    b. seven or eight strong canes per hill.
    c. five or six strong canes per hill.
    d. as many strong canes over 1/2 inch in diameter as possible.

15. Pruning of trailing blackberries during the growing season consists of

    a. shortening the canes to promote lateral growth.
    b. cutting back lateral branches.
    c. no pruning at all.
    d. thinning out weak canes.

16. Fall cultivation of brambles should not be done because

    a. it is not necessary for weed control.
    b. many small roots are destroyed and do not have the time to grow back.
    c. weed seeds are spread over a wide area.
    d. it stimulates late growth and plants do not harden off for winter.

17. Raspberries should be picked every

    a. two or three days.
    b. one or two days.
    c. two days.
    d. any of the above, depending on how hot the weather is.

18. Black raspberries, purple raspberries, and trailing blackberries are propagated by

    a. root cuttings.   c. shoots.
    b. tip layering.   d. suckers.

19. Red raspberries and upright blackberries are propagated by

    a. tip layers.   c. root suckers.
    b. stem cuttings.   d. stem layering.

20. Black raspberries are planted at least 1,000 feet from red raspberries because

    a. red raspberries carry a disease that is harmful to black raspberries.
    b. cross-pollination results in a poor red raspberry crop.
    c. black raspberries carry a disease that is harmful to red raspberries.
    d. none of the above

unit 44
Grapes

OBJECTIVE

To outline a program for the establishment and cultivation of a vineyard.

COMPETENCY TO BE DEVELOPED

After studying this unit, the student will be able to

- select a planting site for a vineyard and give reasons for its selection.
- prepare the soil and plant grapes in the selected planting site.
- describe a fertilizer program for a vineyard.
- list the characteristics of ripe grapes.
- list the procedures for weed control in vineyards.

Materials List

pruners

canes from a grapevine for demonstration of proper pruning techniques

If a hands-on activity involving the planting and care of a grape planting is possible, all planting tools and materials for the construction of a trellis are needed.

Grapes are rapidly becoming a popular homegrown fruit. Grapes have many uses in cooking and are very nutritious. They are consumed fresh, as juices and wines, as raisins, jam and jelly, and as frozen products. Grapes are native to the United States. Many of the domesticated varieties have native wild grape ancestry in at least one parent. Grapes grow in most areas of the United

States, except where the frost-free season is less than 150 days, in acidic sections, and in areas having extremely high temperatures and very humid conditions.

CHOOSING A PLANTING SITE

A level site or a steep slope that is higher than the surrounding area is important to reduce danger of frost damage. The soil should have good drainage capacity and be high in organic matter.

A location to the south or east of a large body of water is favorable since water has a modifying effect on temperature. It protects against fall frost, delays early spring blooming, and thus holds plants until the danger of spring frost is past.

Soil

Grapes grow well on most soil types, providing the subsoil allows deep rooting. *Hardpan* (compacted clayey soil), rock, or wet subsoil close to the surface are not suitable. A deep, fertile, well-drained loamy soil with a moderate amount of organic matter is best. A pH range from 5.5 to 7.0 is satisfactory.

Soils that are overfertilized, rich, or too high in organic matter produce a late maturing crop with a low sugar content. Sandy or light soils tend to produce small crops of early maturing fruit that is high in sugar content. A soil that falls somewhere between these two examples is best.

SOIL PREPARATION

Grapes should not be planted directly into soil that has been in sod for years. Such land should be planted with a cultivated crop such as corn for at least one year before planting grapes. The soil is plowed deeply and disc harrowed until it is in good planting condition.

After a soil test is done, any phosphorus or potash fertilizer needed should be plowed or disced into the soil before planting.

PLANTING

In southern regions of the U.S., grapevines are planted in the fall as soon as they are dormant. North of Arkansas, Tennessee, and Virginia, vines are planted in early spring.

The ideal grape plant at planting time is one year old. The vines are planted at the same depth at which they grew in the nursery. Do not allow roots to dry out during the planting procedure. Immediately after planting, prune to a single stem with two or three buds remaining. The plants are spaced 8 to 10 feet apart in the row, with the rows about 10 feet apart. The plants are placed directly under the trellis wire.

Grapes respond best to shallow cultivation (not over 3 inches deep). The vines should be cultivated soon after planting to kill germinating weeds. Cultivation is repeated as needed for weed control. The grape hoe, consisting of a manually operated blade mounted on a tractor, is an excellent tool for keeping the rows clear of weeds.

TRELLIS CONSTRUCTION

Grapes are trained to grow on a trellis so that the vines are given proper support. Generally, a two-wire trellis is used. Wooden, concrete, or steel posts are placed about 20 feet apart. The end posts must be securely planted at least 3 feet deep to remain firm. Line posts can be set 2 feet deep.

Number 9 steel wire is used for the top wire and smaller number 12 steel wire for the bottom wire. The bottom wire is 30 inches from ground level and the top wire 30 inches from the

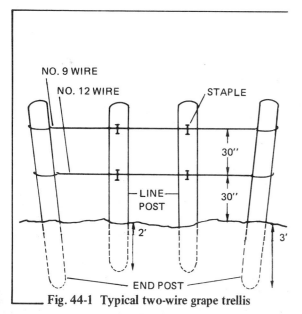

Fig. 44-1 Typical two-wire grape trellis

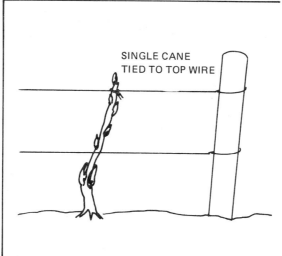

Fig. 44-2 Pruning and training at the end of the first growing season

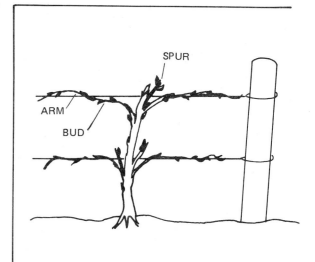

Fig. 44-3 Pruning and training the second and third year

bottom wire, figure 44-1. The wires are fastened to the posts so that they slide through the staples or clips for easy tightening. The wires are tightened each spring before the vines are tied.

TRAINING AND PRUNING

Grapevines are trained to fit the trellis construction described above. Although there are several systems of training, the *four-arm kniffin system* is the most popular method for training bunch grapes and works very well.

After the first growing season, the most vigorous cane is selected and tied to the top wire. It is then cut off above the top wire and all other canes are removed, figure 44-2.

When pruning during the dormant season of the second or third year, select four vigorous canes for arms. Prune each of these canes to eight or ten buds in length. Select four other canes

located as close to the arms as possible and cut each back to two buds each. These short canes are called *spurs*, and the buds on them will develop new canes to serve as the new arms for the vine the following year. All other canes are removed. Each year, this process of pruning is repeated. New arms are formed from spurs each year for the following year's crop. The old arms are removed during the pruning process. Fruit is always produced on new growth from canes that are one year old.

The vines are pruned each year late in winter or early spring before growth begins. Most of the canes are removed each year, with the canes left as arms and renewal spurs greatly shortened, figure 44-3. This may seem to be very severe pruning, but it is essential for continued high yields and quality fruit.

The more vigorous a vine is, the more fruit it is able to produce. The number of buds left to produce fruit on the four arms as described earlier is from thirty-two to forty buds total (eight to ten on each arm). This is considered average. Weak vines have slightly fewer buds and strong vines more. A good rule of thumb is to leave thirty buds and then add ten buds for each pound of one-year-old wood pruned from the vine.

After pruning, the canes are tied to the trellis wire with string. The end of the arm is tied tightly just behind the last bud. That bud is rubbed off so that it does not grow and break the string as the cane grows. Other ties are made as needed to hold the arm to the wire. These are tied loosely so that the cane has room to grow and expand in diameter, figure 44-4.

FERTILIZER

Fertilize grapes in early spring with a complete fertilizer. Apply according to soil test recommendations. The fertilizer should be broadcast because grape roots tend to spread over a wide area.

HARVESTING

Grapes are harvested when they are ripe. This stage is often difficult to detect since some grapes change color and appear to be ripe before they are actually ready to pick. The best test for table grapes is to taste them for sweetness or check the color of the seeds. Seed color changes from green to brown as the fruit ripens.

In grapes used in juices and wines, the percentage of soluble solids or sugar content is important. A refractometer or Balling hydrometer is used to determine ripeness.

It is not always possible to harvest fruit when it is fully ripe. Some grapes crack or split open as ripening progresses and must be picked before they are fully ripe to prevent loss of fruit. Rotting

Fig. 44-4 Grapevine trained to the four-arm kniffin system and tied to a two-wire trellis. A wire fence is shown in the background. (USDA photo)

in rainy weather or danger of early frost may also make early picking necessary.

Table grapes for fresh consumption are generally picked by hand while juice and wine grapes are picked by machine. Figure 44-5, page 542, shows a grapevine with an abundant grape crop.

WEED CONTROL

Weed control is generally accomplished through shallow cultivation and the use of the grape hoe. In one hoeing, the soil is pulled away from the wires toward the center of the rows. About

Fig. 44-5 Grapevine trained to four-arm kniffin system with an abundant crop of fruit. (USDA photo)

two weeks later, the hoe is turned the other way and the soil is pulled back toward the vines and under the trellis. The grape hoe removes most of the weeds and leaves only a few for hand hoeing. Chemical weed killers are also sometimes used. Check local recommendations of the state agricultural experiment station.

INSECTS AND DISEASES

Grapes are sprayed for both insects and fungus diseases. At least three spray applications are necessary for good control. Commercial growers spray up to eight times during the season.

> **Consult the local county agricultural agent for the USDA bulletin on spraying grapes.**

Major diseases which require control are black rot, powdery mildew, and downy mildew. Major insect pests are the flea beetle, leaf hopper, berry moth, and Japanese beetle.

PROPAGATION

Grapes are propagated commercially by hardwood cuttings. See the section of the text on propagation by hardwood cuttings for a description of this technique. If a disease-resistant rootstock is needed for certain varieties, plants are grafted or budded to a disease-resistant seedling rootstock.

STUDENT ACTIVITIES

Complete the following activity as a class project, using a section of school grounds or a local vineyard.

- Plant and prune one-year-old grapevines.

- Fertilize a producing grapevine.
- Construct a grape trellis.
- Prune the grapevine and tie it to the trellis.
- Discuss the site selected.
- Discuss weed control for the planting and carry out, if necessary.
- Observe the application of pesticides.
- Determine when the fruit is ripe and demonstrate proper picking techniques.

SELF-EVALUATION

Select the best answer from the choices offered to complete each statement.

1. Grapes grow throughout most of the United States except in those areas where the frost-free season is less than

   a. 200 days.   c. 100 days.
   b. 150 days.   d. 120 days.

2. A planting site that is _____ the surrounding area helps reduce frost danger to grapes.

   a. lower than   c. higher than
   b. level with   d. protected from

3. Grapes grow well on most types of soil providing that the subsoil allows

   a. deep rooting and good drainage.
   b. nutrients to leach through.
   c. hardpan to hold nutrients and water.
   d. deep cultivation.

4. In areas north of Arkansas, Tennessee, and Virginia, grapes are planted

   a. in the fall.   c. in the early spring.
   b. in August.   d. none of these

5. The proper depth at which to plant one-year-old grape plants is

a. 2 inches deeper than they grew in the nursery.

b. 1 inch deeper than they grew in the nursery.

c. 3 or 4 inches deeper than they grew in the nursery.

d. the same depth at which they grew in the nursery.

6. The two-wire grape trellis has support wires spaced

a. 30 inches apart.   c. 48 inches apart.

b. 36 inches apart.   d. 24 inches apart.

7. In the four-cane system of training grapes, each of the four arms has _____ buds left on it after pruning.

a. five or six   c. eight to ten

b. ten to fifteen   d. twenty to thirty

8. Short two-bud spurs are left on the grapevine after pruning. The purpose of these spurs is

a. to produce extra fruit close to the main stem.

b. to produce canes for new arms the following year.

c. to provide arms in case the selected arms die.

d. to provide foliage to protect the main stem from sunscald.

9. A very vigorous grapevine should be pruned to leave _____ canes and _____ buds than are left on a less vigorous vine.

a. longer, more

b. shorter, fewer

c. the same number of, the same number of

d. none of the above, it depends on the variety

10. A complete fertilizer should be broadcast around grapevines

a. after the fruit is picked.

b. in early fall.

c. after the fruit has set and has started to develop.

d. in early spring.

11. Grapevines are pruned

    a. just after fruiting to remove old canes. c. in early fall.
    b. just after new fruit has set.   d. in late winter.

12. The best test for ripeness in table grapes is

    a. color of the fruit.
    b. sweetness and change in seed color.
    c. softness of the fruit.
    d. ease of separation of the grapes from the vines.

13. Grapes are generally propagated commercially by

    a. hardwood cuttings.   c. budding.
    b. grafting.   d. layering.

14. Grapevines are pruned to remove most of the canes each year because

    a. fruit is produced on current year's wood.
    b. heavy pruning is necessary to produce quality fruit.
    c. heavy pruning is necessary to produce high fruit yields.
    d. all of the above

15. Weed control in vineyards is primarily through use of

    a. shallow cultivation and the grape hoe. c. hand pulling.
    b. weed killers.   d. a disc harrow.

Glossary

Absorb — to take in

Absorption — the process of taking in a liquid or gaseous substance

Accent — an item used to create interest in a landscape

Adventitious buds — buds which occur at sporadic and unexpected places on a vegetative structure

Aeration — movement of air into and through the soil

Aggregation — the clinging together of soil particles to form larger crumblike particles

Annual — a plant which grows, flowers, produces seed, and dies in one year

Anther — saclike structure at the top of the stamen which contains pollen (the male sex cell)

Arm — cane of a vine left after pruning which is tied to a trellis and produces fruit; replaced each year with new wood

Asexually — without the union of male and female sex cells

Banding fertilizer — applying fertilizer in trenches on each side of a row of plants

Basal plate — base of a bulb

Berm — a ridge of soil placed around a newly planted tree to retain water

Biennial — a plant which produces vegetation in one year, flowers the next, and then dies

Biological insect control — control of insects by the use of the natural enemies of certain insects rather than chemical substances

Broadcast — to spread over an entire area

Bud scales — leaves which cover the outside of a bud

Bud scale scar — a scar located where a terminal, or end, bud has been the previous year

Budstick — a small shoot of current season's growth used to cut buds for budding

Bulb — a vegetative structure which consists of layers of fleshy scales overlapping each other, such as the onion

Bulblet — immature bulb which develops at the base of the bulb

Bulb scales — leaflike parts of a bulb which surround the flower bud and are attached to the basal plate

Burlap — a coarse cloth made of jute or hemp often used to protect plants from the weather; holds root ball together on B&B plants

Callus — mass of cells which forms around the wounded area of a plant to start the healing process

Cambium — thin, green, actively growing tissue located between the bark and wood of a plant; in grafting, the cambium of the scion must touch the cambium of the stock

Cane — mature shoot of a vine's previous season's growth

Capillary action — movement of water upward through narrow spaces in the soil

Chemical retardant — a chemical used to slow down, shorten, or dwarf plant growth

Chloroplast — small green particle containing chlorophyll found in leaves

Chlorophyll — the green substance which gives many plants their green color and is necessary for photosynthesis

Cold frame — an outside structure covered on top with glass or plastic used to harden off plants or protect tender plants during the winter

Cotyledons — the first leaves to appear on a plant; seed leaves

Common name — the English name of a plant which may differ in various localities

Compatibility — ability to unite with, grow, and live together as the scion and rootstock in the grafting and budding processes

Complete flower — flower with both male and female parts; sexually complete

Corm — swollen underground stem which grows upright; is a food storage organ and a means of reproduction

Cormel — a new corm produced by a larger corm

Cross-pollination — a process in which pollen (male sex cell) of one plant unites with the egg (female sex cell) of a different plant

Cutting (noun) — a section of stem or root used for propagation of plants

Deciduous plant — a plant which loses its leaves during certain seasons

Dermal — through the skin

Dicot — a plant having two cotyledons, or seed leaves

Dilute — to make weaker or thinner

Direct seeding — planting seeds in a permanent growing site

Division — a method of propagation requiring the cutting and dividing of plants

Dormant — in a resting, or nongrowing, state

Drip line — the imaginary circle which indicates the outer edge or farthest extension of a tree's branches

Embryonic plant — the entire plant before germination; embryo

Endosperm — the food supply for the young developing seedling which is contained in the seed

Environment — the surrounding area

Epidermis — the skin of the leaf

Erosion — washing or blowing away of soil caused by water or strong winds

Evergreen plant — a plant which has leaves or needles throughout the year

Filament — the stalklike part of the stamen, or male part of the flower

Flat — a wooden box with slotted bottom used to start seedlings

Flower stalk — the stem of the plant which supports the flower

Forcing — growing plants to flower at other than their normal season

Fungicide — any substance which destroys or prevents the growth of fungi; usually applied to plants as a spray or dust to control fungus diseases

Germinate — to sprout or begin to grow

Grafting — uniting two different plants so that they grow as one

Grafting wax — a pliable, sticky, waterproof material made of bee's wax, resin, and tallow

Green manure — plants plowed or mixed into the soil to rot and add organic matter to the soil

Guard cells — cells on the underside of the leaf which open and close the stoma, or leaf pores

Hardening off — gradually subjecting plants to more difficult growing conditions by withholding water and decreasing temperature; prepares plants for transplanting

Hardiness — ability of a plant to withstand the minimum temperature of an area

Hardwood cutting — a cutting made from current season's stem tissue which is mature or hard

Harrow — a farm implement pulled behind a tractor which breaks up clods of earth, levels the soil surface, and compacts the seed bed

Herbicide — a substance which destroys weeds

Horticulture — a field which includes growing of fruits, nuts, vegetables, ornamental plants and flowers, and the sale and processing of these items; from the Latin words meaning "garden cultivation"

Host plant — the plant on which a disease or insect lives

Humidity — the amount of moisture in the atmosphere

Hybrid — an offspring of two different varieties of one plant which possesses certain characteristics of each parent

Inconspicuous — small and unnoticeable (as *inconspicuous* flowers)

Inhalation — the process of taking in air through the lungs; breathing in

Initial — the first of something (for example, initial growth)

Initiate — to start, begin

Insert — to place into an opening

Irrigation — the addition of water to plants to supplement natural rainfall

Laminate (tunicate) bulb — a bulb having dry membranous outer scales which protect it from rough handling and drying

Lateral — side branch

Layering — a type of asexual reproduction in which roots are developed on the stem of a plant while it is still attached to the parent plant

Leach — to wash through or out of soil

Leaf axil — the angle made where the leaf leaves the stem; usually contains a bud

Leaf blade — the flat part of the leaf which extends from either side of the leaf stalk

Leaf scale scar — a scar on a stem where a leaf has been attached in a previous year

Lenticels — breathing pores in the bark of woody stems

Lethal — deadly

Limestone — a natural rock used to reduce soil acidity and provide calcium

Lining out — planting cuttings in rows in the field

Loam — soil with approximately equal amounts of sand, silt, and clay; generally considered the best type of soil for plant growth

Major elements — plant food elements required in large amounts: nitrogen, phosphorus, and potash; they must be added to plant media by means of fertilizers

Manure — animal wastes used to improve soil; low in plant food but high in bacteria and other elements used for plant growth

Market pack — plant container which usually holds 6 to 12 plants and is sold as a unit

Maturity — age; plant tissue hardens as it matures

Media — a material which is used to start and grow seeds and plants

Minor elements — elements required by plants in small amounts; they may or may not need to be added to media for the best growth

Misting system — a piped water system used in greenhouses which uses nozzles to spray fine droplets of water on plants

Molluscicide — a chemical used to kill snails and slugs

Monocot — a plant having only one cotyledon, or seed leaf

Mulch — any material used to cover the soil for weed control and moisture retention

Nematocide — a substance used to kill nematodes (small worms)

Neutral — neither acidic nor alkaline; having a pH of 7

Node — the joint of a stem; the swollen place where leaves and buds are generally attached

Nomenclature — a system of naming used to classify a group, such as the botanical names of plants

Nontunicate (scaly) bulb — a bulb which has no outer covering of scales and is much more sensitive to drying or bruising

Nutrients — plant food elements

Oasis — spongelike material used as a base for flower arrangements

Optimum growth — the best possible growth

Orally — through the mouth

Organic matter — dead and decaying plant parts such as green manure crops, peat moss, or animal manures which improve the water-holding capacity of the soil

Organic mulch — material composed of decayed plant life such as straw, peat moss, or wood chips used as a mulch

Ovary — the lower part of the pistil in which the eggs are fertilized and develop; becomes the seed coat or fruit

Ovules — eggs; the female sex cells

Peat pellets — compressed peat moss discs used to start seeds or root cuttings which are transplanted with the plant

Peat pots — pot-shaped plant containers made of compressed peat moss instead of clay or plastic

Perennial — a plant that grows year after year without replanting; plant whose roots live from year to year

Perlite — a white granular material used to help loosen or open up spaces in rooting media

Pest — an unwanted animal, plant, bacteria, or fungus

Petal — one type of leaf on flowers; usually considered the most striking part of the flower

Petiole — stalk or stem which supports the blade of the leaf

pH level — the measure of a soil's acidity or alkalinity on a scale of 1-14; 1-6 indicates an acid, 7 is neutral, and 8-14 indicates a base

Phloem — the bark of stems; conducts food from leaves through the stem to the roots

Photoperiodism — response of plants to different periods of light and darkness in terms of flowering and reproduction cycle

Photosynthesis — manufacture of food by green plants in which carbon dioxide and water are combined in the presence of light and chlorophyll to form sugar and oxygen

Pistil — the female reproductive part of a flower; contains the female sex cells in the ovary

Polarity — the tendency of tuberous roots to grow shoots on one end, or the top, and roots on the other end, or bottom

Pollen — male sex cell in plants

Pollination — the process in which the pollen cell is transferred to the female plant part and the eggs are fertilized

Polyethylene — a clear plastic that holds in moisture but allows the passage of light and a small amount of air

Precooled — cooled for forcing

Propagate — to increase in number, to reproduce

Regenerate — to reproduce or grow anew

Relative humidity — the amount of moisture in the air compared with what it can hold at that temperature; expressed as a percent of the amount of moisture the air could hold if saturated

Renewal spur — a cane of a vine pruned to two buds and left to grow new canes the following year

Reproduce — to increase in number

Resist — to withstand or prevent

Rhizome — underground stem which produces roots on the lower surface, and extends leaves and flowering shoots above the ground

Rodenticide — a substance used to kill rodents

Root hair — a very tiny rootlike extension on small feeding roots; absorbs minerals and water from the soil

Rooting — the development or growth of new roots

Rooting hormone — a chemical, in powder or liquid form, that helps cuttings to root faster and have a greater number of roots

Rootstock — the lower portion of a graft which becomes the stem and roots of the new plant

Scientific name — the Latin name of a plant giving its genus and species

Scion — a short piece of shoot containing several buds which becomes the new top of a grafted plant

Seed coat — the outer covering of the seed

Seedlings — young plants which have been germinated several days

Self-fruitful — able to pollinate itself

Self-pollination — fertilization of a plant by its own pollen

Sepal — green, leaflike part of the flower that covers and protects the flower bud before it opens

Separation — method of propagation that occurs naturally in which reproductive organs of a plant detach from the parent plant to become new plants

Set — to develop fruit

Sexual — reproduction involving male and female sex cells (pollen and egg)

Shallow — not deep, generally less than 2 inches

Shoot — current season's growth of a plant; rapid, new stem growth

Sphagnum moss — a material formed from decayed bog plants; used in growing media

Side-dress — to fertilize along the side of rows near plants

Skew — a tool used to level media in flats before planting

Soluble — able to be dissolved in water

Soluble fertilizer — a fertilizer which will dissolve in water

Stamen — the male reproductive part of the flower containing the male sex cells or pollen

Sterile — free from any living organisms

Stigma — part of the pistil which catches pollen on its sticky surface

Stoma — small pores or holes in the leaf which allow the plant to breathe and give off moisture

Style — the tube in the pistil which leads from the stigma to the ovary and through which pollen reaches the egg

Susceptible — weak, unable to resist (disease)

Symptoms — warning signals of plant growth problems

Taxonomy — the study of plant names and the identification of plants

Tendrils — structures on a grape vine which allow the plant to attach itself to a trellis

Terrarium — a covered or closed container for growing plants

Texture — the coarseness or fineness of plant leaves

Thatch — dead grass clippings and roots which build up in lawns, causing growth problems

Toxic — poisonous

Transpiration — loss of water through the leaves or stems of plants

Transplant — to move plants from one growing location to another thus giving them more space in which to develop

Trellis — a supporting structure made of posts and wire which is used to tie up grapes and some of the bramble fruits

True to seed — reproducing offspring from seed which are nearly exact duplicates of the parent plant

Trunk — the main stem of a plant which lives on for years, supporting new growth each year from arms, spurs, or limbs

Tuber — fleshy root which reproduces by growing roots from an eye, or bud

Tuberous root — a thick root containing large amounts of stored food; has roots located at one end and stem at the other

Turgid — swollen, filled with moisture

Vermiculite — a light mineral with a neutral pH used to increase the moisture-holding capacity of media

Weed — any plant growing where it is not wanted

Xylem — wooden tissue of the plant under the bark and cambium; transports materials from the roots to the stem and leaves

Acknowledgments

W. Atlee Burpee Company, color plates 10-13
Stark Brothers Nurseries, Inc., color plates 14 and 15

Appreciation is expressed to the following for their special technical assistance:

Frances Mary Dewey, Instructor, Junior College of Albany, Division of Russell Sage College, Albany, New York

Jack E. Ingels, Chairman, Department of Plant Science, State University of New York Agriculture and Technical College, Cobleskill, New York

Technical Reviewers

David Manning, Instructor, Wilco Area Career Center, Lockport, Illinois

Ronald D. Regan, Supervisor, Agricultural and Environmental Education, Los Angeles Unified School District

Joseph Huth, Agent, Cooperative Extension Association of Albany County, Albany, New York

David E. Reville, Agent, Cooperative Extension Association of Albany County, Albany, New York

Frank Hegener, New York State Department of Environmental Conservation, Bureau of Pesticides (Section 7, Pesticides)

Thanks are extended to the following for their special assistance:

Dr. Clifford Nelson, Director of Agricultural and Extension Education, University of Maryland

Dr. Conrad Link, Department Head, Horticulture, University of Maryland

Dr. Francis Goin, Professor, Horticulture Department, University of Maryland

Jeffers Nursery, Incorporated, Slingerlands, New York

Hewitt's Lawn and Garden Centers, Latham, New York

Mary Reiley

Judy Shry

The staff at Delmar Publishers

Sponsoring Editor: William W. Sprague
Series Editor, Agriculture: Mary R. Grauerholz
Associate Editor: Mary L. Wright
Cover and Text Designer/Artist: Juanita F. Brown, L.E.P.I. Graphics, Albany, New York

A large portion of the material in this text was classroom tested at Linganore High School, Frederick, Maryland. Unit 2 was classroom tested at Wilco Area Career Center, Lockport, Illinois.

Index

MEAN DATE OF LAST 32° (F.) TEMPERATURE IN SPRING

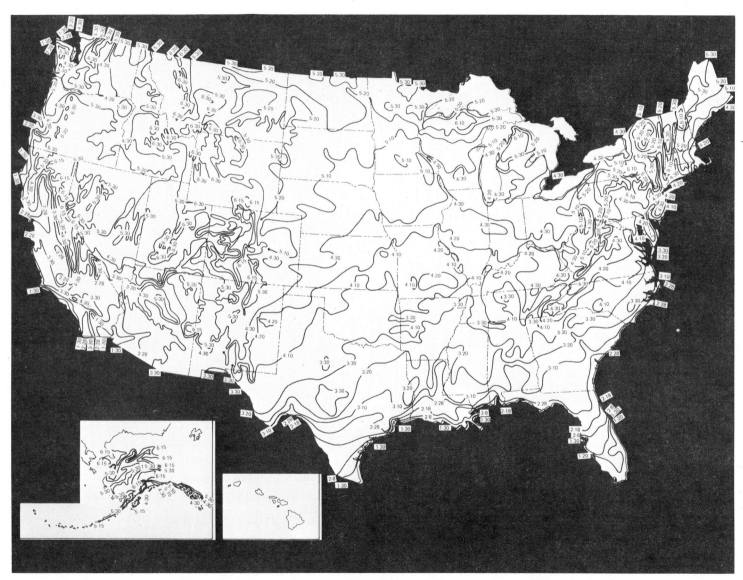

Average dates of the last killing frost in spring. (From USDA *Home and Garden* Bulletin 202)

MEAN DATE OF FIRST 32° (F.) TEMPERATURE IN AUTUMN

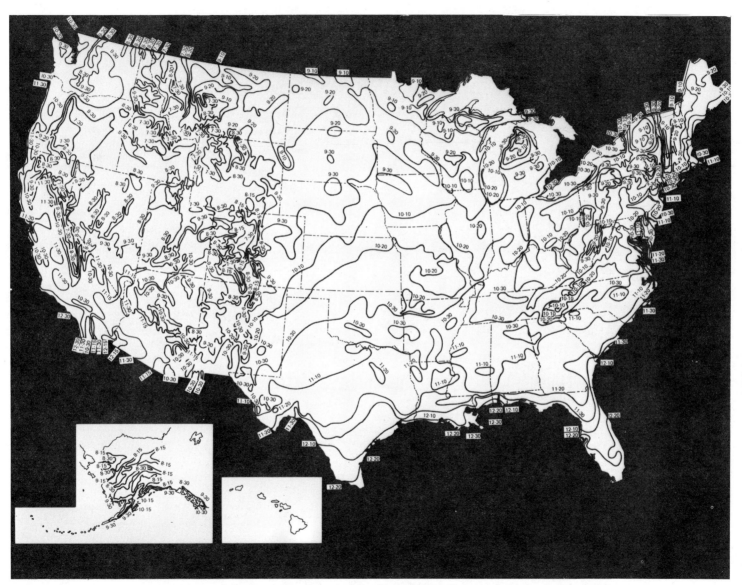

Average dates of the first killing frost in fall. (From USDA *Home and Garden* Bulletin 202)